Profit Strategies for
Air Transportation

Profit Strategies for Air Transportation

Dr. George Radnoti

McGraw-Hill

New York   Chicago   San Francisco   Lisbon   London   Madrid
Mexico City   Milan   New Delhi   San Juan   Seoul
Singapore   Sydney   Toronto

Library of Congress Cataloging-in-Publication Data

Radnoti, George.
    Profit strategies for air transportation / by George Radnoti.
        p.   cm.
    Includes bibliographical references and index.
    ISBN 0-07-138505-3
    1. Aeronautics, Commercial—United States—Finance. 2. Airlines—United
States—Cost of operation. I. Title.
HE9803.A5 R33 2001
387.7'1'0973—dc21   2001051314

McGraw-Hill

A Division of The McGraw·Hill Companies

1 2 3 4 5 6 7 8 9 0   DOC/DOC   0 7 6 5 4 3 2 1

ISBN 0-07-138505-3

The sponsoring editor for this book was Shelley Carr, the editing supervisor was Daina Penikas, and the production supervisor was Pamela A. Pelton. It was set in New Century Schoolbook by Paul Scozzari and Victoria Khavkina of McGraw-Hill Professional's composition unit, Hightstown, N.J.

Printed and bound by R. R. Donnelley & Sons Company.

McGraw-Hill books are available at special quantity discounts to use as premiums and sales promotions, or for use in corporate training programs. For more information, please write to the Director of Special Sales, Professional Publishing, McGraw-Hill, Two Penn Plaza, New York, NY 10121-2298. Or contact your local bookstore.

 This book is printed on recycled, acid-free paper containing a minimum of 50% recycled, de-inked fiber.

Dr. George Radnoti is one of the most intelligent and learned persons I have met in the last forty years. I worked with George at Delta Air Lines in the early 60s and I was always impressed with his rapid grasp of almost any subject especially technical subjects. In addition, he was an enjoyable person with whom to work. I am sure the readers of this book will also be impressed.

HOLLIS L. HARRIS
CHAIRMAN AND CHIEF EXECUTIVE OFFICER
WORLD AIRWAYS, INC.

Contents at a Glance

Contents

Acknowledgments

Forty-four years ago, when I entered the field of airline business, I searched for a book that would serve as a guide for airline operations and economics. This inspired me over these years to produce a book about the subject.

I was lucky to work for great companies in their growing stages, like Delta and American Airlines, way back in the 50s and 60s at the dawn of the jet-age. I had the opportunity to participate in their operations and to be exposed to their vast technical expertise and professionalism. The other significant factor that enabled me to write this book was that I frequently attended courses held for airline operations at Boeing Airplane Company. I am also thankful to Boeing for permitting the usage of some of their materials to be part of this book.

Similarly, I would like to express my thanks to the editors of McGraw-Hill for their dedication and professionalism that contributed to the perfection of this work. My special thanks go to Shelly Ingram Carr, Aviation Acquisition Editor, and Daina Penikas, Editing Supervisor, for their understanding and patience.

I am also in debt to Walt Overend, James Trebes* and Gerald Elikann who were my colleagues, for their encouragement and support.

I want to express my thanks to the 15-year-old, talented cartoonist, Laura Silverstein, for her illustrations in Chapters 10 and 11 (Figs. 10.5 and 11.12).

Finally, I am grateful to my best friend, confidante, and partner, my wife, who was my best critic with a no-nonsense approach. She participated in forging each sentence and scrutinized every aspect of the book with me.

*deceased.

Chapter

1

Introduction to the Economics and Operations of Airlines

The goal of this book is to generate interest, for a financially oriented person, in the technical aspects and, for a technically oriented person, in the accounting and financial outcomes of airline operations. This book will incorporate materials for

- Airline operations, cost analysis, and cost allocation
- Corporate planning
- Aircraft economics and aircraft capabilities (performance evaluation)

This textbook was written for financial and accounting managers engaged and involved in planning, analyzing, and conducting airline operations and for technical managers to be familiar with the financial side of operations. In today's complex world, a thorough knowledge of one side of airline operations also would require a good general knowledge of many other phases of the financial and technical fields. A successful businessman should be fully aware of the technical problems, constraints, and limitations that may occur in airline operations. Similarly, technically oriented managers could improve their job performance by having knowledge about aircraft financing and profitability of operations, as well as accounting problems, cost allocations, etc.

There are numerous textbooks about economics. Nevertheless, airline economy or the so-called transportation economy is a relatively new field, and little literature is available where the technical and economic characteristics are brought together. This book attempts to shed some light on this aspect of analyzing the subject by applying economic principles to airline operations.

Presentations will be shown in graphic form where necessary, but in the great majority, a computerized approach is the solution.

The goal of this book is to bridge the differences between the way financially oriented and technically oriented people think. This will be accomplished by exposing the former to the technical problems of airline operations and the latter to the cost and profit aspects of such operations. Their jobs require interacting in order to enable them to "speak" the same language and have a mutual understanding of the common task to be solved; they need to know more about the other's field.

For example, when an operational manager makes a technically sound decision, he or she also should consider the cost and profit consequences. Perhaps, with an equally sound alternative solution, he or she may improve the cost and profit picture. For each decision he or she makes in his or her mind, a dollar sign should be attached to it. He or she should be familiar enough to appreciate the various depreciation schedules and their effect on profit. Sometimes, considering other alternate solutions may improve the return on investment.

When a financial manager is exposed to, for example, a payload-versus-range curve, he or she may not draw the proper conclusion because he or she never studied it in business school. However, in airline operation, some decisions are influenced by technical presentation having a significant effect on cost and revenue. The preceding examples are just a small part of the many technical-financial presentations. Should the financial staff be familiar with this kind of technical report, the right conclusion and decision could be obtained easily.

It is assumed that the technical knowledge described in this book will be adequate to support a general idea about airplane operations without going into fine details. Similarly, the economic approach is also based on the same principle.

What benefit could be derived from this approach for a technically oriented person?

- Minimize cost and improve productivity
- Gain a better economic perspective for cost-effective decisions
- Gain a better understanding of management decisions
- Use direct and marginal cost for certain decisions
- Arrive at a better buy or lease decision

A financially oriented manager would benefit too:

- By knowing the capabilities and limitations of a given airplane
- By planning attainable goals
- By better understanding of the operational limitations

After studying the contents of this book, each individual may select a specific area to extend and further his or her knowledge by learning more about that subject of his or her choice.

The charts, tables, and cost and other information were prepared for the purpose of showing methods and procedures only. They are not meant to be used for current operation because costs, taxes, depreciation, etc. are constantly changing.

For emphasis, I repeat: The purpose of this book is merely an introduction to today's business practices where operations and economics go hand in hand.

2

Aviation Industry

2.1 Introduction

Today, the aviation industry plays a significant role in our country's economy. It consists of general aviation and commercial air transportation. There are several trades connected with the airline industry, such as aircraft and engine manufacturers, vendors, and suppliers. Aircraft maintenance by itself is another by-product of aviation, developed mainly in airports or nearby areas. There are also about 1 million persons involved in aviation and the connected trades.

The aviation industry started about 90 years ago, and its development shows an exponential growth. Several books describe the history and development of air transportation. In this book, the present state of the aviation industry will be described briefly.

2.2 U.S. Major Air Carriers

Major air carriers generally serve larger cities with a significant volume of traffic and use large commercial aircraft. These types of air carriers have each a reported revenue of over $1 billion per year. The following airlines are members of this group:

- Air Alaska
- American Airlines
- American Eagle
- American Trans Air
- Continental Airlines

- Delta Air Lines

- DHL

- FedEx

- Northwest Airlines

- Southwest Airlines

- United Airlines

- UPS

Note: American Airlines is in the process of absorbing TWA. However, United Airlines and U.S. Airways merger has been called off.

2.3   U.S. Nationals

The airlines in this group have a revenue range between $100 million and $1 billion per year. Presently, there are 31 airlines in this group, but a few of them face economic hardship that may lead to an insolvency in the near future.

2.4   U.S. Regional Airlines

The revenue range for this group is up to $100 million per year. For the larger regional airlines, this range varies between $10 and $100 million per year. By code sharing with larger air carriers (see below), regional airlines created a useful service to the flying public, and they also enjoyed the benefits of associating themselves with a larger carrier.

Presently, there are about 500 regional jets operating, and this number is expected to triple within the next 10 years. The higher speed, better utilization, and marketing advantages accelerated the widening and growth of the regional air carrier industry. The economics of the higher speed contributed to additional time available for more flights and, as a result, increased revenues. Meanwhile, the service became more appealing to the public.

When two airlines have the same identification code, it is called *code sharing*. Major airlines benefited from this association with their regional partners. It opened up for them the small community market they could not service before with their larger equipment. The smaller airports with shorter runways could not support their operations, and the larger aircraft were not economically feasible.

The regional airline-partners market share also was improved by the major airlines' advertising campaigns. Similarly, the convenience of flying passengers to a hub or direct to their destination over a shorter distance contributed to a rapid growth. The introduction of regional jet service and the technical and sometimes financial help that came from the major airline strengthened the position of the regional code-sharing airlines against the local competitor non-code-sharing airlines.

2.5   General Aviation

General aviation is a large industry. The public is only familiar with a few services within this category, such as police helicopters, traffic-reporting news helicopters, ambulance services, and fire fighters using aerial methods of spreading water or chemicals over endangered areas. However, these constitute only a small segment of general aviation. In the following, a few sectors of general aviation are presented.

2.5.1   Business and corporate aircraft

One section of general aviation consists of business and corporate aircraft. These airplanes are not used for paying passengers. The types of airplanes used in the business sector could be, for example, the Cessna 172, where the owner is piloting his or her own aircraft. Such usage is basically for sales, contacting customers in a few nearby states, flying from point to point, and on weekends, maybe for pleasure rides.

Corporations that have more than one airplane also may have a separate flight department (or one person designated to handle the flights), and they generally employ professional pilots. The commercial airlines' network covers mostly larger cities, whereas smaller communities have low-frequency service or none at all. Industrial development is seeking to establish plants in smaller communities, where the cost of labor is lower and/or materials are obtained more easily. Approaching these communities by commercial airlines would require too much traveling time, particularly if the airplane must go through a hub spoke. Trip time would be prohibitive. By using business and corporate airplanes, the savings in time and cost are significant.

2.5.2   Various general aviation activities

There are about 200,000 airplanes engaged in general aviation. Half this number is used for personal flying. The other half is used in various fields of activity.

Agriculture uses airplanes for crop dusting, spreading seeds, fertilization, etc. Aerial operations are conducted for land survey, inspection, photography, and topography. This category, as mentioned earlier, includes the daily use of small airplanes by traffic reporters, fire fighters, ambulance services, and law enforcement agencies.

2.5.3 General aviation airframe and engine manufacturers and maintenance

There are numerous airframe and engine manufacturers that build general aviation aircraft. The range includes from Beech to Bombardier and from piston to turboprop and jet aircraft. Boeing is considering entering this market with a scaled-down B717-100 airplane that would be heavier than the CRJ900. Actually, this is a model MD-90 series with a comfortable, large jet cabin, but with the operational and economical characteristics of a small jet.

In the United States, business and corporate airplanes have a 25 percent share in general aviation, whereas instructional flying amounts to only about 10 percent. The approximately 200,000 general aviation type airplanes require maintenance, repairs, and parts installation that have to be performed by Federal Aviation Authority (FAA)–licensed maintenance personnel. Similarly, planned inspections must be completed by qualified mechanics. In the airports or nearby, fixed-base operators conduct these tasks.

Additional functions connected with general aviation aircraft would be the sale or resale of airplanes, fueling, and flight instructors for student pilots. Since presently in the United States there are about 300,000 licensed pilots, pilot education by flight instructors is an essential part of aviation.

General aviation is a large and diversified industry and includes activities from a one-person shop to thousands of workers in airport hangers.

Another aspect of this industry is airport development. Funds were provided by the U.S. government based on the Airport-Airways Development Act in 1970 and the Airport and Airways Improvement Act of 1982. The growing number of aircraft used by general aviation also contributed to these airport improvements.

2.6 Fractional Sector of Air Transportation

The low utilization of corporate aircraft made their operation expensive and the cost of ownership prohibitive. Some companies specialize in part ownership of corporate aircraft. This permits a more economical operation and permits ownership even for companies on a smaller budget.

Another approach to ownership is that two or more corporations come to an agreement to use a corporate airplane and share their ownership and expenses based on a prearranged quota system.

Both approaches have the same goal: to have an airplane at their disposal with reasonable ownership and maintenance costs. Meanwhile, this arrangement enables each corporation to have fast transportation available for their managers and executives that they could not afford otherwise.

3

Airline Revenues and Operating Costs

3.1 Revenue Categories

The revenue-generating ability of an airline depends on many factors. Examples of normal revenue accounts and the related activities that generate the revenue are outlined in Table 3.1. The analyst must be able to relate these accounts to the forecast statistics of the airline industry in order to obtain operating revenues.

The sources of revenue can be classified as passengers, cargo, excess baggage, mail, and miscellaneous revenues from nonoperative sources.

3.1.1 Passengers

Yield is defined as air transportation revenue per unit of traffic carried in air transportation. It is shown in cents per passenger miles or cents

TABLE 3.1 Normal Airline Revenue Accounts

Revenue account	Generating medium
Passenger	Passenger traffic
Freight	Freight traffic
Mail	Government contracts
Excess baggage	Passenger traffic
Charter	Available aircraft time
Duty-free sales	On-board sales
Services performed	Maintenance handling for other airlines
Leasing income	Lease of equipment to other airlines

per passenger kilometers. Generally, it is obtained for the whole system. It may be broken down to regions or city-pairs.

If significant differences exist in yield between regions served, then yield should be applied by regions. When individual routes are examined, yield should be applied by city-pairs.

Yield generally is applied to all revenues obtained from scheduled revenue-passenger miles flown, including nonrevenue passengers.

3.1.2 Freight

Freight or cargo revenue has become increasingly more important in airline revenues. The larger airplanes now have good load-carrying capabilities, and cargo forecasts indicate a higher growth percentage than passenger growth. This revenue is measured in cents per ton-mile or cents per ton-kilometer. Mail revenue is measured in the same way as cargo revenue.

3.1.3 Excess baggage revenue

This is more significant in the operation of international airlines. It may be expressed in percentages of passenger revenues.

3.1.4 Other revenues

In this category, the following miscellaneous revenues are included: charter operations, royalties, ground transportation and handling services, rents, in-flight sales, contract maintenance, training, and other revenue-generating services.

3.2 Cost Categories

3.2.1 Historical background of direct-operating-cost (Air Transport Association) cost formula

In 1944, the Air Transport Association (ATA) introduced a cost formula for estimating airplane direct operating cost. A few revisions followed in 1948, 1955, 1960, and most recently in 1966.

With the title, "Standard Method of Estimating Direct Operating Cost," 1966 was the last time the ATA published a method defining direct operating costs. This is the basis from which today's costs were developed. The object was to have a standardized method of estimation of direct operating cost for various types of airplanes. The formula was based on actual cost data as reported to the Civil Aeronautics Board (CAB, now the Department of Transportation) on

Form 41. At that time, it presented an "industry average" value. Since then, it has been used, updated, modified, and changed to conform to the changed conditions. It could be expressed in dollars per mile or dollars per hour.

To review quickly the influencing variables of this ATA formula, Table 3.2 is provided.

3.2.2 Historical background of indirect-operating-cost (IOC) formula

A standard method for estimation of indirect operating costs was developed in conjunction with the ATA's DOC method. At the request of the FAA way back in 1964, the two major manufacturers, Boeing and Lockheed, jointly developed the indirect-operating-cost (IOC) method, which was finalized in 1970. Like the DOC method, the IOC method is now obsolete. Cost categories and their parameters are established and listed in Table 3.3.

IOC is sometimes defined as a percentage of DOC, expressed in dollars per block-hour, dollars per trip, or dollars per mile.

TABLE 3.2   Direct-Operating-Cost (DOC) Cost Categories

Cost category	Variables affecting cost
Crew	Number of flight crew Max. takeoff weight Speed
Fuel	Block fuel Fuel price Trip distance
Maintenance	Airframe price Engine price Airframe weight Engine static thrust Number of engines Labor rate Utilization Overhead rate
Insurance	Airplane purchase cost Annual utilization Insurance rate
Depreciation	Airplane purchase cost Depreciation schedule Investment spares ratio Annual utilization

TABLE 3.3   Indirect Operating Cost (IOC) Categories

Cost categories	Parameters influencing cost
Passenger service	Load factor
Aircraft servicing	Trip length
Traffic servicing	Block speed
Reservation and sales	Landing weight
Advertising and publicity	Tons of freight/mail
Ground property	Maintenance labor cost
Maintenance/depreciation	Etc.
General and administration	

3.2.3   Present methods for defining operating costs

Aircraft manufacturers developed in-house direct and indirect operating cost data by using selected airline cost information as their basis. These cost data were obtained by the U.S. Department of Transportation (DOT). Through careful evaluation of these data, they developed typical operating cost figures. Statistics and curve fits were applied, and to present their airplanes to airline customers successfully, it was necessary to provide realistic operating cost figures.

3.2.4   Methods of cost allocation

There are many ways to allocate costs. Tables 3.4 through 3.7 allocate costs based on fleet requirements, which may not be identical or even similar to the classic accounting systems shown in textbooks.

Some airlines divide costs into the following four categories:

1. Cash operation cost (COC)
2. Direct operating cost (DOC)
3. Indirect operating cost (IOC)
4. Total operating cost (TOC)

Cash operating costs might be better described as out-of-pocket cost to the company in order to get an airplane from point A to point B. They include flight crew pay, fuel and maintenance costs, and landing fees. Direct operating costs include cash operating costs, to which is added insurance (paid on an annual basis) and depreciation. Indirect operating costs include all the essential supporting items necessary to run the airline, such as administration, office rentals, advertising, and

TABLE 3.4   Airplane-Related Costs

Account	Related factors
Cockpit/cabin crew	Crew compliment Block hours Crew salaries
Fuel and oil	Fuel price/aircraft performance Mile/trips flown
Maintenance	Labor rates Aircraft design and age Number of flight hours
Aircraft service	Aircraft size (weight/seats) Number of departures Salaries/contract rates
Landing fees	Aircraft weight Airport rate schedules Number of departures
Navigation fees	Miles flown Government/airline cost schedules

TABLE 3.5   Traffic-Related Costs

Account	Related factors
Traffic servicing and reservation and sales	Passengers/tons transported Salary rates Fixed expenses Variable expenses
Food, liability insurance	RPK/RPM* produced Food cost Insurance rates

*RPK/RPM = revenue-passenger kilometer or mile.

TABLE 3.6   System-Related Costs

Account	Related factors
Advertising and publicity	Airline size
General and administrative	Management philosophy Salaries/overhead rates
Taxes	Government policies Tax laws

TABLE 3.7   Annual (Period) Related Expenses

Account	Related factors
Depreciation (amortization)	Original asset cost Useful life Salvage value
Hull insurance	Asset insured value Insurance rates
Lease/rental	Original asset cost Lease rates Supplies/demand (Short-term rentals)
Finance (interest) charges	Amount borrowed Term of loan Interest rates Company financial rating Lenders involved

all other items not included in the preceding categories. Total operating costs are a combination of all the preceding categories.

Other classifications are listed below:

- Direct operating costs

  Flying operations
  Fuel cost
  Insurance
  Maintenance and maintenance burden
  Depreciation/rents/interest

- Indirect operating costs

  Landing fees
  Navigational fees
  Station costs

  - Passenger-related
  - Airplane-related
  - Cargo-related

  System overhead

  - Administrative
  - Advertisement
  - Promotion
  - Other

  Start-up costs (if any)

In the case of a new airline or the addition of airplanes to an existing airline fleet, start-up cost should be considered in the budget.

3.2.5   Flying operations

Cockpit crew cost can be classified into six categories:

1. Gross weight pay
2. Mileage pay
3. Benefit pay
4. Training pay
5. Personal expenses (hotels and meals)
6. Payroll taxes

The actual expenses vary for domestic or international flying, low- or high-cost airlines, and from country to country.

Another way of establishing cost figures is to take the budget of the flight department and divide it by the hours flown in order to obtain dollar per block-hour values. *Caution:* This approach is only applicable if the flight hours are relatively constant without any significant fluctuation for a given time period. When this is not the case, adjust the budget based on aircraft type, schedule, flight hours, and market projections.

Cabin crew assignment and cost are based on the number of passengers the airplane is able to carry and are defined by the applicable regulations. For example, Federal Air Regulation (FAR) 91 requires "for airplanes having more than 19 but less than 51 passengers, one flight attendant; for airplanes having more than 50 but less than 101 passengers on board, two flight attendants," etc. Depending on the air carrier's contract, the range of cabin crew cost is between $35 and $55/h.

For example, a budget for a flight department is $312,500 per month (fuel costs not included), and flying time is 825 block-hours. By dividing the dollar amount by the block-hours, a figure of $378/block-hour is established for this operation. This approach presupposes that during the near future (e.g., month) no new and/or additional aircraft are introduced into the existing fleet and the budget and block times are fairly constant from one month to the next. For example, if and when the next month is known to have increased traffic and the budget figures are planned to be $423,200 and flying time is 918 block-hours, the new value will be $461/block-hour, reflecting the anticipated changes.

Another approach is to take the budget apart and classify it into constant expenses such as rent and electricity and variable expenses such as volume of passengers and/or cargo:

Constant expenses $153,200

Variable expenses $159,300

The application of 825 block-hours for this time period would result in

$185/block-hour for constant expenses

$193/block-hour for variable expenses

$378/block-hour total

This means that the second approach furnishes the same result.

3.2.6 Fuel cost

Fuel consumption depends on airplane-engine combination, airplane operating policy, and maintenance. The operating policy is set by the flight operations department, and it affects the following:

- Derated takeoff (contributes to lower engine temperature, extends engine life, and lowers maintenance costs)
- Climb schedule
- Cruise policy (LRC, Mach)
- Descent schedule
- Optimal altitude
- Reclearance policy
- Tankering
- Fuel flow degradation based on airplane age
- Conduct performance audit
- Airport characteristics and payload

The maintenance department is responsible for keeping the airplane in such a condition that drag is minimal while engine performance is maximal. Fuel savings will be discussed in greater detail in Chapter 7.

The best way to calculate fuel cost is to obtain fuel costs for each station listed in the schedule. For example:

Trip: JFK to LHR

Trip fuel: 100,000 lbs

Fuel price at departure station at JFK: 55¢/gal

Fuel cost = (trip fuel/density) × fuel price

$$\$8208 = (100,000/6.7) \times 0.55$$

Another, less accurate method for defining fuel cost is based on average fuel flow of the engines in pounds per hour or gallons per hour. For example, one could use an average cruise weight at a reasonable cruise altitude (depending on the judgment of the performance engineer) and establish a gallons per hour consumption. Another approach would be to collect total fuel requirements and total time from the company's records. The calculation becomes more complex if more types of aircraft are flown. From total fuel and total time, a fuel/time figure could be established in gallons per hour and multiplied by an average fuel cost in cents per gallon. A dollars per hour value could be established for the whole system.

The following example shows how to apply this average gallons per hour figure for a given trip:

Trip time: 6.0 h

Average fuel consumption: 2200 gal/h

Average fuel price: $0.60/gal

Fuel cost = trip hours × average fuel consumption × average fuel price

$$\$7920 = 6.0 \times 2200 \times 0.60$$

3.2.7 Insurance

There are three types of insurance, and each is described briefly below.

Hull insurance. Hull insurance may be related to the replacement cost of airplanes. Generally, it is a certain amount of the airplane price (less engine costs) and is expressed in percentages.

War-risk insurance. War-risk insurance is around 3 to 5 percent of the airplane hull cost, but depending on the circumstances, it may be higher.

Passenger insurance. This is based on mileage and is in the range between 10 and 30 cents per passenger.

3.2.8 Block-hours and flight-hours

Since maintenance costs are calculated and expressed in flight-hours, there is a need to understand the differences and relationship between flight-hours and block-hours. Picture an airplane at a terminal building. As soon as the engines are started and the airplane proceeds to move away from the terminal building, the time is recorded immediately in the cockpit. If there is no other traffic as the airplane taxis to the assigned runway and the airplane is cleared for takeoff, this takeoff time is recorded. Thus taxi-out time represents the time elapsed between the start of the engine and takeoff. If the airplane is in a queue for this runway due to traffic, additional time is necessary for takeoff.

This additional waiting time is included in taxi-out time. When the airplane is occasionally assigned to another (further) runway due to traffic, wind, or any other reason, it has to travel to the newly assigned runway, and this will represent an additional taxi-out time.

Taxi-out time depends on factors such as the time of day, day of the week, season, weather conditions, traffic, and airport layout (distance between terminal building and runway). Once the airplane lands at the destination airport, the time is recorded, and the airplane proceeds to the terminal building. When engines are shut down, the time is recorded anew.

Block time is defined as time elapsed between engine start at departure and engine shutdown at the destination airport. *Flight time* is defined as time elapsed between takeoff time at the departure runway and time of landing at the destination airport.

The difference between block time and flight time is the taxi-out and taxi-in time:

Block time = taxi-out time + flight time + taxi-in time

Each airline tries to minimize taxi time and fuel because they do not contribute to profit.

Cargo airplanes take off during night hours or when there is no heavy passenger airplane traffic. This shortens taxi-out time and block time. Block time is sensitive to airport layout and airplane traffic. Flight time is sensitive to en-route winds and/or traffic congestion at the airports (in the case of a holding pattern).

3.2.9 Maintenance cost

The figures for maintenance cost are given in flight-hours, whereas direct operating cost is calculated in dollars per block-hour. In order to incorporate maintenance cost into the direct operating cost, dollars per flight-hour have to be converted into dollars per block-hour. Taking an average figure for the relationship between flight-hour and block-hour, a ratio of 1.20 could be used, or another one could be selected based on the operator's experience.

1.0 flight-hour = 1.20 block-hour

For example, a budget contains $518,300 for direct maintenance cost and a total of 912 flight-hours:

$518,300/912 = $568.3/flight-hour

912 flight-hours = 1.20 × 912 = 1094 block-hours

$518,300/(912 × 1.20) = 473.7 block-hours

This example shows how to switch from flight time to block time, and vice versa.

Maintenance cost consists of airframe maintenance and engine maintenance. Airframe (including APU) maintenance is divided into labor and materials.

Airframe direct labor cost per trip:

$$\$ \text{ (Airframe labor per trip)} = LR \ (ACL + AL \times FH)$$

where LR = direct labor rate, \$/man-hours
  ACL = man-hours/cycle
  AL = man-hours/flight-hours
  FH = flight-hours/trip

Note: Cycle consists of a takeoff and a landing.

Airframe direct material cost per trip:

$$\$ \text{ (Airframe material cost per trip)} = ACM + AM \times FH$$

where ACM = material\$/cycle
  AM = material\$/FH

Example of maintenance cost calculation:

Trip = 3000 miles

Flight time = 6.55 hours

Given cost elements:

Labor rate (LR) = \$24/man-hour

Labor (AL) = 3.08 man-hours/flight-hours

Cycle (ACL) = 8.75 man-hours/cycle

Material (AM) = 83.28 man-hour dollars/flight-hours

Cycle (ACM) = 359.00 AM \$/cycle

Labor = $LR \ (AL \times FH + ACL)$

  $\$ = (\$/\text{man-hours}) \times [(\text{man-hours}/FH) \times FH + (\text{man-hours/cycle})]$

$\$694 = 24 \times 3.08 \times 6.55 + 24 \times 8.75$

  Material = $AM \times FH + ACM$

    $\$ = (\text{Material}\$/FH) \times FH + (\text{material}\$/\text{cycle})$

    $\$904 = 83.28 \times 6.55 + 359$

Total airframe cost:

$$\$904 + \$694 = \$1598$$

Engine maintenance cost is based on a similar approach; engine manufacturers provide the basic cost values (which depend on the type of engine):

Labor (EL) = 2.31 man-hours/FH

Cycle (ECL) = 1.62 man-hours/cycle

Material (EM) = 181.44 EM$/FH

Cycle (ECM) = 135.00 EM$/cycle

Labor = $LR \times (EL \times FH + ECL)$

$ = ($/man-hours) $\times$ [(man-hours/FH) $\times$ FH + (man-hours/cycle)]

$402 = 24 \times 2.31 \times 6.55 + 24 \times 1.62$

Material = $EM \times FH + ECM$

$ = (material$/FH) $\times$ FH + (material$/cycle)

$1323 = 181.44 \times 6.55 + 135$

Total engine cost:

$$\$1323 + \$402 = \$1725$$

When labor is given in dollars per flight-hour, the equation changes:

Airframe:

Labor (ADF) = 73.92 $/FH

Cycle (ADC) = 210.00 $/cycle

Flight-hours (FH) = 6.55 flight-hours

Airframe labor:

$/trip = $ADF \times FH + ADC$

$ = ($/$FH$) $\times$ FH + $/cycle

$694 = 73.92 \times 6.55 + 210$

Engine:

Labor (EDF) = 55.44 $/FH

Cycle (EDC) = 39.00 $/cycle

Flight-hours (FH) = 6.55 flight-hours

Engine labor:

$/trip $= EDF \times FH + EDC$

$\$ = (\$/FH)(\times FH + \$$/cycle

$\$402 = 55.44 \times 6.55 + 39$

Total direct maintenance cost is the sum of airframe labor, airframe material, engine labor, and engine material, which gives a total of $3323.

Examples. Let us take the budget of Omega Airlines and calculate maintenance cost of a mixed fleet. The fleet consists of three types of airplanes:

Three "Alpha" airplanes

Nine "Beta" airplanes

One "Gamma" airplane

For the Alpha-type airplane with a scheduled 1055 flight-hours this month, the following maintenance jobs have to be performed, according to the maintenance program:

1. "A" check after each 15.5 flight-hours

   "A" check planned 1055/15.5 = 68 times

   "A" check requires 58 man-hours

$$58 \times 68 = 3944 \text{ man-hours/month}$$

   3944 man-hours/1055 flight-hours = 3.73 man-hours/flight-hour

2. "B" check after each 330 flight-hours

   "B" check planned 1055/330 = 3.19 times

   "B" check requires 700 man-hours

$$700 \times 3.19 = 2233 \text{ man-hours/month}$$

   2233 man-hours/1055 flight-hours = 2.11 man-hours/flight-hour

$$3.73 + 2.11 = 5.84 \text{ manhour/flight hour}$$

For the Beta-type airplane schedule for 3316 flight-hours, the following maintenance work is applicable:

3. "D" check after each 1500 flight-hours

   "D" check planned 3316/1500 = 2.2 times

   "D" check requires 1150 man-hours

$$1150 \times 2.2 = 2530 \text{ man-hours/month}$$
$$2530 \text{ man-hours/3316 flight-hours} = 0.76 \text{ man-hours/flight-hour}$$

The Gamma-type airplane, which has a schedule for 405 flight-hours, is programmed with the following maintenance jobs:

"B" check after each 100 flight-hours
"B" check planned 405/100 = 4.0 times
"B" check requires 200 man-hours

$$200 \times 4.0 = 800 \text{ man-hours/month}$$
$$800 \text{ man-hours/405 flight-hours} = 2.0 \text{ man-hours/flight-hour}$$

According to the budget figures, the following cost items are known for this month:

Labor	$422,510
Overtime	53,905
Fringe benefits	72,000
Pensions	24,000
Total	$572,415

During this month, outside contractors are hired to conduct other maintenance functions at a cost of

910 man-hours (@ $55/h)	$50,500
115 man-hours (@ $110/h)	12,650
Total	$63,150

Subtracting the expenses paid to the outside contractors, the in-house cost will be

Labor	$372,010
Overtime	41,255
Fringe benefits	72,000
Pensions	24,000
Total	$509,265

Thus man-hours = (man-hours/flight-hours) × monthly flight hours.

	Man-hours/flight-hours		Flight-hours		Man-hours
For Alpha airplane	5.84	×	1055	=	6160
For Beta airplane	0.76	×	3316	=	2520
For Gamma airplane	2.00	×	405	=	810
				Total	9491

Dividing total labor by man-hours will provide dollars per man-hour:

$$\$509,265/9491 = 53.65 \ \$/\text{man-hour}$$

For each airplane type, the following is calculated:

Airplane type	Labor × Man-hours		Dollars
Alpha	53.65 × 6160	=	330,537
Beta	53.65 × 2520	=	135,198
Gamma	53.65 × 810	=	43,450
Total			$509,105

With the preceding exercise, the original budget was reconstructed. The next step is to develop the dollars per flight-hour figures:

Airplane type	Labor/flight-hour		$/flight hr
Alpha	330,537/1055	=	313.30
Beta	135,198/3316	=	40.77
Gamma	43,456/405	=	107.30

The preceding calculation thus indicates that for the Alpha-type airplane this month, the cost of each flight-hour is $313.30.

The preceding examples dealt with direct maintenance cost only. Similarly, the maintenance burden cost figures have to be defined. The following example shows how to apply dollars per flight-hour and obtain maintenance costs.

Example

Trip: 3000 miles

Flight time: 6.55 hours

Airframe and engine labor and material, in dollars per flight-hour: $394.16

Airframe and engine cost per cycle: $743.00

Total direct maintenance cost: $/FH × FH + cycle

$$394.16 \times 6.55 + 743 = \$3324$$

Example   An airline has a monthly budget for direct maintenance cost of $343,000. Total flying time for the given time period is 875 flight-hours. Direct maintenance cost is 343,000/875, or 392 $/FH. Internal maintenance cost also will be influenced by the learning curve and the newness of the airplane.

The preceding examples dealt with various maintenance costs, but other cost elements have to be considered when establishing a maintenance budget. Line stations maintenance cost is one such element.

Spare cost also should be considered. Spares can be obtained within a very short time from the manufacturers, but sometimes even that short time is too long to satisfy the need and proves to be costly. The best solution is to have spares on the hangar floor, but this ties down the airline's funds.

Another solution is to have spares owned and handled by a supplier(s) on the hangar floor so that the airline can pay for the spares as they are used. This enables the airline to use its funds somewhere else.

Another approach is to consider maintenance cost on or off the airplane. Each airline has its own maintenance procedures and work schedule for various airplane types.

It is difficult to compare maintenance procedures and maintenance costs because each airline has a different method of maintaining its airplanes. Some use outside maintenance services only, some use in-house maintenance, and some use a combination of both. The choice of maintenance procedure is based mostly on cost-effectiveness.

Cycle/flight-hour effect. There is a definite relationship between maintenance cost and flight time; the maintenance cost per flight hour decreases with the increased flight time, as shown in Fig. 3.1.

Learning factor. Figure 3.2 shows learning-curve improvement as a function of airplane time expressed in years. Note that the man-hour requirement is decreasing and the learning factor is increasing with time and that the vertical scale goes up from 1.0 to zero.

This is in direct correlation with the newness factor of maintenance cost. The time spent maintaining the airplane, based on the learning curve and newness of the airplane, is inversely proportional to its maintenance cost.

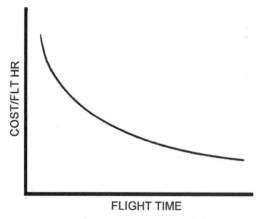

FIGURE 3.1 Cost per flight-hour versus time.

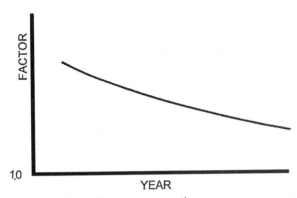

FIGURE 3.2 Learning curve versus time.

Newness factor. The curve in Fig. 3.3 depicts the typical effects of the newness of a standard-body aircraft related to maintenance costs. The factor is applicable for both labor and material costs for the engines and airframes.

In conclusion, as maintenance personnel get more knowledgeable about their work and gain more experience, maintenance cost has a tendency to decrease as a function of time.

3.2.10 Maintenance-Burden Costs

Maintenance burden or overhead consists of all costs of maintenance except direct labor and direct materials. Overhead costs are related to the maintenance work and should be associated with a specific type of airplane. Many of the burden costs are allocated to purchasing, facilities, etc. These costs could be allocated to maintenance activities. However, because of their nature, it is difficult to execute the allocation precisely. The allocation should be considered on a "fair-share" basis. Since maintenance burden is related to maintenance cost, it is expressed frequently as a percentage of direct maintenance cost. Burden cost consists of vacation pay, sick pay, various employee benefits, buildings, pensions, record keeping and statistical personnel, payroll taxes, trainees, instructors, and supervisory personnel.

Totaling the other cost elements such as rent, electricity, telephone, heating, and administrative work, the sum in this case is $438,103. Dividing this amount by the total man-hours, the result is $46.10 per man-hour. In this way, the indirect maintenance or burden is defined by airplane type:

$$\$438,103/9491 \text{ man-hours} = \$46.10/\text{man-hours}$$

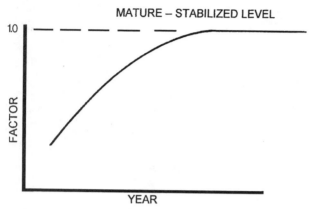

FIGURE 3.3 Newness factor versus time.

Airplane type	$/Man-hour × man-hour		Dollars
Alpha	46.10 × 6160	=	284,391
Beta	46.1 × 2520	=	116,324
Gamma	46.1 × 810	=	37,389
Total			$438,103

After these figures have been obtained, the cost in dollars per flight-hour can be defined for either type of aircraft. For example, for the Alpha airplane type:

$$\$284{,}391/1955 \text{ flight-hours} = 269 \text{ \$/flight-hour}$$

The purpose of showing these calculations is to indicate how to handle man-hours, flight-hours, and the budget when the fleet consists of a mixture of airplane types and dollars per flight-hour has to be obtained.

The addition of a new aircraft type to the fleet may increase the maintenance burden. In this case, a careful analysis has to be conducted to define the magnitude of the maintenance burden.

Note: Cost allocation can be done in many ways. Each airline, based on its structure, organization, and philosophy, approaches the maintenance budget in a way that is logical, convenient, and leads this complex task to a correct solution.

3.2.11 Methods of depreciation

Depreciation is an expense that reflects the diminishing value of a capital asset such as airplanes, engines, spare parts, ground equipment, or any other equipment with a useful (economic) life longer than 1 year and with a value over the amount determined by company policy. The useful life of each object is established by Internal Revenue Service rulings, and it changes from time to time whenever tax laws are amended. This time period, however, is not necessarily identical with the actual time of use; sometimes the objects can be used for an extended period of time. In some cases, their usefulness is shorter, or they become obsolete with technical advances. This is the reason why companies in most instances establish two different time periods for the duration of an object's depreciation. After the useful time period is over, the object may still have some value if sold or further used with reduced capacity. This value is called *salvage* or *residual value*.

Some of the customary terms used for depreciation are defined below:

Depreciation is the decrease in value of objects because of use and time.

Life is the time period established as the useful life of an object.

Residual is the salvage value received for an object after expiration of its useful life.

Base cost is the original cost of an object when acquired, including taxes, freight charges, import duties, or any other expenses related to the acquisition, e.g., engine, buyer-furnished equipment, spare parts, cost of training, etc.)

Yearly depreciation is the yearly cost of use, calculated by various methods (see below), considering the base cost plus capital improvements, salvage value, and useful life.

From the various depreciation methods applicable to a given object, companies choose the most convenient one allowed by law. The most commonly used methods are listed below:

1. Linear
2. Sum-of-the-year method
3. Double-declining balance
4. Double-declining balance and linear

Linear depreciation

$$DEPLIN = (Al - S)/n$$

where $DEPLIN$ = yearly depreciation
      Al = amount of airplane investment, e.g., \$10,500
       S = salvage or residual value, e.g., \$500
       n = economic life of airplane, e.g., 6 years

Note that economic life is not necessarily identical with the physical life of an airplane.

Inserting the preceding figures into the *DEPLIN* equation, a yearly depreciation of $1666.67 is obtained.

An item purchased on January 1 costs $10,500, and after 6 years, it has a residual value of $500. Table 3.8 shows the calculation of yearly and total depreciation.

These figures were obtained by applying the equation shown above. The yearly depreciation and the undepreciated portion or book value at the end of each year is

$$BOOK_t = Al - [(Al - S) (n/t)]$$

where $BOOK_t$ is book value at year t. Here, $t = 4$.

At the fourth year,

$$3833.33 = 10500 - [10,000/(6/4)]$$

Sum-of-the-year method. This method adds up consecutive numbers of years from year 1 to the last year of the economic life of the equipment. If the economic life is 5 years, the denominator is $1 + 2 + 3 + 4 + 5 = 15$. The numerator is the remaining years of the economic life. In the first year, the numerator is 5; in the second year, it is 4; etc.

Example   The economic life of an object is 6 years. It was purchased at the beginning of the year for $10,500 and will have a residual value of $500. Table 3.9 shows the calculation of yearly and total depreciation. Depreciation allowance is at the year t:

$$DEP_t = \frac{n - (t - 1)}{[n (n + 1)]/2} (Al - S)$$

TABLE 3.8   Linear Depreciation

Year	Yearly depreciation	Book value (balance)
0	0	$10,500.00
1	$1,666.67	8,333.33
2	1,666.67	7,166.67
3	1,666.67	5,500.00
4	1,666.67	3,833.33
5	1,666.67	2,166.67
6	1,666.67	500.00

TABLE 3.9 Sum of Year-depreciation

End of year	Denominator	Yearly depreciation	Book value
0	0	0	$10,500.00
1	6/21	$2854.17	7642.86
2	5/21	2380.95	5261.90
3	4/21	1904.76	3357.14
4	3/21	1428.57	1928.57
5	2/21	952.38	976.19
6	1/21	476.19	500.00

At the fourth year using the preceding equation, $t = 4$, and

$$\text{Depreciation} = \frac{6 - (4 - 1)}{[6\,(6 + 1)]/2}\,(10{,}500 - 500)$$

Thus, in the fourth year,

$$1428.57 = (3/21)10{,}000$$

and the book value is

$$BOOK_t = \frac{(Al - S)(n - t)(n - t + 1)}{n(n + 1)} + S$$

Let's calculate the book value for the same year:

$$\text{Book value} = (10{,}500 - 500)\,\frac{(6 - 4)(6 - 4 + 1)}{6(6 + 1)} + 500$$

Thus, in the fourth year,

$$1928.57 = 10{,}000[(2 \times 3)/42] + 500$$

Double-declining balance. The advantage of this method is the acceleration of depreciation in the early years of use. This is especially useful because of the rapid changes in technological domains and objects quickly becoming obsolete. Table 3.10 shows the calculation of yearly and total depreciation using the same example as with the previous method.

The equation for the yearly depreciation is

$$DEP_t = (2/n)BOOK_z$$

where $z = t - 1$. Thus

$$\text{Depreciation} = (2/6)3111.11 = 1037.04$$

at the fourth year.

The book value is

$$BOOK_t = Al[1 - (2/n)] \quad \text{at the t power}$$

Book value $= 10,500(1 - 0.3333)$ at the fourth power $= 2074.07$

at the fourth year where $t = 4$.

Double-declining balance and linear from ($n/2$ + 1). This method is a combination of the two methods described before. The switch from double declining balance is at year $n/2 + 1$. In this case, the residual is defined. Again, tabulation reflects both methods side by side (Table 3.11).

In the fifth year, the straight-line method results in a larger amount of depreciation than the double-declining one. This is the year when the combined method switches to linear.

TABLE 3.10   Double declining balance-depreciation

End of year	Yearly depreciation	Book value
0	0	$10,500.00
1	$3500.00	7000.00
2	2333.33	4066.67
3	1555.56	3111.11
4	1037.04	2074.07
5	691.36	1382.72
6	460.91	921.81

TABLE 3.11   Double declining balance and linear from $n/2$ + 1-depreciation

End of year	Depreciation, double declining	Depreciation, linear	Book value	Depreciation combined
0	0	0	$10,500.00	0
1	$3500.00	$1667.67	7000.00	10,000.00/6 = 1666.67
2	2333.33	1300.00	4666.67	6500.00/5 = 1300.00
3	1555.56	1041.67	3111.11	4166.67/4 = 1041.67
4	1037.04	870.37	2074.07	2622.22/3 = 870.37
5	691.36	787.04	1287.03	1574.07/2 = 787.04
6	460.91	787.04	500.00	787.04/1 = 787.04

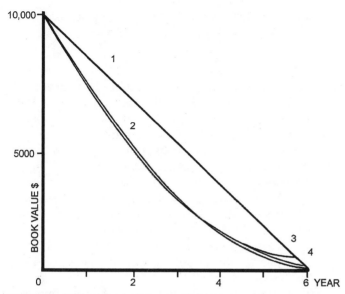

FIGURE 3.4   Various depreciation methods.

Figure 3.4 presents the various methods in graph form:

1. Linear
2. Sum of the year
3. Double declining
4. Double declining and linear after $n/2 + 1$

For economic analysis, the straight-line method is used. Airlines treat depreciation annually by airplane type. The depreciation schedule has a great importance in any cost calculation in order to achieve a realistic cost and earnings figures. It is also calculated in block-hours per year, as shown in the following example for airplane type A:

Total investment	$36,073,500
Economic life, years	16
Residual (salvage) value (10%)	$3,607,350
Total amount to be depreciation	$32,466,250
Depreciation per year	$2,029,134
Total block time per year	5130
Depreciation in dollars/block-hour/year	$395

The total price of an airplane should include all airplane-related expenses such as the cost of the engine, buyer-furnished equipment, spare parts, and training. When airplanes are delivered at different intervals, escalation indices issued by the Bureau of Labor Statistics should be used for future deliveries. An airplane's economic life for depreciation purposes is established for x years, with a y percent residual value. Federal tax laws control the values of x and y.

3.2.12 Airplane leasing

Many airlines turn to leasing when they plan to extend their fleet or try to establish a new airline. Leasing has advantages over purchasing; e.g. it would not tie down capital. Leasing companies obtain loans at lower rates, and this lowers lease payments. In addition, it may be a faster approach to execute a lease agreement than to go through the banking circuit. Furthermore, leasing has certain tax advantages.

Airlines generally consider leases for a short time period because payments get larger with time. The airline does not own the airplane.

In the case of a true or regular lease, the owner of the airplane can deduct depreciation. The airline deducts the lease payments.

In the case of a capital lease, the airline buys the airplane at a lower price before the lease period expires, and the leasing period is about 75 percent of the airplane's economic life.

There are a few other types of lease arrangements. For example, one company can take advantage of depreciation, while another pays lower lease payments. Terms of lease and lease rates are expressed as percentages of airplane and spares costs—generally at an annual rate between 10 and 11 percent. Leases can be calculated either yearly or monthly.

$$\text{Yearly lease} = \frac{TIN_v}{K(1/i)}$$

where $K = \{1 - [1/(1 + i)^{pp}]\}$
TIN_v = airplane and spare parts costs
i = interest rate
PP = number of payment periods

3.2.13 Interest costs

Interest expenses should be considered when cost is analyzed. Interest and depreciation are considered as total ownership. The loan life will equal the depreciation period.

No depreciation is used with airplane leases. Interest is calculated with the declining-balance method on a constant principal payment. Here is the equation to be used:

$$TOTINT = 0.5 \times i \times TTL \times (LNP + 1)$$

where $TOTINT$ = total interest over loan life
TTL = total amount of loan
LNP = total number of loan payments
i = interest rate

Example

Cost of airplane, engines, spare parts, etc.	$20,000,000
Loan life	20 years
Financing	100%
Interest rate	10%
Number of payments	20
Total block-hours per year	3000

By using the preceding equation, the total interest paid is divided by the total block time (at 3000 hours per year), yielding the dollars per block-hour.

Another approach is to define interest cost for each year, and by considering 3000 block-hours per year, a yearly variable interest cost is obtained.

Interest is based on the amount of money invested in airline operations, such as the airplane, ground support equipment, and spares. Airline interest cost is an annual charge and will vary by year depending on prevailing interest rates and the debt repayment schedule.

3.2.14 Financing cost

For non-aircraft-related investments, payment schedules for loans on other capital investments (i.e., buildings, sales, offices, hangars, maintenance facilities, computers, communications equipment, and other non-aircraft-related facilities) should be considered.

3.2.15 Landing charges and navigation fees

Charges for landing fees are assessed in various ways. Many times landing fees are established in proportion to the maximum takeoff or landing weight and sometimes as a monthly charge. In cases of high volume, fees can be discounted. The amount also could depend on the time of the day, whether the flight is international or domestic, and whether it is a training flight, etc. Sometimes stage 2 aircraft have higher fees. Congested airports charge higher fees, too. Such fees may include air navigational charges and actual ground handling charges by station and aircraft types. There is a large fluctuation in landing fees. Large airplanes have lower costs per passenger-seat or ton than smaller airplanes. Landing fees are published in the International Air Transport Association (IATA) manual.

Landing charges sometimes include other associated charges such as rent fees, parking fees, terminal fees, and handling facilities. Table 3.12 shows a few landing fees.

TABLE 3.12 Landing Fees

Station	B747	DC10	AIRBUS
FRA	$3900	$2500	$1600
LHR	729	729	729
JFK	2600	1750	1180

Air navigation fees are charges for the use of landing and navigation aids. Navigation fees could be considered part of the landing fees. Generally, the charges are related to the airplane weight (AW).

$$AW = \sqrt{\text{max takeoff weight}/50} \quad \text{and} \quad Z = DISTF \times AW$$

where $DISTF$ is the distance factor in hundreds of kilometers.

$$NAVFEES = K \times Z$$

where K is the unit rate in U.S. dollars (varies in each country). The relevant ICAO and ATA publications usually contain the proper navigation fees.

3.2.16 Station costs

Ground handling expenses include the cost of several services, such as ramp handling, cleaning the airplanes, servicing galleys, water, lavatories, catering food and other passenger services, passenger airport charges, and subcontractors for passenger handling charges.

For cargo, handling is done by subcontractors or by station personnel. The typical items included in station costs are

Flight processing

Weighing

Supervision

Transit in and ramp handling

Passengers handling:
- Catering (food and other passenger services)
- Liability insurance
- Passengers sales commissions
- Subcontracted passengers handling

Cargo handling:
- Cargo liability insurance
- Cargo sales commission
- Subcontracted cargo handling

Airplane cleaning:

- Cleaning airplane
- Servicing water
- Lavatories

Galleys

Fueling

Online maintenance

Unscheduled maintenance

Miscellaneous services not listed above

Passenger cost can be broken down for each passenger, and cargo cost can be broken down for each ton.

Passenger handling cost. The handling of ticketing and sales per passenger and baggage can be expressed in dollars per passenger. This cost is associated with the trip length; for an airline with short trips, this cost would be relatively more than for an airline having long trips. Southwest and Qantas airlines are good examples. While Qantas transports 300 passengers on a 10-hour flight and pays a one time handling fee, Southwest handles maybe two or three times as many passengers during the same time period and pays twice (or more) the handling costs.

Airline sales department. The airline's sales expenses are budgeted yearly and added to the station cost. For example, if Golden Star Airline has a station in San Marino where an airline office is functioning, the cost of maintaining this office has to be included into the station cost.

Advertising and publicity. Similarly, using the preceding example of Golden Star Airline, if the airline will advertise in San Marino's *Giorno* newspaper, the cost of the advertisement should be included as a station cost.

Defining station cost can be done in many ways, two of which are shown here. The first method requires that each cost item be examined and classified into constant expenses, such as the renting of facilities and salaries, or variables in conjunction with the traffic fluctuation in and/or out of the station. This, however, is a very tedious approach because it is time-consuming and the precise data are not always available. Particularly in the planning stage where no past experience is available, the following assumption is considered: The traffic into and out of the station is known and does not fluctuate excessively. The budget for this station, divided by the number of transits, would furnish a fairly close estimate. Below is an example involving Alpha Airlines:

Station	Budget	Frequency	Transit cost
BOS	$34,683	13	$2,667
FRA	403,215	31	13,006
JFK	585,950	54	10,850
LHR	91,397	27	3,385

When a schedule is generated from these figures, the budget could be reassembled. In other words, when a monthly budget is allocated and the airline operation is simulated, the sum of the cost elements should be equal to the original amount of the budget shown above.

3.2.17   Systems overhead

So far the costs have been allocated into different categories, such as crew cost, maintenance, depreciation, and landing fees. Nevertheless, there are certain cost elements that by their nature cannot be classified into clear and well-defined categories. Some of these expenses are listed below:

- General and administrative expenses
- Advertising and publicity
- Headquarters building and office maintenance
- Salaries and wages of personnel not classified in any of the previous categories

These costs, called *overhead,* are added up and divided into the block-hours flown (total block-hours flown within a year is relatively stable). For example:

System cost per year	$4,000,000
Total block time (hours)	3,335
System cost (overhead)	$1,200/block-hour

3.3   Start-Up Costs

Start-up cost is an additional cost when a new airplane is added to an existing fleet or when establishing a new airline. The following items can be listed as start-up costs:

- Facilities
- Spares
- Ground equipment

- Training:
  Flight crews
  Cabin crews
  Maintenance personnel
  Other personnel
  New hires
  Others

A significant part of start-up cost consists of personnel training, ground support equipment, and initial spare purchases. To mention a few expensive ground equipment items:

Tow tractors

Passenger stairs

Toilet trucks

Potable water trucks

Galley trucks

Cabin-cleaning trucks

Ground power units

Air-conditioning units

Air-start units

Deicing/washing trucks

A few additional items for freight handling:

Belt conveyors

Lower-lobe container loaders

Transporters

Main-deck cargo loaders capable of loading/unloading pallets and/or containers

Miscellaneous tools

Note: These items are needed at the main base and at certain stations, too. Management has to examine the best economic solution, such as whether to buy or rent this equipment or engage a service company or another airline. Such ground equipment can cost a few million dollars, so this matter needs very careful evaluation.

Personnel training includes crew training and training of maintenance and other personnel affected by the introduction of the new airplane type. Flight crew is the costliest item, and it is broken down into five categories:

- Ground training
- Simulator training
- Aircraft time
- Out-of-production time (while in training)
- Route training

To train a captain of a smaller-sized airplane (e.g., DC-9 or B737) to fly a larger type of airplane (DC-10 or B767), the following expenses should be considered (figures are approximate, varying from airline to airline and from year to year):

Ground school (6 days)	$900
Simulator (18 hours)	9,700
Flight training (5 hours)	18,000
Salary and expenses (6 weeks)	6,500
Total	$35,100

These figures refer to only one crew member, so they have to be multiplied by the number of crew members designated for training. If the smaller airplane mentioned before is not retired from the fleet, another set of crew has to be trained for their operations. Training cost could easily reach into the millions of dollars. Cabin crew training is less costly, but it also has to be considered. Similarly, maintenance personnel also should be prepared for the handling of the new airplane type to acquire the necessary knowledge for the new procedures required. On a smaller scale, personnel of other departments also have to be trained. All these training expenses should be part of the investment and accounted for.

3.4   Aircraft Commonality and Economics

Fleet expansion is an important part of economic studies. When there is a choice between adding airplanes of the same type into the fleet or having a new type of airplane with no commonality within the existing fleet, a detailed cost analysis is required.

The commonality could be an advantage for an airline operator. Boeing has a set of airplane families. Airbus established two airplane families, namely, the single-aisle airplane and the wide-body (which can be two- or four-engine airplanes). A smaller airplane manufacturer, Embraer, established a regional jet airplane family. It built the ERJ-145, which has 45 seats, and it built the ERJ-135, with seats for 37 passengers. These small jets can fly at Mach 0.78 and have a range of about 1300 miles.

What is important is that they have 90 percent commonalties (Fig. 3.5). The benefit could be classified into three different groups:

- Same type of ratings means the pilot is permitted to fly any one of the airplanes that has the same rating.
- The high degree of commonality permits a pilot convert to fly another type of airplane by attending a "difference training" class only.
- A multiple qualification permits a pilot to fly any type of aircraft in the same "family," assuming that it satisfies the prescribed requirements.

The following design philosophy made this approach possible:

- Commonality of cockpit layout
- Identical primary flight controls, navigation display, auto pilot, flight director, etc.
- Normal operating procedures, abnormal and emergency procedures, communications, etc.

A full-transition training requires about 25 days. With the preceding approach, it could take from 1 to 9 days only. This cost savings is a significant part of crew training. The cost differences could be large enough to reverse the economic consequences.

Let us examine the benefits of expanding an existing fleet versus purchasing a new type of airplane. There are some benefits when the same aircraft type is added to an existing fleet:

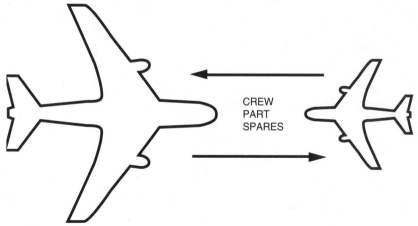

CREW
PART
SPARES

FIGURE 3.5 Commonality.

- Flight operations:
  Reduced crew training
  Less training time
  Less simulator time
  Less flight training
  Less cabin crew training

- Only new hires have to go to a full training session; the existing personnel need a partial course only

- Maintenance:
  Less training time
  Fewer new tools
  Less spares cost
  Slightly increased maintenance burden, if any

There are some estimates that about 30 percent in airframe and about 40 percent in avionics training can be saved.

For a new type of airplane, all these items are more costly and require more spending. The family approach permits a mechanic with two or three licenses to work on all types of airplanes, resulting in reduced management costs, higher productivity, and a more flexible work schedule.

- Depreciation:
  Less ground support equipment to purchase
  Fewer initial spares
  Fewer spares to store
  Fewer personnel to hire

As a percentage of DOC, the introduction of a new type of airplane with no commonality with the existing fleet has the disadvantages shown in Table 3.13. As a function of DOC, the magnitude of penalties is approximate only and would vary for each airline.

TABLE 3.13   DOC penalties for new type of airplanes

Item	Penalty in % DOC
Flight crew	2.6
Crew utilization	3.2
Maintenance + burden	1.2
Initial training	2.1
Spares	1.0
Depreciation	1.3

3.5   Stockage Level

This depends on spare inventory reorder level requirements. Spares planning philosophy has an effect on spare expenses. Expenses are based on

- Main base (rotables, repairables, and expendables)
- Line stations or flight kits
- Insurance items (landing gear, flight control equipment, cowlings)
- General supplies
- Spare engines
- Spare modules
- Engine parts
- QEC kits
- Engine overhaul/repair location and flow time
- Number of months supply of expendables stocked
- Component overhaul flow time
- Spares overhead cost, their definition, and their allocation in accounts
- Spares pooling agreement with other airlines
- Value of spares currently in stock per airplane type

Reorder time defines the level of stockage, and this affects the amount of money tied down in spare parts. This is the area where improvements can be achieved.

3.6   Spares Policy

The type of airplane has a significant effect on spares. For an existing fleet, adding the same type of airplane requires a few additional spares and parts. By having a new type of airplane in the fleet, the number and cost of spares are increased drastically. Figure 3.6 shows the number of aircraft versus spares cost. Route network also has an effect on spares, since each station has to have a certain amount of spares available.

The maintenance program adopted by an airline has an effect on spares when hard-time replaceable components are involved. The level of contract maintenance also has an influence on spares. In order to share common parts, some airlines join pooling agreements. This is a useful arrangement for small airlines and for line station operators.

The best way to apply Fig. 3.6 would be through an example. Providing initial spares for four new airplanes would cost 17 percent

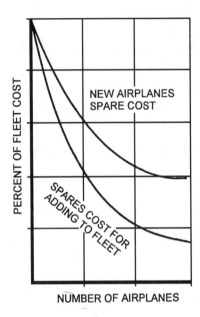

FIGURE 3.6   Spares versus number of aircraft.

NUMBER OF AIRPLANES

above the cost of this fleet. A year later, adding eight similar airplanes to the fleet would cost only 7 percent of fleet cost. In the first case, the reading is done on the line labeled "New Airplanes Spare Cost." Entering the figure a second time on the horizontal scale at 8, one has to move up to the line "Spares Cost for Adding to Fleet," and moving left, the figure indicates 7 percent of fleet cost on the vertical scale.

Below, a typical spares provisioning plan is listed. The quantity is not defined because it will be based on the available and newly purchased aircraft.

Engine	Airframe
Spare engines	Rotable/airline
Engine maintenance spares	Expendable items/airline
Power plant kit for engine buildup	Rotable/vendor
Accessory items for engine buildup not included in power plant kit	Expendable/vendor
Turbine reversers	
Engine nose cowl	
Spare APU (including turbine kit and accessory items)	

3.7   Spares Availability

As a general rule, airlines invest 14 to 16 percent of the value of their fleet in inventories. Airlines are focusing their attention on reducing the value of their inventory in such a way that it would not jeopardize their operation because an airplane on the ground loses money and the goodwill of its passengers and/or shippers in the case of cargo operations. A few problems are listed below:

1. *When should an item be repaired or discarded?* Experience dictates that the cost of repair should be below 62 to 67 percent of the purchase price of the part in question. If the cost of repair is above this limit, the item should be discarded.

2. *Parts are categorized into go or no-go items.* A go item per the airworthiness document is an item where, when takeoff is permitted, the airplane can fly to its maintenance base. In case of a no-go item, the part has to be changed; otherwise, the airplane is not authorized for takeoff. The acceptable dispatch reliability requires at least 85 percent spares availability (Fig. 3.7). There is a certain relationship between the amount of money spent on spares and spares availability that can be expressed in percentages. A typical curve shows this relationship in Fig. 3.7.

At the low-availability percentage area, a given amount of money could increase the spares availability by a relatively high percentage, but the same amount of money at a higher-availability region would increase the spares availability by only a few percentages. This indicates a diminishing return of the invested money because parts have high reliabilities and are expensive. Airlines try to minimize their spare parts inventories and at the same time increase their dispatch

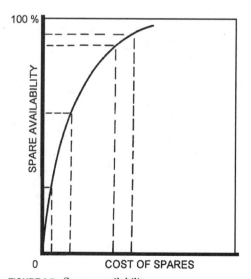

FIGURE 3.7   Spares availability versus spares cost.

reliability. Since 100 percent availability of spares is the ideal solution, airlines are striving for new approaches to solve this problem. The following solutions are possibilities:

- *International Airline Technical Pool. (IATP).* For example, if American Airlines needs a part in Frankfurt, Lufthansa will supply it. There is no charge for this, but it has to be returned within a specified time.
- Manufacturers, in an attempt to make their product more desirable, are helping their customers reduce their inventory. For example, the inventory is shipped in proportion to the aircraft delivery schedule, and this results in cost savings.

Boeing main part distribution center is in Seattle, and other Boeing distribution centers are located in various parts of the world. Furthermore, for airplane-on-ground (AOG) critical parts, Boeing promises to ship parts in between 2 and 24 hours.

3. A spare parts provider offers airlines 100 percent dispatch reliability. It can satisfy this requirement because it has contracts with several airlines and can enjoy the economies of scale. It saves significant amounts of money for the airlines.

4. Another factor that could benefit airlines is their own automatic test equipment. In the case of a false diagnosis, a given part is removed, it stays for a long time in the repair shop, and then is sent back to the airlines without repair. This automatic test equipment contributes to a better use of the parts, lowers failure rates, saves time, and lowers inventories.

The Coordinating Agency for Supplier Evaluation (CASE) is an agency that establishes standards for purchasers. One important feature is 100 percent traceable quality inventory because, unfortunately, there are bogus parts floating around.

3.8 Spares Commonality

The beneficial effect of commonality results from having airplanes be as identical as possible. The amount of incremental spares investment gets smaller for each additional identical airplane added to a fleet. This incremental benefit can be quite substantial (Fig. 3.8) when comparing the addition of a new airplane type with adding to an existing fleet. An example of main-base repairables is shown below.

The fleet consists of 30 type Y airplanes. Main-base repairables for airplane types Y and X are 80 percent common. Incremental investment is $3.0 million. The original investment in spares for five type X airplanes should be $7.0 million.

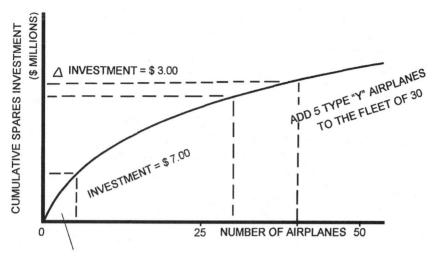

5 NEW TYPE "Y" AIRPLANES

FIGURE 3.8 Spares commonality.

Common parts + uncommon parts = total investment

($3.00 × 0.80) + ($7.0 × 0.20) = $3.80 million

Commonality savings = $3.20 million

Note: For five type X airplanes, the spares cost would amount to $7.0 million. Because of commonality, a savings of $3.2 million is achieved.

Parts can be identified as identical or similar. Similar parts cannot be exchanged. Their advantages include the use of common test equipment, maintenance and shop overhaul procedures, and some identical subparts. This improves troubleshooting capabilities and reduces mechanical delay time.

3.9 Ground Support Equipment

The following applies when additional airplanes are added to a fleet:

- Additional ramp equipment (if needed)
- Tools
- Training equipment
- Evaluate additional workload and decide to do it in-house or contract it to outside company

3.10 Marginal Cost

Previously, direct and indirect operating costs were discussed in great detail. Now, a quick discussion will focus on marginal cost. The basic question is as follows: There is a given size of airplane that, for the sake of argument, has a 100-passenger capacity. Breakeven load is 40 percent, and operational load factor is 60 percent. How would revenue and cost change if an additional 20 passengers were to be carried on this flight?

On the revenue side, it would increase by the price of 20 additional tickets. On the cost side, each element has to be examined carefully and separately. Let's take cost elements individually.

Flying operations. No additional expenses are necessary. The flight crew flies the airplane anyway, and the additional 20 passengers have no effect on crew expenses.

Fuel cost. The additional 20 passengers are equivalent to an additional weight of $[20 \times (165 + 35)] = 4000$ lb. This additional 4000 lb would require an additional fuel burn of $\{[(4000 \times 0.25)/6.7] \times .60\} = \$89$, where 4000 lb is the additional passenger weight, 0.25 is the required fuel factor at the 3000-mile distance, 6.7 is fuel density in pounds per gallon, and 0.60 is the fuel cost in dollars per gallon.

Depreciation. There is no effect on depreciation.

Landing fees. When landing fees are based on airplane weight, this would have a minimal effect on landing fee in the magnitude of a few dollars.

Station costs. Passenger handling, insurance, and food are the expenses that should be considered. Assuming $15 per passenger for these items, the cost would be about $300 (for these 20 additional passengers), and thus a total expense of $400 should be appropriate for these additional 20 passengers. Knowing the additional revenues and the associated cost would furnish the marginal dollar amount.

3.11   Facilities

Management has to know with the introduction of additional airplanes the effect on all phases of operations. Is the hangar large enough for housing a larger sized airplane? Can existing facilities handle the additional workload? Is the hangar door large enough, or does the airplane fit into the hangar and/or ramp area? Is enough parking space available for the aircraft? A chart was developed to present various parking space requirements for the nose in parking, parallel parking, and 45-degree angle-in and angle-out parking. Fifteen feet of building clearance for the nose in parking is required.

Figure 3.9 shows the parking space requirements for a given type of airplane. Would the maintenance department have enough personnel, spares, and tools? Is there a need for a new simulator? Everything has

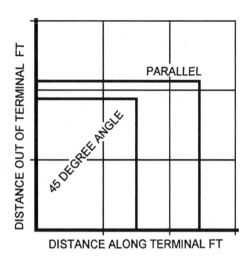

FIGURE 3.9 Airplane parking space requirements.

to be planned up to the smallest detail because when the airplane is delivered, no delay for any reason should be tolerated. Generally, airlines prepare readiness programs where individual managers present the solutions to questions asked by higher management or bring up new problems and their solutions. There are meetings at department levels and progress reports on the problems. The department heads have regular meetings with higher-level management to solve problems when introducing additional airplanes to a fleet.

It should be mentioned briefly that the passenger waiting area has to be planned in such a way that it is not overcrowded and should be able to handle passenger movement in a satisfactory manner.

It is interesting to note that when passenger terminals were designed in the early days of operations, the passenger areas could satisfactorily accommodate the passengers they carried. Now, with higher passenger volume and with the introduction of larger airplanes, passenger areas are overcrowded, and a good many passengers have to stand because not enough seats are available. This makes air travel difficult for elderly people who cannot stand for long periods of time. Airlines are now recognizing this problem because the aging population is slowly increasing.

3.12 Cost Savings by Volume

Purchase prices generally are discounted on the basis of volume. These purchases may include fuel, spare parts, passenger food, and numerous other items.

3.13   Cost Escalation—Inflation

Expenses have the tendency to increase over time. One of the causes is inflation. Planning into the future realistically always should account for the inflation factor. One has to study the various economic indicators prepared by the government and/or other sources projecting inflation rates into the future.

3.14   Foreign Currency Exchange

Airlines dealing with customers paying with foreign currency have to consider the significant effect of exchange fees and fluctuations of foreign currency against the dollar. One of the reasons for a small U.S. international airline's bankruptcy was that its (small) profit was absorbed by the losses suffered from the fluctuation of foreign currency and paying high exchange fees.

3.15   Graphic Presentation of Direct Operating Costs

Before the introduction of computers, airlines developed various charts for the determination of direct operating cost. Table 3.14 is based on the ATA-defined calculation of DOC. The constants in the

TABLE 3.14   Direct Operating Cost based on ATA method

Direct operating cost of transport airplanes	Units to be used on chart	Example
This diagram is used to calculate the direct operating in dollars per mile.	Two sets of units may be used: 1. Block fuel, lb	DC-8 Trip length: 900 statute miles
Any aircraft may use the cost formula as follows:	2. Block speed, mi/h	Block fuel: 28480 lb
DOC = $K_1/V + K_2(F/D) + K_3$ = $/mile	3. Trip length, statute mile	Block speed: 452.2 mi/h
Where K_1, K_2, and K_3 depend on type of airplane, F = block fuel in lb, D = distance in statute miles, V = block speed, mi/h	and/or 1. Block fuel, lb 2. Block speed, knots	K_1 = 468.7 K_2 = 0.0223
Entering the graph with the known values, DOC can be obtained.	3. Trip length, nautical miles	K_3 = 0.044429 DOC = 1.79 $/statute mile
Calculation is based on ATA method of DOC calculation		

ATA formula are functions of airplane characteristics. Logarithmic scales had to be used, and for a quick determination of DOC, this table was satisfactory at that time.

Entering Fig. 3.10 with speed and trip length and moving through several subcharts, the DOC is furnished in dollars per mile. It was a tedious chart reading.

Next, a nomograph is introduced (Fig. 3.11) to define DOC for a specific airplane. This is a quick and simple way to present DOC. There are three parallel lines. One is calibrated for fuel, and the other one for time. For a given trip, connecting the fuel and time values with a ruler, the DOC figure is furnished on the third (middle) scale.

3.16 Revenue and Cost versus Time

So far revenue and cost have been discussed independently of each other. Now consider them together as a function of time (in years). The figures are tabulated in Table 3.15 and are presented in Fig. 3.12.

Typically, in the early stage a company (point AB) operates with low revenues and high costs. This is the time when a company tries to establish itself and builds up a clientele/passenger base.

At point B, the stage is finally reached where revenues are equal to expenses. In the next stage (at C), the company is growing and makes a profit. The time period during B and D could be months or years or many years or decades (Golden Ram Airline, months; PanAm, decades).

This stage actually shows the management capabilities keeping up with the times. Management has to discover the point of diminishing returns when profit starts to decrease. In this case, maybe new management has to be introduced with fresh ideas and new ways to conduct business. One way to interpret the graph just presented is shown in Table 3.15.

In the last few years (around point D), revenue has been leveling off, and cost figures are increasing above revenue. This may indicate that capacity reduction may be the proper way to bring the company back into profitability (Fig. 3.12).

Many airlines made mistakes when business was good and profit was growing; they overextended themselves, and their investment generated only diminishing returns. Not realizing this trend, many companies were driven to make mistakes when business was good and profit was growing; the result frequently was bankruptcy.

Sometimes, after a certain time of operations, the revenue curve dips below the revenue = cost curve. Depending on the circumstances, the company could file for bankruptcy (Chapter 11). Later, it may recover, like America West, or it could file for Chapter 7.

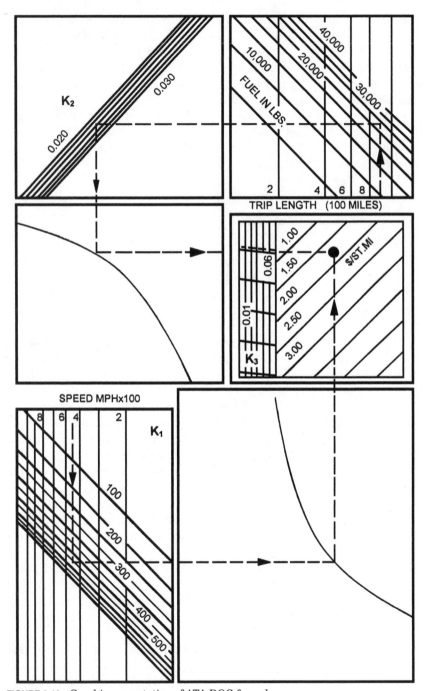

FIGURE 3.10 Graphic presentation of ATA DOC formula.

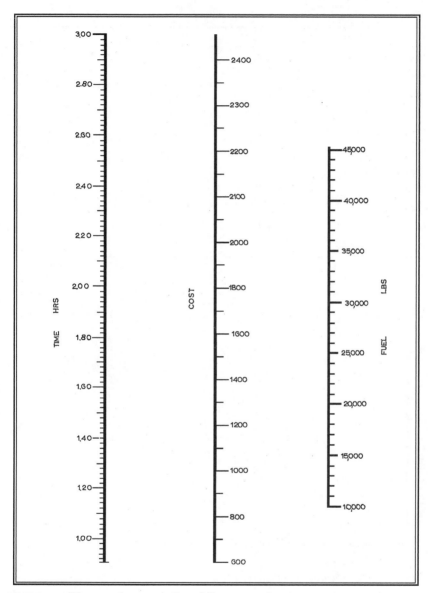

FIGURE 3.11   Nomograph presentation of direct operating cost.

TABLE 3.15   Revenue and cost in million dollars

Year	Revenue	Cost	Remarks
1985	2.0	4.2	Start-up
1986	2.8	4.3	
1987	3.4	4.4	
1988	4.2	4.5	
1989	4.8	4.7	Profitable yr
1990	5.6	4.9	Profitable yr
1991	6.3	5.2	Profitable yr
1992	7.0	5.6	Profitable yr
1993	7.7	6.3	Profitable yr
1994	7.4	7.7	Year with loss
1995	6.5	8.7	Year with loss

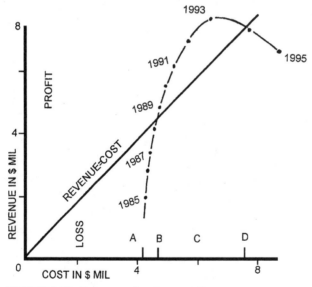

FIGURE 3.12   Revenue and cost presentation.

Chapter 4

Aircraft Economic and Performance Evaluation

4.1 Introduction

Economical and performance evaluations for aircraft were conducted even before the computer age; the presentations were either tabulated or in graph format. Decisions were based on past experience, instinct, and "gut feelings." It was partly science, partly art.

With today's complex computer programs, we can prepare information in numerous sophisticated ways. A complete airline operational and financial simulation will be discussed later. Tables and graphs are still useful in aircraft evaluation, and they deserve to be discussed in more detail.

In this chapter various methods are presented to cover the many aspects of airplane characteristics. There is no need to study all the methods. It is more important instead to have the knowledge to select the best method to do the evaluation you intend.

Evaluation can be divided into two different procedures:

1. Evaluation of an airplane over its range, defining its payload versus range capabilities, cost expressed in various units (dollars/mile, cents per passenger-mile or ton-miles or replacing miles with kilometers), profit potential presented for the range of the airplane, and cost figures for passenger and/or cargo operations.

2. Evaluation of each individual trip. This involves defining various cost figures for passenger and/or cargo operations and defining estimated profit before taxes for certain selected trips.

Here, both methods will be applied to show the many sides of the evaluation process.

4.2 Aircraft Operational Weights Presentation

In Chapter 8, airplane weights are described and defined in great detail. Going one step further, a graphic presentation of weights is shown in Fig. 4.1.

It is very important to know all weights, their definitions, and their limits because this is the basis for any further analyses. In this example, the maximum zero fuel weight of 241,500 lb, the maximum landing weight of 257,000 lb, the maximum takeoff weight of 281,000 lb, and the ramp weight of 283,500 lb are the legal limits of this airplane and are listed (generally) in Section 1 of the Federal Aviation Administration (FAA)–approved flight manual. These limits cannot be exceeded. On the vertical weight scale, the first weight shown is the operating weight empty. Adding load, the maximum zero fuel weight limit could be reached. So far, the airplane has zero fuel on board; from this point on, only fuel can be added. The reserve fuel is the fuel located between landing weight and zero fuel weight, and it is shown as 15,500 lb. In the case that the reserve fuel is 23,000 lb, it will decrease the zero fuel weight and load because the landing weight limit cannot be exceeded. The trip fuel (block fuel = trip fuel + taxi fuel) is located between take-off weight and landing weight. When more trip fuel is required (for a longer-range)

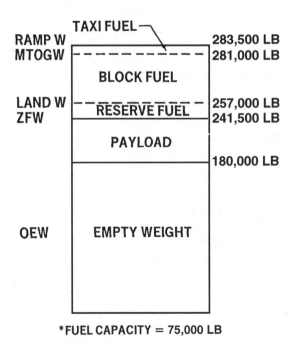

*FUEL CAPACITY = 75,000 LB

Figure 4.1 Aircraft weight definition.

the maximum takeoff weight cannot be exceeded. By adding more trip fuel, the landing weight has to be lowered to accommodate the increasing amount of trip fuel. This would decrease zero fuel weight and load. The block fuel is defined between landing weight and ramp weight. (To be precise, the taxi-in fuel is included in the block fuel, but it is taken from the reserve fuel. Once the airplane has landed and is on the ground, there is no need for the reserve fuel.) When the trip fuel is, for example, 35,000 lb, the takeoff weight will remain unchanged (the maximum of 281,000 lb). The landing weight will be changed (reduced) to 246,000 lb, zero fuel weight (23,000 lb reserve fuel) to 223,000 lb, and the load to 43,000 lb only. In conclusion, the legal limits were observed.

4.3 Load

When the word *load* is mentioned in connection with an aircraft, it may mean either cargo or passengers.

4.4 Passenger Load

This load is based on the number of seats in the aircraft multiplied by the sum of the passenger weights and baggage weights. Generally, the industry uses an average weight of 165 lb per passenger and 35 lb for baggage per passenger; however, each airline can establish its own averages in its load calculations.

Figure 4.2 shows the increase in the number of passengers carried by each airplane as a function of range. It shows the rapid growth of air transportation in the past few decades.

4.5 Cargo Load

This load consists of two elements: the cargo itself and the tare weight. Tare weight is the weight of pallets and nets and/or containers. For example, a 90,000-lb load could consist of an 85,500-lb payload and 4500-lb tare weight, assuming (DC-8-73) 18 pallets at 250 lb each. The part of the load that the shipper (customer) pays for is called the *payload*. Load includes payload and tare weight:

$$\text{Load} = \text{payload} + \text{tare weight} \qquad (4.1)$$

If tare weight is 0,

$$\text{Load} = \text{payload} \qquad (4.2)$$

Note: Evaluating an airplane's load potential in cargo configuration, the full load is considered with zero tare weight. Airlines then make

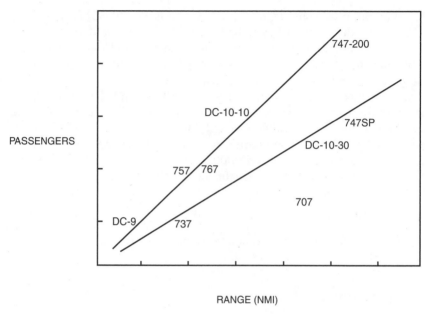

RANGE (NMI)

Figure 4.2   No passengers versus range for a few selected airplanes.

adjustments for the applicable tare weight, and this may vary per airline, per airplane, and per type of load and various combinations of pallets and/or containers.

4.6   Required Data for Conducting Aircraft Economic and Operational Studies

Airline higher management reviews from time to time its existing fleet from the standpoint of profit, marketing, equipment, and efficient use. For example, an operator having a fleet of B767 aircraft would like to explore the economics of a larger airplane (B747) and its economic and operational consequences. At the conclusion of the studies, the operator may decide to buy the B747 aircraft or additional B767 aircraft, buy some other type of aircraft, or just keep the status quo.

After evaluation of the economic and operational factors, the study could be expanded or terminated. Frequently, an airline's management conducts economic studies for various types of aircraft to familiarize themselves with their economics without any intention to buy any aircraft. Sometimes it is necessary to evaluate the competition's equipment as well as its economics and performance.

In order to prepare these studies, middle management has to supply upper management with economic, technical, and marketing information in a certain sequence:

1. Economic and operational assumptions applicable to the study
2. Aircraft general information and weight summary
3. Payload versus range curve
4. Direct operating cost in dollars per mile
5. Direct operating cost:
   a. Cents per passenger-mile
   b. Cents per ton-mile
6. Breakeven load factor presentation
7. Miscellaneous items:
   a. Trip data for selected routes
   b. Marketing forecast
   c. Forecast for ground equipment development
   d. Additional miscellaneous information

In the following pages, airplane performance and airplane cost figures will be discussed in greater detail. When management contemplates fleet expansion and a choice has to be made between adding the same type of aircraft to the existing fleet or acquiring a new type of aircraft, various airplane performance characteristics and economic factors are compared.

In order to prepare such studies, operational and economic factors must be defined. Before any calculation is made, the basis has to be established, spelling out all the conditions applied. Below, a list of the assumptions for such studies is presented:

Date of study: Feb. 2002

Aircraft: Name (B767 or Airbus, etc.)

Engine: Name (P&W, GE)

Year of delivery: 2002

Engine thrust: lb

Airplane price: $

Cost of buyer-furnished equipment: $

Engine price: $

Cost of spare parts: $

Spare allowance:
   airframe: %
   engine: %

Prepayment schedule: Years before delivery

Amount of prepayments per schedule: $

Prepayment interest rate: %

Yearly utilization: Trips per year or block-hour per year

Interest rate: %

Duration of loan: Years

Payment schedule: Years

Lease rate: % (if applicable)

Lease rate per month or year: $

Depreciation: x years to y% residual

Fuel price: Cents per gallon

Weight summary:

  Max. ramp weight: lb
  Max. takeoff weight: lb
  Max. landing weight: lb
  Max. zero-fuel weight: lb
  Operating weight empty: lb
  Engine weight: lb

Aircraft configuration:

  Passenger
  Convertible
  Freighter
  Combi

Passenger operation:

  Max. payload passenger operations:

    Passengers: Number
    Total passenger and baggage weight: lb
    Cargo weight (if any), upper and/or lower deck: lb
    Tare weight: lb

  Freight operations:

    Max. payload: lb
    Max. cargo weight: lb
    Tare weight: lb
    (Pallets and/or igloos and/or containers): Number
    Volume upper level: ft^3
    Volume of lower level: ft^3
    Bulk volume (if any): ft^3

Cost factors:

  Direct operating cost:

    Flying operations: $/block-hour
    Fuel and oil*: $/block-hour

---

*This can be based on system averages or calculated for each individual trip based on fuel price at the departure station.

Maintenance and burden: $/block-hour
Depreciation/rents: $/block-hour
Insurance: $/block-hour
Interest: $/block-hour
Indirect operating cost:
Passenger service
Flight and traffic service
Reservation and sales
Ground equipment maintenance and depreciation
Nonflight items
Transport-related items
Advertising and publicity
General administration

Note: Sometimes indirect operating cost is expressed as a percentage of direct operating cost, in dollars per block-hour or only in dollars per trip. The indirect operating cost items are usually not used in the evaluation process.

Operational factors:
Cruise policy: Mach number
Reserve: International/domestic/island
Alternate distance*: Nautical miles
Enroute wind: Percentage of wind probability
Temperatures:
Airport[†]: 84°F
Enroute: Standard for winter, standard + 20°F for summer[†]

Note: These variables indicate the complexity of this report and should be well established before any presentation takes place. For example, in the financial section, assuming a 10 or 12 percent interest rate could change the outcome of the financial conditions. Similarly, in the operational section, changing the cruise policy from Mach number 0.80 to 0.83 could result in significant changes. Dating the report would indicate the applied tax and depreciation laws. It is important to clearly define the basic assumptions because the final conclusions are based on these basic assumptions.

Looking back into the past, it is interesting to note the changes in economic assumptions based on various events, such as political effects (fuel prices), changes in tax laws (depreciation), changes in insurance policies (reliability), annual utilization (improved scheduling), and inflation. Table 4.1 examines three time periods.

---

*Generally, 200 nautical miles international and 150 nautical miles for domestic operations, when not specified.
†Higher in the tropics.

TABLE 4.1   Historical Data

Items	Before 1973	After 1980	After 1990
Spare allowances	9.70%	6.00%	6.00%
Annual utilization	3650 block-hours	3780 block-hours	3900 block-hours
Insurance	3.00%	0.50%	0.7–0.85%
Depreciation/residual	12 yrs/5%	15 yrs/10%	20 yrs/10%
Fuel price	10.71 cents/gal	90 cents/gal	50–60 cents/gal

4.7   Introduction to Payload versus Range Curve

The payload versus range curve is presented in Fig. 4.3. It is prepared by calculating the load for various distances and plotted as a function of range. The importance of this chart lies in the fact that the payload versus range curve reflects the airplane's economic potential because each pound of payload is generating revenue. This chart is used to compare various types of airplanes.

The maximum payload is constant up to point 1. This is the load between maximum zero-fuel weight (MZFW) and operating weight empty. This load is limited by the structural limitations of the maximum zero fuel weight. Point 1 is the maximum range of the maximum payload. For trips between point 1 and point 2, the load slightly decreases.

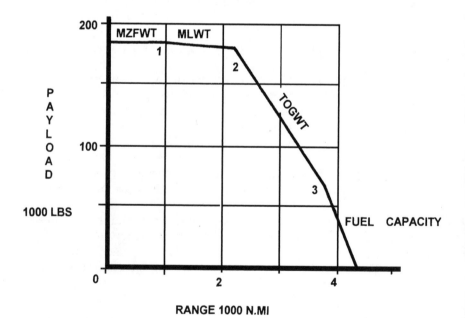

Figure 4.3   Payload versus range.

This is caused by the maximum landing weight (MLW) limit. As the reserve fuel is slowly increased with range, it will exceed (in this example) the weight difference existing between maximum landing weight and maximum zero-fuel weight. Landing weight cannot exceed its maximum limit, and the legal amount of reserve fuel has to be carried. The zero-fuel weight and the load are the only two items that can be decreased when the reserve fuel is more than the weight differences between the maximum landing and maximum zero-fuel weight. When the reserve fuel is less than the weight difference between the landing weight and the zero-fuel weight, the payload is maximum and constant until it reaches point 2. In other words, point 1 moves toward point 2. Figure 4.4 shows this situation when point 1 collapses into point 2.

Point 2 becomes the range of maximum payload. At point 2, the airplane takeoff weight (TOW) has reached its maximum limit. Beyond point 2, payload is traded for fuel in order to move the airplane further to its destination. Beyond point 3, fuel capacity limits airplane performance until the payload decreases to zero (lower payloads require less trip fuel). In addition, the tank capacity limit has to be observed.

Note: Sometimes the maximum payload range point is called the *knee point* (2). Generally, the point on the payload range chart is also called *knee point* at number 3 in Fig. 4.3, where the (taxi-out fuel, trip fuel, and reserve fuel) total fuel is equal to the fuel tank capacity.

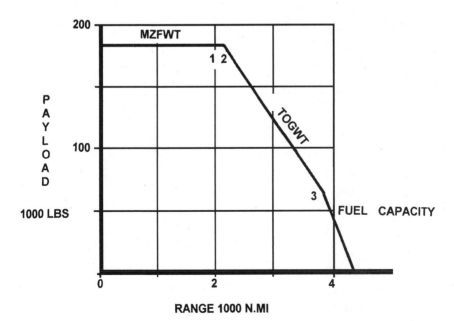

Figure 4.4   Payload versus range.

When the payload range curve is presented, the following information is needed:

Weight summary, containing
  Max. ramp weight
  Max. takeoff weight
  Max. landing weight
  Max. zero-fuel weight
  Operating weight empty

For passenger operations:
  Max. seat capacity
  Lower-level load capacity, if used

Freight operations:
  Max. load potential
  Payload
  Tare weight

Airplane performance:
  Cruise type:
    Long-range cruise
    Mach 0.77 up to 0.88
    320 kias, etc.
  Reserve fuel:
    Domestic
    International
    Island reserve

Note: When no suitable alternate is available, a reserve fuel equivalent to 2 hours of fuel flow is applied.

  Alternate:
    Based on alternate distance
    Or (if alternate not specified) 150 or 200 nautical miles

Engine:
  Engine type
  Number of engines
  Thrust per engine
  Temperature at takeoff airport:
    Standard
    Standard plus 10, etc.
  Enroute temperature:
    Standard
    Standard plus 10, etc.

Enroute wind reliability
85%
75%, etc.
Yearly/seasonally/monthly

Note: Date of preparation is an integral part of the whole report, and without it, the report could lose its validity.

Any change in any of the input data would invalidate the payload range curve and require the preparation of a new one. For example,

- Operating weight empty may change for various reasons.
- Seat capacity could become lower or higher.
- In freight operation, tare weight could change due to, e.g., going from palletized to containerized operation or into a mix of pallets and containers in any kind of combination.
- Cruise policy could change from, e.g., Mach 0.80 to Mach 0.83.

In Fig. 4.5, three small charts are developed to illustrate graphically the different phases of the payload versus range chart shown in Fig. 4.3. From zero distance up to point 1 in Fig. 4.3, the corresponding picture is presented in Fig. 4.5A.

- Range is increasing to point 1.
- Payload is constant and maximum.

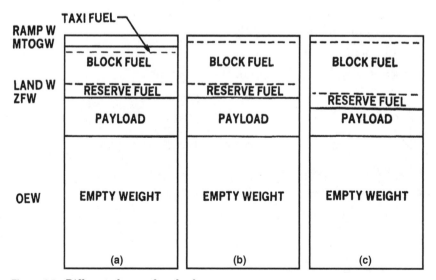

Figure 4.5 Different phases of payload versus range.

- Reserve fuel increases slightly.
- Block fuel increases.
- Takeoff weight increases (still below structural limit).
- Landing weight is at maximum or below.
- Zero-fuel weight is at maximum.

Moving from point 1 to point 2, the following situation exists (Fig. 4.5*B*):

- Range is increasing to point 2.
- Payload decreases slightly.*
- Reserve fuel increases slightly.
- Block fuel increases.
- Takeoff weight approaching structural limit.†
- Landing weight is at maximum or below.
- Zero-fuel weight is at maximum or below.

Note: When the spread between landing weight and zero-fuel weight can accommodate the reserve fuel, then point 1 moves to point 2, and payload constant is up at this point. This is the maximum range with full payload.

Moving from point 2 to point 3, the following situation exists, as shown on Fig. 4.5*C:*

- Payload decreases (to exchange payload for fuel to obtain additional range).
- Reserve fuel increases slightly.
- Block fuel increases.
- Takeoff weight is at maximum and constant.
- Landing weight decreases (as block fuel increases).
- Zero-fuel weight decreases (as block fuel increases).

Moving from point 3, the following changes take place:

- Payload decreases to zero.
- Reserve fuel is constant or decreases slightly.
- Block fuel increases.

---

*If reserve fuel is more than the difference between max. landing and max. zero-fuel weights.
†At point 2, the takeoff weight reached its structural limit (Fig. 4.5*B*).

- Takeoff weight is at maximum and constant.
- Landing weight decreases (as block fuel increases).
- Zero-fuel weight decreases (as block fuel increases).

Note: When the amount of block and reserve fuel is at the limit of tank capacity, the trip cannot be continued until the payload is reduced. Less payload reduces airplane landing weight. For the same trip length, the reduced airplane landing weight requires less block fuel. Finally, the payload will be reduced to zero.

The preceding detailed explanation is necessary because many decisions are made based on the information furnished by payload versus range charts. The next chart presents takeoff weight versus range in conjunction with payload versus range charts. As range starts to increase (see Fig. 4.5A), so do trip fuel and takeoff weight. At a certain range, the takeoff weight reaches its maximum, legal, certified limit defined in the FAA-approved flight manual, and from this point on, the takeoff weight is constant. This point is the range of the maximum payload. As range increases, takeoff weight has to be constant, but trip fuel increases and gradually takes the place of payload. Let us say that the airplane is carrying the maximum of 273 passengers and baggage and a certain load in the lower deck. As the range is increasing from the maximum

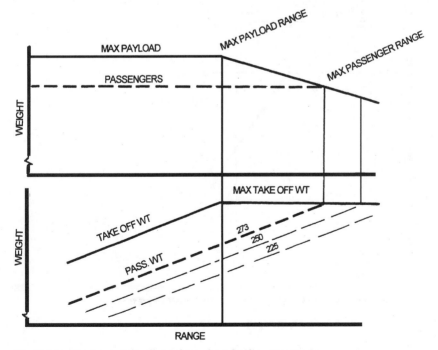

Figure 4.6 Maximum takeoff weight and payload versus range.

payload range, first the lower-level cargo is decreased down to zero. This is the point where the 273-passenger line hits the takeoff-weight line. Beyond this range, the passenger load starts to decrease in order to have enough fuel to reach the destination (Fig. 4.6).

Let us say that for a given aircraft, the maximum payload range point is 2243 nautical miles and the trip has to go from JFK to London (from JFK, 3000 nautical miles). The plane has to fly an additional 757 nautical miles further from the maximum payload range point. The required trip fuel (from JFK to LHR) loaded on the airplane lowers the landing weight, zero-fuel weight, and payload. It should be mentioned that the operating weight empty cannot be changed, and the maximum takeoff gross weight cannot be exceeded. See Fig. 4.5C for this condition. This causes the load to decrease, and the fuel replaces this part of the load in order to move the airplane to its destination (beyond the maximum payload range point).

In this case, a graphical presentation is very useful (Fig. 4.7 and Table 4.2).

There is another limitation, namely, the fuel tank capacity. When the amounts of ramp or taxi-out fuel, trip fuel, and reserve fuel are the same as the tank capacity, the airplane is at the end of its range. This is called the airplane's *maximum range*. This range can be extended by adding auxiliary fuel tanks.

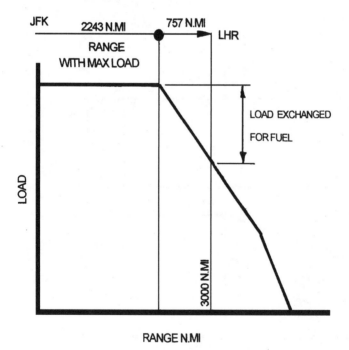

Figure 4.7 Payload versus range for JFK to London trip.

TABLE 4.2   Weights Affected By Range

Range	Up to 2243 nautical miles	Beyond 2243 nautical miles
Takeoff weight	Increasing	Constant*
Payload	Constant	Decreasing
Block fuel	Increasing	Increasing†
Landing weight	Constant	Decreasing
Zero-fuel weight	Constant	Decreasing

*When constant, it has reached its maximum limit.
†Tank capacity has to be observed.

4.8   Payload Range Curve for Regional and General Aviation Aircraft

A payload versus range curve is presented in Fig. 4.8. It is for a small, short-range passenger airplane. The airplane's maximum range is 2400 nautical miles.

This curve indicates that the airplane is able to carry 60 passengers and, for example, 11,000 lb of cargo up to 1800 nautical miles. The range could be extended to 2200 nautical miles, and the airplane is capable of carrying the original 60 passengers without any additional cargo load. This 11,000 lb of load was exchanged for fuel in order to fly the airplane beyond the 1800 nautical miles. The number of passengers is decreased, and their weight is exchanged for fuel in order to move the airplane further to the end of range that is, in this case, 2400 nautical miles. At that point, the payload is zero.

Reviewing the payload range curve of this airplane, the maximum cargo potential is 34,000 lb that the airplane can carry up to 750 nautical miles. Assuming that the airline operates on a route network with an average trip length of 800 nautical miles but no trip length beyond 2200 nautical miles, this airplane (Fig. 4.8) would be a good candidate for this range. Naturally, further study would be needed to evaluate other operational and economical characteristics.

Note: Introducing a one-stop operation would extend the range of the airplane, but it would require two landing fees, two station costs, etc., and lost time and may not make it economically feasible.

4.9   Range Consideration for Regional and General Aviation Aircraft

Data for two selected regional aircraft are shown side by side in Table 4.3. One is a short-range airplane, and the other is a long-range airplane.

As far as the range of a given route structure is considered, the following can be stated: The cents per seat-mile value is the lowest for a turbo-prop airplane at the range of up to 300 to 350 nautical miles versus a

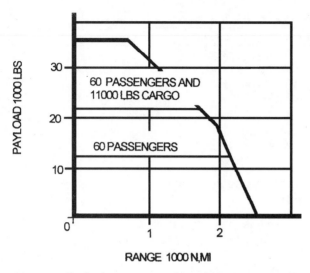

Figure 4.8 Payload versus range for a short-range aircraft.

TABLE 4.3 Regional aircraft data

Airplane type	Range	Takeoff weight	No. of passengers
Embraer ERJ-190-100LR	2300 nautical miles	105,500 lb	98
Fairchild 328-110	455 nautical miles	14,100 lb	19

high-speed turbo-prop or a (regional) jet airplane. The best economical range for a high-speed turbo-prop is around 300 to 450 nautical miles. A longer range is economical for a (regional) jet airplane. The reason for this is the higher productivity of the jet airplane; it can carry more loads (passengers and/or cargo) with higher speed. The average route length has an important influence on the type of airplane that will be selected. On a route system, for example, where the average sector length is 170 nautical miles and the range of the trip length lies between 95 and 290 nautical miles, no one would buy a B757 airplane. For this range, an ATR-72 or SAAB-340 would be considered. A typical example is presented in Fig. 4.9.

The next presentation shows a route system where routes are grouped into short- and longer-range routes. For the short-range routes, a small turbo-prop airplane would be ideal, but for the longer-range routes, a different longer-range-capability airplane would be satisfactory. This indicates that the route structure by itself has a certain effect on the type of airplane selected. There is not one type of airplane that could satisfy both types of route systems economically (Fig. 4.10).

RANGE 100 N.MI

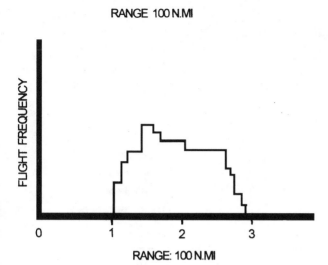

Figure 4.9   Route length distribution.

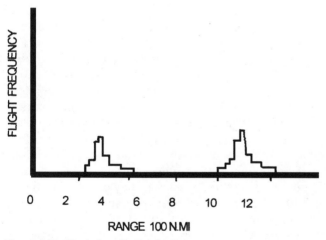

Figure 4.10   Route length distribution.

This example presents the relationship between range and types of airplanes that exist and should be considered.

4.10   Large Aircraft Payload versus Range Curve

The next type of payload versus range curve was prepared for large cargo airplanes. The B747 airplanes are powered with JT9D-7A and JT9D-70A engines (Fig. 4.11).

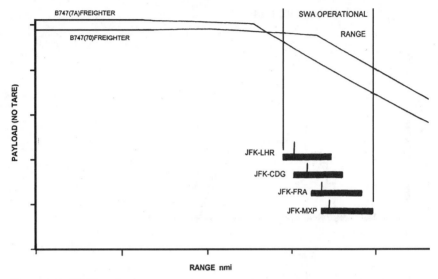

Figure 4.11   B747 payload versus range.

Figure 4.11 was prepared by the now-defunct Seaboard World Airlines to present the payload capabilities on its four main routes. For this operation, the B747-200 aircraft was selected, powered by JT9D-70A engines. The payloads for these trips are shown in Table 4.4.

In Fig. 4.11 for the B747 powered with JT9D-70A engines, the maximum payload range is around 2000 nautical miles. Payload is decreased slightly beyond this point until it reaches about 3400 nautical miles. This indicates limitations due to maximum landing weight. Reserve fuel starts to exceed the spread between maximum landing and maximum zero-fuel weights around 2000 nautical miles. Beyond 3400 nautical miles (see Fig. 4.3 and point 2) the decrease in payload is significant because payload is traded for fuel to extend the range (see Fig. 4.5C). This figure shows the full potential payload, since tare weight is considered zero. Once the tare weight is established for the airline's particular operation, the true profit potential could be defined.

Note: Subtract tare weight from payload.

In domestic operations or up to about 2600 nautical miles, this airplane carries around 10,000 lb (cargo) less weight due to its higher operating weight empty than the B747 powered with JT9D-7A engine.

Since Seaboard's revenues were derived from long-haul trans-Atlantic operations, the choice of aircraft was the B747-200 with the JT9D-70A engines due to better performance in the operational range of the aircraft.

Table 4.4 and Fig. 4.12 show the payload-carrying capability for the selected operational range.

TABLE 4.4   Payload for Selected Trips

	Payload, in lb	
	Eastbound	Westbound
JFK–LHR	240,000	228,000
JFK–CDG	239,000	220,000
JFK–FRA	238,000	207,500
JFK–MXP	232,000	200,000

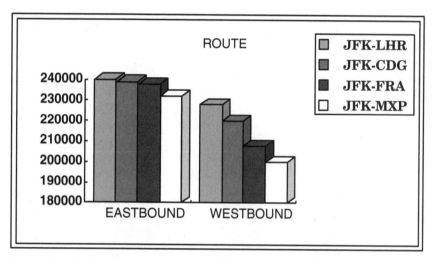

Figure 4.12   Graphical presentation for selected trips.

Eastbound loads are higher than westbound loads due to the enroute wind effect. The trips from JFK to LHR and JFK to CDG are located on the payload range curve before the payload is traded for fuel (see Fig. 4.3 before point 2). The trips for JFK to FRA and JFK to MXP are beyond the range of the maximum payload.

Table 4.4 and Fig. 4.12 are based on a yearly 85 percent wind relia-bility. The now-defunct Seaboard World Airlines had four major trips to Europe, as indicated on the payload versus range curve (Fig. 4.11). Each heavy black horizontal strip shows the range of operational pay-load values of each trip with wind effect. The JFK to LHR trip is described in more detail. The distance at zero wind is 3000 nautical miles, as indicated by the short vertical line on the top of the heavy horizontal black strip on the payload versus range curve (Fig. 4.11).

The magnitude of the wind effect is shown in Fig. 4.13. Wind effect is considered with 85 percent probability of not exceeding the value of 54 knots headwind for the London to New York trip and 21 knots tailwind for the New York to London trip on a yearly basis. (In case of headwind, more fuel is burned than for zero wind, and in case of tailwind, less fuel is burned than for zero wind.) This wind effect can be translated into "equivalent distances," as shown in Fig. 4.13. At 54 knots headwind and 21 knots tailwind, a wind scale can be positioned on the wind-effect strip. On the left side, tailwind is shown, whereas headwind is shown on the right side. Due to the characteristics of the payload versus range curve, the break point is located at 35 knots headwind for the LHR to JFK trip. Beyond this point, payload is traded for fuel in order to continue the trip. The operational wind values will be between these two extreme points of 54 knots headwind and 21 knots tailwind with an 85 percent probability of not exceeding these wind magnitudes.

One day the trip from JFK to LHR could have an 8-knot tailwind, whereas from LHR to JFK a 17-knot headwind may be experienced. On another day, different wind values will be observed. Eastbound trips generally have tailwinds, whereas westbound trips experience headwinds.

To illustrate the usage of this chart, an example is presented:

Trip: London to New York

Enroute wind: 45-knot headwind

What payload can be expected?

Enter Fig. 4.13 and move to the wind scale of 45 knots of headwind. Move up to the payload versus range curve chart and on the left-hand side the payload scale reads 232,000 lb. Figure 4.13 is a detailed, enlarged section of Fig. 4.11.

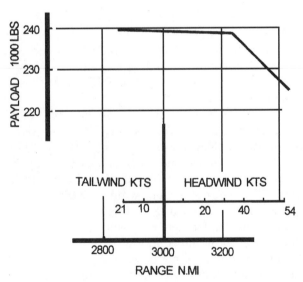

Figure 4.13 Wind effect on range and payload.

Note: JFK-LHR zero-wind distance is 3000 nautical miles. In case of a 45-knot headwind, the fuel burn would be equivalent to a trip length of 3340 nautical miles at zero wind. At this distance, the reading would be a payload of 232,000 lb.

To calculate equivalent distance, the following equation can be used:

$$eqd = \frac{tdz \times as}{as + wr} \tag{4.3}$$

where eqd = equivalent distance
tdz = trip distance at zero wind
as = airplane speed at zero wind
wr = selected percentage reliability equivalent (enroute) wind

Note: As an example, take a trip distance of 3000 nautical miles and an airplane average cruising speed and headwind of 45 knots (at 85 percent yearly reliability). The result is an equivalent distance of 3340 nautical miles.

By using Eq. (4.3) for any other distance and wind, the equivalent distance can be calculated.

For passenger operations, a similar presentation (Table 4.4) can be used, as shown in Fig. 4.14. Here, besides passengers, an additional cargo capability is available. A typical round trip is presented here.

On the eastbound flight, the full number of passengers and a certain amount of cargo are carried. On the westbound flight in this case, the full number of passengers is still carried, but the additional cargo volume is small due to the high headwind.

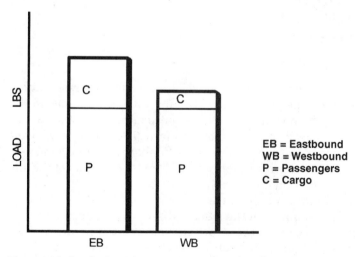

Figure 4.14   Load comparison-passenger operations for a given trip.

This could be a flight for JFK to MXP. Preparing similar charts for another type of airplane, a comparison could be presented easily. The airplane with the highest payload has the best load-carrying capability.

Studies, comparisons, and/or evaluations frequently are conducted for individual trips. The assumptions, conditions, and any other basic airplane performance data and, if necessary, economic and financial data should be furnished:

Date of study: Dec. 20, 2001

JFK to/from Milan

Distance: 3440 nautical miles

Standard day performance

Cruise: Mach .84

Reserve fuel: International

Alternate: 200 nautical miles or any suitable alternate airport

Enroute wind considered

Reliability: 85 percent yearly

Engine name

The following weights should be listed:

Ramp weight

Takeoff weight

Landing weight

Zero-fuel weight

Operating weight empty

Note: These and similar notes should be part of any presentation. Without these notes, the presentation loses its value. Any change in the assumptions has a significant effect on the outcome of these studies. For example, a cruise speed of Mach 0.80 or Mach 0.84 has a significant effect on fuel flow and the cost of operations. Similarly, enroute wind reliability and/or various operating empty weight or other weights of the airplane have an effect on fuel flow.

Figure 4.15 presents the departure airport's temperature effect on payload. The vertical scale (on the left-hand side) is the takeoff weight, and the horizontal scale is the departure airport temperature. The heavy horizontal line on the graph represents the maximum takeoff weight (Max TOWT), i.e., the certified maximum takeoff weight per

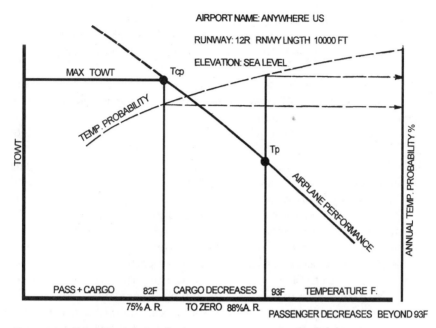

Figure 4.15   Takeoff weight versus departure airport temperature.

the FAA-approved flight manual and the airplane performance limit line (slanted line on the right-hand side). On the (vertical) right-hand scale, the temperature probability of the temperature not being exceeded is shown in percent. At a colder temperature than T_{cp} (where T_{cp} is the temperature associated with the maximum load capability; beyond this temperature, the load is decreasing), the runway length (e.g., Anywhere Airport USA and runway 12L) would permit a takeoff weight that is above the certified takeoff weight, but the certified limit must be observed. For example, up to T_{cp} this airplane may carry 263 passengers and 30,000 lb of cargo. Assuming 205 lb per passenger and bags, a total of 83,915 lb (53,915 + 30,000 lb) is carried.

Beyond temperature T_{cp}, the takeoff weight decreases, and so does the payload. As temperature increases to T_p (T_p is the temperature where the load is limited to the full load of passengers and baggage only), the cargo is diminished to zero. At that temperature, the airplane is still carrying the full load of passengers. Beyond T_p temperature, the passenger load is decreasing. The probability of reaching T_p is small. In domestic operations, the passenger-carrying capability would not be affected significantly by high temperature (see the probability scale).

These temperatures, T_{cp} and T_p are variables and depend on airport characteristics and airplane performance.

At certain airports (mostly in the tropics), it is worthwhile to study the temperature effect on operation because the daily high temperatures of these airports are close to the airplane maximum operating temperature limit.

Preparing similar charts for other types of airplanes, a comparison would show the takeoff performance for each airplane for the same airport-runway combination.

To show the significance of the graph in Fig. 4.16, a trip is selected, e.g., New York to San Francisco. The airplane is able to carry its maximum load, which consists of 250 passengers and 10,000 lb of cargo. The airplane is carrying this load between the temperature range of 0 and 82°F. As the departure airport temperature rises (beyond 82°F), the airplane takeoff weight decreases (so does cargo), as shown in Fig. 4.15. When the temperature reaches 93°F, absolutely no cargo can be carried on the airplane. Beyond the high temperature of 93°F, the number of passengers must be decreased. This is not a problem in our domestic operations due to the low probability of reaching this critical high temperature. However, in the tropics, it would place some limitations on airplane performance. As temperature increases, the airplane can carry less cargo or fewer passengers, as shown on the temperature and passenger scales.

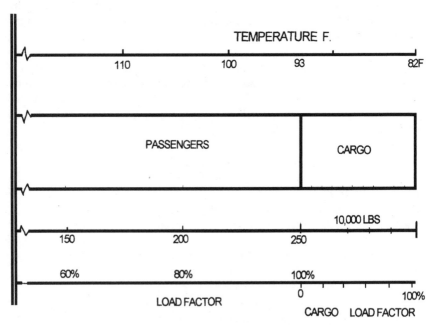

Figure 4.16 Temperature effect on cargo and passenger load for a selected trip.

To sum up, as temperature increases from 82 to 93°F, cargo load decreases to zero. At 93°F, the airplane is able to carry the maximum passenger load of 250. The number of passengers that can be carried decreases beyond 93°F. As temperature increases, the probability of not exceeding these temperatures increases. In other words, it is less probable that the airport (departure) temperature would climb that high.

Let us assume that the airplane is carrying 10,000 lb of additional cargo and 250 passengers. It can be calculated that for each degree beyond 82°F the cargo decreases by 909 lb. For example, at 87°F, the load would be decreased by (87 − 82) × 909 lb = 4545 lb, or the remaining cargo on the airplane is only 10,000 − 4545 = 5455 lb. Its load factor is 54.55 percent, and this amount of cargo will generate revenue.

Note: To be very precise, assuming that one pallet is necessary to carry these 10,000 lb of load, the pallet weight of 250 lb should be considered and the revenue should be adjusted accordingly.

Let's consider 100°F. What can the airplane carry at this temperature?

The chart indicates that the airplane cannot carry cargo at all and is not carrying the maximum number of passengers either.

It was established that each degree Fahrenheit corresponds to a load reduction of 909 lb. The difference in temperature between 93°F and, in our case, 100°F is 7°, and this multiplied by 909 lb comes to a total of 7 × 909 = 6363 lb, which is the decrease in weight due to the higher temperature. Considering 210 lb for each passenger and luggage, 6363/210 = 30.3, or 31 passengers should be subtracted from the maximum of 250, and as a result, 219 passengers can be carried at 100°F.

This exercise indicates the importance of knowing how to get the most out of the airplane cargo/passenger-carrying capability. Where a ticket price of $300 and 312 trips per year is considered, just losing one passenger per trip could amount to $300 × 312 = $93,600. For more airplanes, this loss in revenue could add up to a large amount.

Note: Payload versus range curves for the same aircraft may vary from airline to airline based on specific operational conditions. For a domestic airline, the reserve fuel requirement is different from that for international operations. The selected cruise policy could vary from airline A to airline B (e.g., long-range cruise versus Mach 0.84). One airline may have a heavier operating weight empty (Fig. 4.17) due to its higher buyer-furnished equipment. Different engines on the same type of aircraft (PW versus GE) may have different weights and fuel flows. All weight changes have an immediate effect on the weight of the payload. Furthermore, in many instances, the same aircraft type manufactured a few years later will have a higher manufacturer weight empty due to additional technical improvements. This growth could be different for

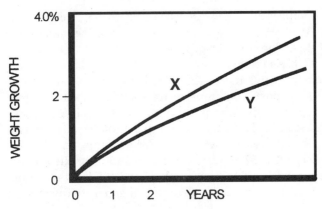

Figure 4.17 Manufacturers' empty weight growth versus time in years.

aircraft x or aircraft y. The same type of aircraft may have different configurations, since one could be palletized and the other containerized, and the tare-weight differences could be up to some 20,000 lb. Similarly, identical passenger airplanes could have different numbers of seats available based on company policy.

4.11 Airplane Range Capabilities Presentation

Frequently, operators are interested in the range of an airplane and ask questions such as how far can an airplane fly from Newark Airport? By presenting a chart with the action radius of the airplane, a good approximation is furnished. The two radii, namely, from, for example, Newark and Los Angeles, are not the same because headwind (Newark) and tailwind (Los Angeles) effects are considered (Fig. 4.18).

Figure 4.19 shows a similar presentation. Here, the arrow from EWR shows a scale where various payloads and their ranges are indicated. It can carry passengers a certain distance. This airplane carries 175 passengers to Kansas City, but it can only carry 79 passengers to Seattle, Los Angeles, or San Diego. The advantage of this presentation is to have a good graphic picture of the airplane's capability as far as range is concerned.

A very similar presentation could be prepared for cargo operations, where, for example, the figures 175, 122, and 79 could stand for cargo in thousands of pounds.

Another presentation permits a quick evaluation of the range capability of the airplane. The runway length for a given airport elevation and temperature define takeoff weight, and these, in turn, define the

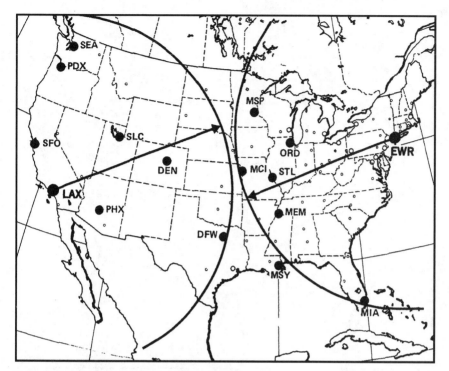

Figure 4.18 Aircraft maximum range capabilities.

airplane's range capability. This is shown in Fig. 4.20. For example, a 10,000-ft runway permits a takeoff weight of 390,000 lb, and with this takeoff weight, the airplane is able to carry a certain load to a range of 4200 nautical miles.

Modifying the preceding chart, a new chart can be developed where runway length versus range is shown (Fig. 4.21). Runway length available always should be longer than runway length required.

The takeoff weight is the one that changes the most. Takeoff weight is affected by many conditions, such as runway length, runway slope, airport elevation, airport temperature, obstacles, trip length, and enroute wind. All these factors are considered when establishing runway length required for the preceding conditions, which are based on the FAA-approved flight manual.

In the upper left-hand corner of Fig. 4.21, the runway length available is indicated for Amsterdam Airport. The required runway length for this type of airplane is less than the runway available. This trip is not limited by runway length.

Note: The required runway is a function of airplane and engine combination. The runway available is the actual (concrete) length of runway.

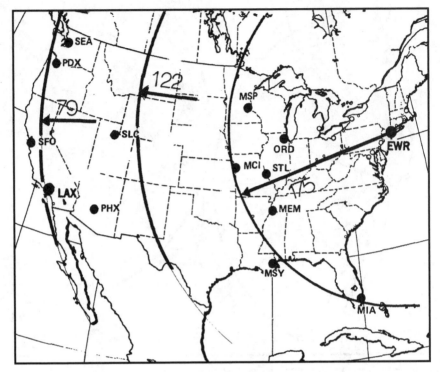

Figure 4.19 Aircraft range capabilities with various loads.

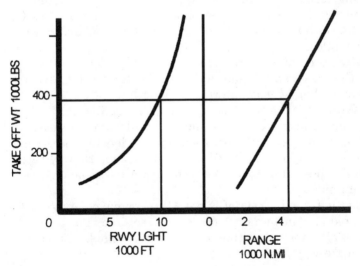

Figure 4.20 Runway length and takeoff weight versus range.

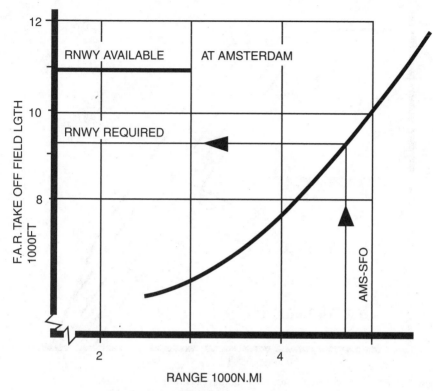

Figure 4.21   Takeoff field length versus range.

For any runway length, the range is available and is easily extracted from Fig. 4.22 for a given airplane-engine combination on this payload versus range chart.

In Fig. 4.20, a relationship exists between runway length and take-off weight. Figure 4.22 presents the effect of variation in takeoff weight = runway length on the payload versus range chart. The payload curves are generally prepared for the maximum takeoff weight, but they do not show the influence of the various takeoff weights due, for example, to shorter runway length or the available lower load. The purpose of this figure is to show the effect of runway length on the payload curve.

4.12   Airplane Fuel Burn Characteristics

At this point in the discussion, fuel burn should be examined more closely. An airplane consumes the most fuel in climb configuration, followed by cruise and descent. The fuel consumption in pounds per hour

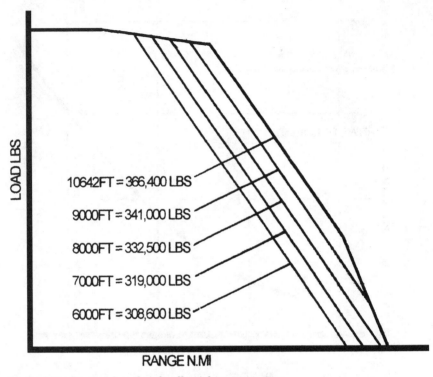

Figure 4.22   Runway length–takeoff weight versus range.

is the highest on short trips. Here, half the trip time is spent in climbing, and the relationship between trip time and climb time is about 50 percent. On a long trip, the high pounds per hour climb fuel flow becomes a small part of total fuel. As trip distance increases, the total fuel flow (pounds per hour) decreases. Beyond 2000 nautical miles, the fuel flow curve is relatively flat. Higher fuel flow shows up in the airplane operational cost, where an airplane flying on shorter range has a higher cost (Fig. 4.23).

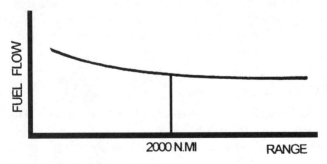

Figure 4.23   Fuel flow versus range.

4.13   Aircraft Engines

Airframe manufacturers offer various types of engines. A few major engine manufacturers are named here: Pratt & Whitney, General Electric, and Rolls Royce. Similar engine sizes are offered with different fuel consumptions and different engine weights. For smaller airplanes, Pratt & Whitney offered its JT8D series of engines. The Boeing 747 is powered by the JT9D series, with, respectively, JT9D-3, -3A, -7, -7AH, -7A, -7Q, -7F, -7J, -70A, and -7R4 engines, and General Electric has its CF6 series engines, and Rolls-Royce has its RB Series, just to mention a few.

As the engines were updated, engine thrust went up, and fuel consumption was improved. Selecting the proper airframe-engine combination, management was able to select engines with the best performance for their operations.

Considering fuel cost savings, another item also should be thought about carefully when selecting engines, namely, engine weight. One type of engine having the lowest fuel flow could be a good candidate for a given airplane. If it is heavier than another type of engine with higher fuel flow, the weight difference should be translated into fuel because carrying the extra weight requires additional fuel. In the final analysis, it may cause the burning of more fuel, thus resulting in higher operating costs.

Engine economics have an effect on the selection of an engine for a given airplane. Engine cost is expressed in engine flight hours (EFT). There are many factors affecting engine cost, and a few are listed next:

Shop visit. Shop visit rate is expressed in the number of shop visits per each 1000 hours (engine flight hours).

Flight cycle (one takeoff and one landing). This varies by the airline route structure. Longer flights have fewer flight cycles. Qantas flies long routes, and domestic airlines such as Southwest or Japan Airlines domestic operations have many short flights, to mention two extreme cases.

Engine flight hours on wing. Longer engine hours are more economical.

Work scope. The expenses are based on heavy or light work procedures. This partly depends on the design, e.g., high or low engine gas turbine temperature (EGT).

The deterioration in different modules varies, and the replacement of life-limited parts has a profound effect on maintenance costs.

The evaluation of engine cost is a very complex procedure because it is based on many factors, some of which are airline operational mode, number of cycles, engine thrust, application of assumed temperature method during takeoff and climb, schedule characteristic, and other considerations not mentioned here.

From a maintenance point of view, engine design should be analyzed in terms of easy maintenance, availability of spare parts, and duration of the required maintenance jobs. Similarly, the manufacturer's experience, past records, and support play a major role in the decision process.

4.14   Direct Operating Costs in Dollar per Statute Mile or Nautical Mile versus Range

Figure 4.24 shows the direct operating cost (DOC, $/statute mile) versus range for different sizes of airplanes. A method of calculation will be shown for this figure. For the time being, it is assumed that the direct operating cost for a given airplane is expressed in dollars per block-hour. The same goes for flying operations, maintenance, maintenance burden, and depreciation/rents/lease (whichever is applicable), as mentioned in the preceding chapter.

Example Let's say a trip length of 3500 statute miles has been selected. Direct operating cost is given in this example as $718.21 per block-hour, and for fuel price, 50 cents per gallon is considered. All costs will be expressed in dollars per statute mile.

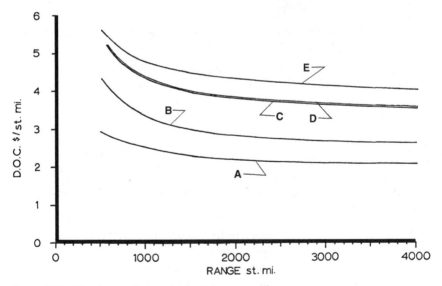

Figure 4.24   Direct operating cost comparison per mile.

The following data are available:

Block time: 6.705 hours
Block fuel: 98,800 lb/14,746 gal
Block speed: 522 mi/h
Distance: 3500 statute miles
Fuel price: $0.50/gal
Zero wind

To make it easier to follow the calculations, the dimensions are shown at each step. DOCLF= DOC less fuel cost

$$
\begin{array}{ccccc}
\text{DOCLF} & / & \text{speed} & = & \text{dollar/statute mile} \\
718.21 & / & 522 & = & 1.3757 \quad (4.4) \\
\text{\$/block-hour} & & \text{statute mile/block hour} & & \text{dollar/statute mile}
\end{array}
$$

$$
\begin{array}{ccccc}
\text{DOCLF} & \times & \text{distance} & = & \text{dollars} \\
1.3757 & \times & 3500 & = & 4.815 \quad (4.5) \\
\text{\$/statute mile} & & \text{statute miles} & & \text{\$}
\end{array}
$$

Alternate method:

$$
\begin{array}{ccccc}
\text{DOCLF} & \times & \text{block-hours} & = & \text{dollars} \\
718.21 & \times & 6.705 & = & 4.815 \quad (4.6) \\
\text{\$/block-hour} & & \text{block-hours} & & \text{\$}
\end{array}
$$

$$
\begin{array}{ccccc}
\text{Partial trip cost} & / & \text{distance} & = & \text{dollars/statute mile} \\
4815 & / & 3500 & = & 1.3757 \quad (4.7) \\
\text{\$} & & \text{statute miles} & & \text{\$/statute mile}
\end{array}
$$

Note: trip fuel cost is not included.

Trip fuel 98,800 lb @ 6.7 lb/gal (density) = 14,746 gallons (4.8)

$$
\begin{array}{ccccc}
\text{Fuel requirement} & / & \text{distance} & = & \text{fuel/mile} \\
14,746 & / & 3500 & = & 4.213 \quad (4.9) \\
\text{gallons} & & \text{statute mile} & & \text{gal/statute mile}
\end{array}
$$

$$
\begin{array}{ccccc}
\text{Fuel/statute mile} & \times & \text{fuel price} & = & \text{dollars/statute mile} \\
4.213 & \times & 0.50 & = & 2.1065 \quad (4.10) \\
\text{gal/statute mile} & & \text{\$/gal} & & \text{\$/statute mile}
\end{array}
$$

Alternate method:

$$\text{Fuel requirement} \times \text{fuel price} = \text{fuel cost}$$

$$
\begin{array}{cccc}
14,746 & \times & 0.50 & = & 7373 \\
\text{gallons} & & \$/\text{gal} & & \$
\end{array}
\tag{4.11}
$$

$$\text{Total DOC in dollars} = 4815 + 7373 = 12,188 \tag{4.12}$$

$$
\begin{array}{ccc}
\text{Total trip cost} & / & \text{distance} & = & \text{dollars/statute mile} \\
12,188 & / & 3500 & = & 3.482 \\
\$ & & \text{statute mile} & & \$/\text{statute mile}
\end{array}
\tag{4.13}
$$

$$
\begin{array}{c}
\text{Total \$/statute mile} = \$/\text{statute mile (DOCLF)} + \$/\text{statute mile (fuel)} \\
3.4822 = \quad\quad 1.3757 \quad\quad\quad + 2.1065
\end{array}
\tag{4.14}
$$

The shorter the trip is, the higher the DOC becomes; when range increases, DOC decreases. A larger airplane with a higher purchase cost has a higher DOC.

Conducting similar calculations for a set of different trip lengths, the cost curve can be established.

4.15 Direct Operating Costs in Cents per Passenger-Mile and/or in Cents/Ton-mile versus Range

Another cost chart is developed where cost is expressed in cents per passenger-mile or cents per ton-mile. This chart is based on the value of the previous DOC ($/statute mile). This is obtained by dividing the dollars per statute mile value by the number of available seats or, in the case of a freighter airplane, by the number of available tons. These specific cost figures are used when comparing various types and sizes of airplanes. In this way, the so-called unit cost is established in cents per passenger-miles for passenger operations and cents per ton-mile for cargo operations (Fig. 4.25).

There is a lowest point on each curve, and this point corresponds on the payload versus range capability chart to the maximum payload range. This is the lowest-cost point. Beyond this point, the payload decreases rapidly and the cost increases significantly.

Since airplanes come in various sizes with different fuel consumptions and costs, there is a definite need to be able to compare them. Cents per passenger-mile or cents per ton-mile cost satisfies this requirement. It is a good indicator, for example, of how much it costs in cents to move one passenger or one ton of cargo a distance of 1 mile. These specific cost figures can then be compared for the various air-

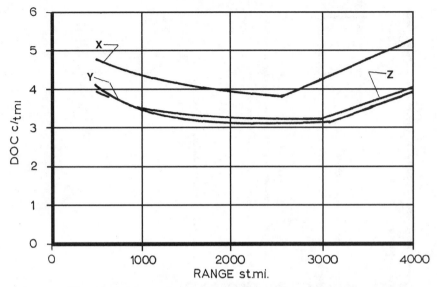

Figure 4.25 Direct operating cost per passenger-mile or ton-mile.

planes with various engine configurations and sizes. This would show what type of airplane proves to be more economical for carrying passengers or cargo (expressed in cents per passenger-mile and/or cents per ton-mile).

For specific trips, a bar-chart comparison is presented. Taking the reference airplane cost figure for a given trip length such as 1000 miles, the other two airplanes cost figures are compared. This chart indicates the airplane that has the best economy (Figure 4.26).

For this individual trip, the same basic data should be furnished that were mentioned previously in terms of the payload range curve. These data should include a weight summary, type of operations (passenger or freight), and airplane performance (cruise type, reserve fuel, etc.).

4.16   Cargo Density

By definition, *density* is the weight or load divided by the volume expressed in pounds per cubic feet. When an airplane carries 100,000 lb and the volume is 8200 ft^3, the density is 12.1 lb/ft^3. This brings up the question of what the designated density should be when an airplane is in the design stage. The actual experience varies between 5 and 23 lb/ft^3. Generally, it is observed that the average cargo density decreased during recent decades to about 8 lb/ft^3. The ideal airplane, however, should be able to carry the maximum load without any restrictions (Fig. 4.27).

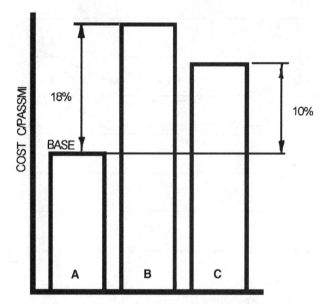

Figure 4.26 Direct operating cost cents per passenger-mile comparison.

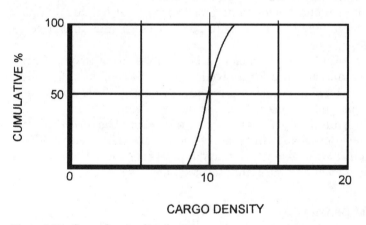

CARGO DENSITY

Figure 4.27 Cargo density distribution.

Note: In 1972, the Civil Aeronautics Board (CAB) and 11 U.S. airlines cooperated in an extensive survey of airfreight density. This survey was conducted as part of the CAB's domestic airfreight rate investigation. The survey included actual observations of palletized and containerized loads for 4 weeks of freighter and passenger aircraft departures.

Boeing studies were conducted of actual airfreight carried by air cargo carriers on over 10,500 flights of B707 size airplanes.

The weight, volume, and density data obtained for more than 100,000 pallets and containers contributed to the following conclusion: For international operations, the projected average cargo density for a single B747-200F will range from 8 to 12 lb/ft³, with a mean average of approximately 9.8 lb/ft³. Average domestic cargo density was about 1 lb/ft³ lower.

The next density chart (Fig. 4.28) presents the density versus payload. The slanted line indicates the space limit. Payload and density can increase until they reach the limiting volume of the airplane. When payload is constant and density is increasing, then volume is decreasing. This density chart has a relationship with the payload and density versus range chart. For let's say a type A airplane, any point on the slanted line indicates a constant volume.

Payload, space limited

$$25{,}000 \text{ lb/5 lb/ft}^3 \text{ (density)} = 5000 \text{ ft}^3 \text{ volume}$$

$$50{,}000/10 = 5000$$

$$90{,}000/18 = 5000$$

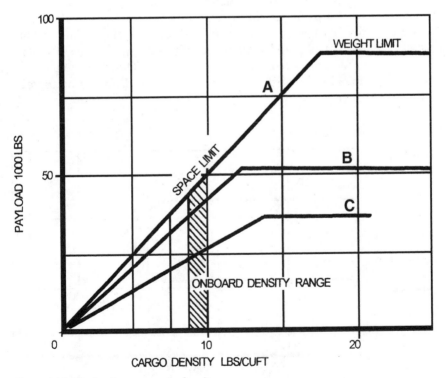

Figure 4.28 Payload versus cargo density.

Payload, weight limited

$$90,000/20 = 4500$$
$$90,000/23 = 3900$$

The average densities of selected commodities are listed below:

Furniture: 3.5 lb/ft^3

Cut flowers: 5.2 lb/ft^3

TV sets: 8.6 lb/ft^3

U.S. domestic mail: 10.5 lb/ft^3

Auto parts: 12.4 lb/ft^3

Figure 4.29 shows a comparison between type A and type B airplanes. It presents the percentages when both types of airplanes are volume-limited. Type A is volume-limited around 35 percent of the time, as opposed to 75 percent of the time in the case of the type B. As the range increases, the weight decreases, and the airplane becomes increasingly less limited by volume (Fig. 4.29).

Another presentation for cargo airplanes compares direct operating cost per cubic foot per trip, as shown on Fig. 4.30. Here, the airplane labeled "Base" is compared with other different types of aircraft. This presentation is considered for a 700-mile trip. The base airplane is more economical than airplane type B, but type C is the better airplane for this trip.

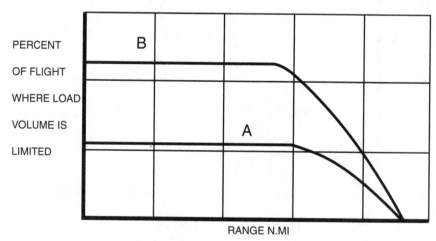

RANGE N.MI

Figure 4.29 Flights limited by volume versus range.

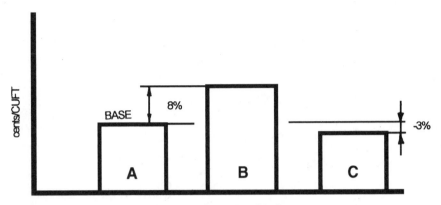

Figure 4.30 Direct operating cost per volume, in cubic feet.

4.17 Cargo Operations in Cents per Ton-Mile versus Density

The next chart (Fig. 4.31) presents DOC in cents per ton-mile versus density in pounds per cubic foot for cargo operations. A flight from JFK to LHR has been selected here. When payload is maximum, density is constant. When payload is below maximum, density is variable. For a type C aircraft, the cost for a partial load is shown with the associated load factor.

By entering the chart with any given density value, payload and load factors can be found. With a low load factor, the cents per ton-mile figures are practically off the chart, indicating a very costly operation (Fig. 4.31).

For comparison purposes, airplane types A and B are shown. They are larger than type C aircraft, and their specific costs in cents per ton-mile indicate improved economic operations.

4.18 Density versus Range

Similar to the payload versus range curve, a density versus range curve can be developed. The maximum payload range point is at the same location as the one on the payload versus range curve. Up to this maximum payload range point, payload is constant and so is density. Once beyond this point, load and density are decreasing (Fig. 4.32).

4.19 Partial Load Presentation as a Function of Range

The payload versus range curve in Fig. 4.3 presents the full potential of the airplane. Many times, the airplane flies with partial loads only, and this should be presented in an easy-to-read format. (Fig. 4.33)

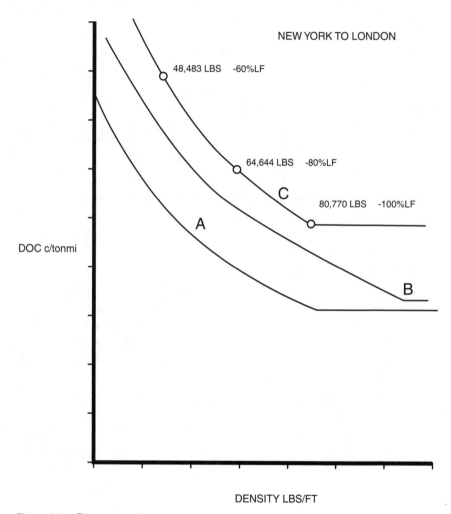

NEW YORK TO LONDON

48,483 LBS   -60%LF

64,644 LBS   -80%LF

C

80,770 LBS   -100%LF

A

DOC c/tonmi

B

DENSITY LBS/FT

Figure 4.31   Direct operating cost in cents per ton-mile versus density.

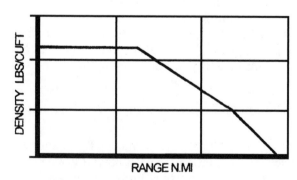

DENSITY LBS/CUFT

RANGE N.MI

Figure 4.32   Cargo payload versus density.

The lines indicate the number of passengers this airplane is able to carry. Partial loads and their maximum range are also shown in the figure. The vertical scale shows the number of passengers and the payload the airplane is able to carry. The airplane maximum load capability is 62,000 lb. On the main deck, there is space for 225 passengers. The total weight of passengers and baggage at 210 lb per passenger is 225 × 210 = 47,250 lb. Subtract this amount from the airplane maximum load of 62,000 lb, and 14,750 lb of cargo potential could be expected. The airplane range with 62,000 lb is 2650 nautical miles.

For example, when the airplane carries only 135 passengers or a load factor of 60 percent, or an equivalent load 135 × 210 = 28,350 lb, the range of the airplane is 3800 nautical miles.

In another example, what would be the maximum range of load carrying 135 passengers and 10,000 lb of cargo? Convert the number of passengers to equivalent load by multiplying 135 passengers by 210 lb to get 28,350 lb. Add the available 10,000 lb of cargo to the passenger weight to get 38,350 lb. Enter the chart with this load, move horizontally until you reach the payload scale, and read the range for this load. It is 3400 nautical miles.

Another approach is to convert the load into equivalent passengers. Converting 10,000 lb into 10,000/210 = 47.6 passengers and adding the original 135 passengers, a total of 182 passengers is now considered. The range with this load also would be the same—3400 nautical miles.

In conjunction with the preceding chart, a DOC (cents per passenger mile) versus range chart can be prepared (Fig. 4.34). For example, 135

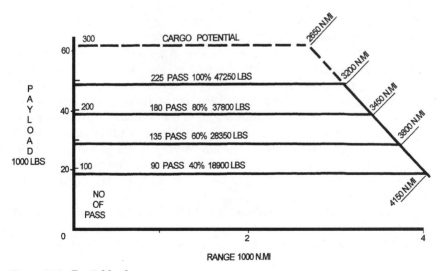

Figure 4.33 Partial loads versus range.

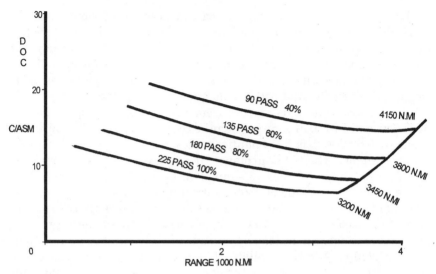

Figure 4.34 Direct operating cost in cents per passenger-mile for partial loads versus range.

passengers are on board for a given trip of 3000 nautical miles. This is equivalent to a 60 percent load factor, and the cost is 13 cents per passenger-mile. The trip cost equation (Eq. 4.14, based on DOC only) would furnish an operating cost of $52,650 for this particular trip.

Cents/passenger-mile × miles × passengers = cost

$$0.13 \times 3000 \times 135 = \$52,650 \tag{4.15}$$

The purpose of this chart is to furnish cost data for partial loads of the airplane and for comparing the economics of two (or more) airplanes.

The next chart is prepared for a fictitious airplane operating in a cargo configuration. It has a maximum payload range point at 3000 nautical miles. It can carry 177,000 lb (100% Load factor) and has a density of about 13 lb/ft³. With an 80 percent load factor, it can carry 141,600 lb, has a density of 10.1 lb/ft³, and a range of 3400 nautical miles. With a 60 percent load factor or 106,200 lb of load, it has a range of 3900 nautical miles. The sales department has to be fully aware that the airplane range is a function of the available load. A scale on the right-hand side on this chart (Fig. 4.35) indicates densities. In this way, payload and density are available simultaneously.

For another fictitious airplane (Fig. 4.36), the unit cost for cargo operations in cents per ton-mile is presented for various load factors. A linear interpolation between load factors is permitted. The importance

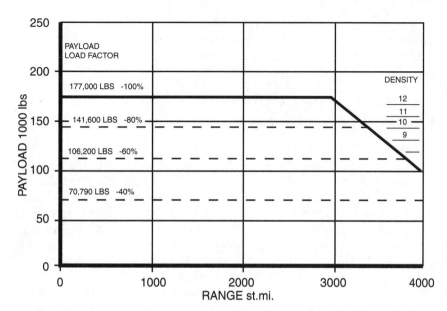

Figure 4.35   Partial payload versus range.

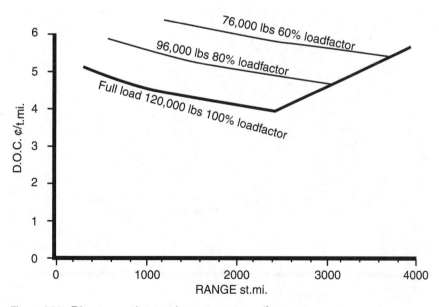

Figure 4.36   Direct operating cost in cents per ton-mile versus range.

of this chart is to represent partial load and cost figures in cents per ton-mile over the range of the airplane.

This graph furnishes answers to the following questions:

- For a given trip length at a given load factor, what is the payload, and what is the operating cost in cents per ton-mile?

- What is the trip cost for a given load at a given trip length?

Note: Trip cost is calculated by applying Eq. (4.14).

For cargo operations, cents per ton-mile versus density is presented for the range of the airplane in Fig. 4.37. This chart presents information for cargo operations based on the furnished density figures. Cargo tariff is based on weight and density. To see its importance, an example is presented for an airplane that has a volume of 10,770 ft³. In addition, it is able to carry a maximum of 100,000 lb. The load to be carried is mercury and the density of the mercury is 845 lb/ft³. Considering the airplane's volume of 10,770 ft³, a total of 10,770 ft³ × 845 lb/ft³ = 9,100,650 lb should be carried. However, due to airplane weight limitations, only 100,000 lb can be carried. The available volume is 10,770 ft³, but only 118.3 ft³ is used. The airplane is weight limited, and if 100,000 lb of cargo yields 30 cents per pound, the revenue would be $30,000.

Let us examine the other end of the spectrum. Let's say the load consists of feathers and is packed at 2 lb/ft³. To carry 100,000 lb of feathers would require 50,000 ft³. Due to limitations of available

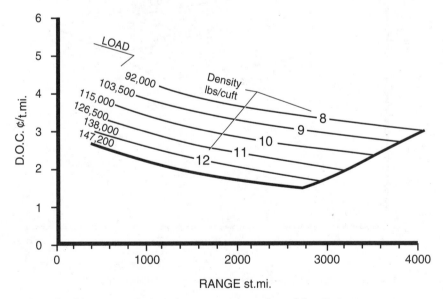

Figure 4.37 Direct operating cost in cents per ton-mile and density versus range.

cube volume of 10,770 ft^3, only 10,770ft^3 × 2lb/ft^3 = 21,540 lb can be carried. The airplane is cube or volume limited. The revenue at 30 cents per pound would be $6462. These two (extreme) examples indicate the airplane's weight or volume limitations and their relationship.

4.20   Airplane Load Factor Definitions

Load factor is the measure of utilization of an airplane for passenger and/or cargo operation. It represents, for the whole system, the relationship between the number of passengers flown and the total number of seats available. For example, if an airplane has 100 seats available and carries 60 passengers, the load factor for this flight is 60 percent. There is another variable, namely, the *breakeven load factor,* which is a number expressed in percentage where operating costs and generated revenues are equal. For example, when this airplane has to carry 40 passengers to generate revenue that equals the cost of operating the airplane, this means that the breakeven load factor is 40/100 = 40 percent.

The load factor can fluctuate significantly between flights; the next day may be much lower or higher. As with any other means of transportation, load factor has its peaks and valleys. Certain flights have high load factors in one direction and very low load factors on the same route for the return flight.

These principles are also valid for a cargo airplane. When a cargo airplane is able to carry 100,000 lb and carries only 60,000 lb, it has a 60 percent load factor. If this airplane has to carry 40,000 lb of freight to generate revenue that equals the operating cost, a (40,000/100,000) 40 percent breakeven load can be established. If the breakeven load factor is higher than the operating load factor, this trip generates significant losses. If an unsatisfactory load factor is observed systemwide, management has to take steps to alleviate this situation. If an airline has 600,000 seats available per year and sells only 320,000 seats (or tickets), this would be a 53.3 percent load factor.

4.21   Breakeven and Operating Load Factor

Breakeven load is an important element in the evaluation process. If an airplane can carry 300 passengers and requires 156 passengers to generate enough revenue to cover operating costs, the breakeven passenger load factor is 52 percent. This is an index by which different-sized airplanes may be compared. The average industry load factor is now around 60 percent or more. A high breakeven load factor is not preferred because it is harder to fill the airplane.

TABLE 4.5   Operating and Breakeven Loadfactors

Items	Airline A	Airline B	Airline C
A/C operating load factor, %	52.75	60.01	59.70
Breakeven load factor, %	45.70	58.40	61.60

Note: Now airlines are flying with higher load factors by decreasing flight frequencies and charging high penalties for changing trips or no shows. This situation changed drastically after the September 11, 2001 tragedy.

To illustrate the importance of breakeven load factor in relationship to airplane operating load factor, the data of three carriers are presented side by side. One is a low-cost carrier. The other two are trunk airlines, and one of them has economic difficulties. Table 4.5 contains the data.

Airline A is a low-cost airline. It has a profitable operation. Its breakeven load factor is relatively low compared with its operating load factor. Airline B also has a profitable operation, but there is a narrow band between breakeven and operating load factors. This means that if a certain magnitude of fluctuation exists in its operating load factor, it could result sometimes in a negative cash flow. It is evident that airline C is experiencing economic difficulties. The relationship between the operating and breakeven payloads shows the airline's problem. A graphic presentation between the industry operating and breakeven payloads is illustrated in Fig. 4.38.

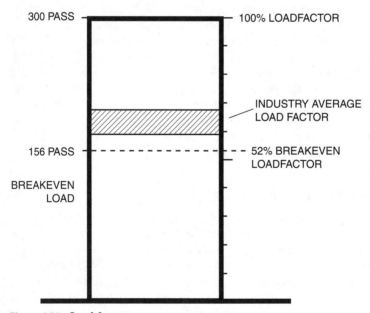

Figure 4.38   Load factors.

4.22 Breakeven Passenger Numbers, Load Factor, and Profit—Passenger Operations

A multitude of questions are raised when evaluating breakeven figures, load factors, yields, and profits. To find the proper answers to these questions, a chart is developed. Typical questions include

- What is the breakeven number of passengers, and at what yield and load factors does it occur?
- What is the breakeven number of passengers at a given yield?
- What is the dollars per passenger value at a given yield?

A practical example will explain the use of these types of charts. The following conditions exist:

Airplane type: Alpha 755

Trip cost: $25,000

Trip distance: 3000 nautical miles

Maximum number of passengers: 250

The breakeven number of passengers, yields, and load factors are presented in Table 4.6.

It is easy to conclude that at low yield, the breakeven number of passengers is quite high with a high breakeven load factor. By increasing the yield, the breakeven number of passengers decreases in a nonlinear relationship (Fig. 4.39) under the following conditions:

Trip cost: $25,000

Trip distance: 3000 nautical miles

Maximum number of passengers: 250

TABLE 4.6   Yield Breakeven Number of Passengers and Load Factors

Yield in cents per passenger-mile	5.0	10.0	15.0	20.0
Breakeven number of passengers	166	83	55	42
Load factors	68%	33%	21%	18%

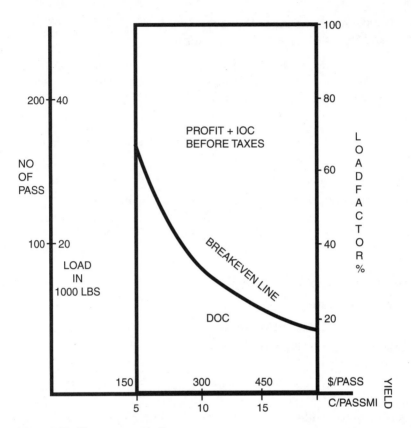

Figure 4.39   Passenger operations.

For a given yield, the breakeven number of passengers could be obtained from Fig. 4.39. For example, at 10 cents per passenger-mile, the breakeven number of passengers is 83, and the breakeven load factor is 33.6 percent. There is one breakeven value for each yield at 15.0 cents per passenger-mile. The breakeven number of passengers is 55, and at 200 lb per passenger and baggage, the weight is 11,150 lb and the associated breakeven load factor is 23.2 percent.

For the relationship between cents per passenger-mile and dollars per passenger, Table 4.7 is presented.

The revenue is tabulated as a function of yield in cents per passenger-mile in Table 4.8.

TABLE 4.7   Relationship between cents per passenger-mile and dollars per passenger

Yield in cents per passenger-mile	5.0	10.0	15.0	20.0
Yield in dollars per passenger	150	300	450	200

TABLE 4.8   Yield, Revenue and Profit-Passenger Operations

Yield in cents per passenger-mile	5.0	10.0	15.0	20.0
Revenue at full load*	$37,500	$75,000	$112,500	$150,000
Profit + IOC before taxes	$12,500	$50,000	$87,500	$125,000

*Full load is 250 passengers.

Note: In establishing profit, a trip cost of $25,000 is subtracted from the revenue at full load. Actually, the trip cost of $25,000 is slightly variable because the number of breakeven passengers changes with yields and affects the required amount and cost of trip fuel (variable number of passengers slightly changes amount of trip fuel needed and ultimately its cost). In this example, trip fuel is considered constant.

Another example is shown here: The airplane carries only 145 passengers at 12.0 cents per passenger-mile. What profit could be expected for this trip? The following conditions prevail:

Trip distance: 3000 nautical miles

Maximum number of passengers: 250

Passengers carried: 145 (58 percent load factor)

For the solution, we note the following:

Breakeven number of passengers at 12.0 cents per passenger-mile: 69

Difference: 145 − 69 = 76

Profit (applying yield in dollars per passenger): 76 passengers × 360 $/passenger = $27,360

Profit (applying yield in cents per passenger-mile): 76 passengers × 3000 nautical miles × 12.0 cents per passenger-mile = $27,360

Note: Breakeven number of passengers of 69.48 is rounded to 69.

The next chart (Fig. 4.40) shows the relationship between number of passengers carried and load factors versus cost and revenue. Any number of passengers carried before the intersection of the yield and cost lines are carried at a loss. Beyond this point, the airplane shows profit. There is a simple relationship to calculate profit, as shown in Eq. (4.16):

$$Rev = actualpass \ (maxprft/maxpass) \qquad (4.16)$$

where rev = revenue
$actualpass$ = number of passengers carried
$maxprft$ = revenue at maximum number of passengers carried
$maxpass$ = maximum number of passengers

Substituting the numbers from the previous example and reading data from the chart, the same result could be obtained (Fig. 4.40). For example:

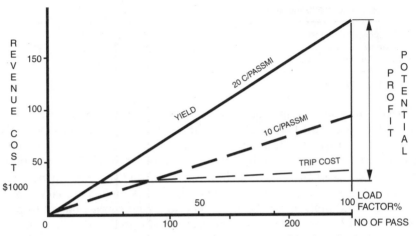

Figure 4.40   Passenger operations: profit/loss presentation for selected trip.

Revenue at maximum number of passengers

$$= \$90{,}000 = (250 \times 3000 \times 0.12)$$

Revenue at 145 passengers

$$= \$52{,}200 = [(145/250) \times 90{,}000]$$

$$\text{Less trip cost} = \$25{,}000$$

$$\text{Profit (before taxes)} = \$27{,}200 = (52{,}200 - 25{,}000)$$

Note: There is a $160 difference between the two profit figures because of rounding the number of passengers (see above).

Note: The trip cost is increasing slightly because carrying more passengers requires an increased amount of tripfuel. This is indicated on the dashed line of the trip cost on theis chart. To simplify the calculation, use constant trip cost data.

4.23   Breakeven Load, Load Factor, and Profit—Cargo Operations

For freight operations, another chart was developed (Fig. 4.41 for a given type of aircraft and for a given trip length. The following conditions prevail:

Trip cost: $25,000

Trip distance: 3000 nautical miles

Maximum payload: 150,000 lb

The following example (Table 4.9) shows the usefulness of this chart. What profit can be expected at

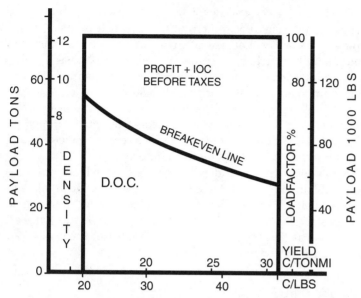

Figure 4.41   Cargo operations.

TABLE 4.9   Yield, Revenue Profit Cargo Operations

Yield in cents per ton-mile	15.0	20.0	25.0	30.0
Breakeven in tons	55.5	41.6	33.3	27.7
Breakeven in pounds	111,000	83,200	66,666	55,400
Yield in cents per pounds	22.5	30.0	37.5	45.1
Revenue at maximum load	$33,750	$45,000	$56,250	$67,500
Profit + IOC before taxes	$8,750	$20,000	$31,250	$42,500

Load factor: 68 percent

Trip distance: 3000 nautical miles

Yield: 25.0 cents per ton-mile

The solution according to Fig. 4.41 is as follows:

At 68 percent load factor, the load is:	102,000 lb
Breakeven load is:	66,000 lb
Difference:	36,000 lb

Profit (applying yield in cents per pounds): 36,000 lb × 37.5 cents/lb = $13,500

Profit (applying yield in cents per ton-mile): (36,000/2000) × 3000 nautical mile × 25 cents/ton-mile = $13,500

The next chart (Fig. 4.42) was developed for cargo operations. The result from this example could be read from this chart. It is based on the same principle shown in Fig. 4.40 except that here the load is expressed in tons or pounds and the other chart has it in number of passengers.

In conclusion, charts for passenger and cargo operations indicate that there is a minimum load factor for each yield. Below this load factor, the operation is not profitable. As yield decreases, the minimum load factor has to increase, and vice versa. Both charts—passenger and cargo operations—should be reviewed periodically, since economic factors such as inflation can have drastic effects on these figures.

For leased airplanes, Fig. 4.43 indicates the breakeven load. Entering the chart with yield and the amount of monthly lease rate, the breakeven values can be obtained. For passenger operations, the left-hand scale indicates the breakeven number of passengers and the breakeven load factor. For cargo operations, the right-hand scale indicates the breakeven payload, breakeven load factor, and density.

4.24 Lower-Lobe Cargo Effect on Breakeven Load—Passenger Operations

For any passenger-carrying airplane, the cargo in the lower lobe contributes to increasing the revenue. This means that the breakeven number of passengers will be less. This relationship is presented in Fig. 4.44, where the breakeven number of passengers is plotted versus range. There are certain lines on the chart labeled "Cargo Load Factor" for various percentages of the load. Entering the chart with the range, proceeding up to the proper "Cargo Load Factor" lines, and then turning to

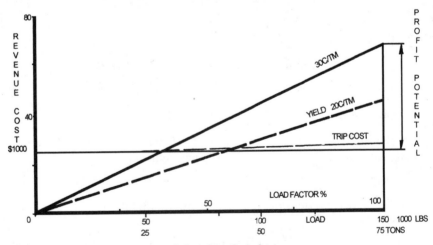

Figure 4.42 Cargo operations: profit/loss for selected trip.

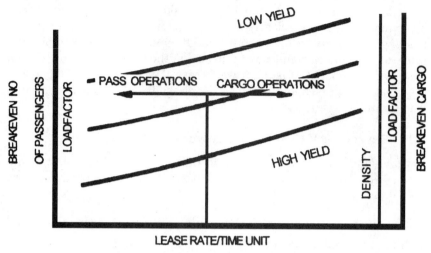

Figure 4.43   Breakeven load versus lease rate.

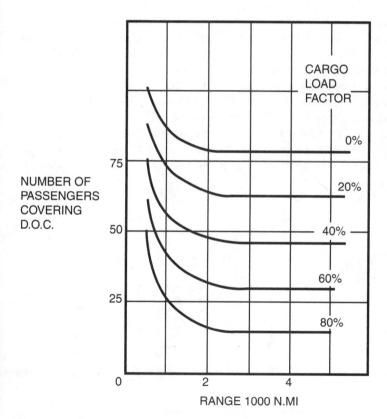

Figure 4.44   Passenger breakeven number and cargo versus range.

the left on the vertical scale, the (lower) breakeven passenger number can be obtained. This chart clearly indicates the improvement in breakeven passenger numbers brought about by additional cargo. For example, at a range of 2000 nautical miles with no load in the lower level, a breakeven passenger number of 77 is required. In case of a load factor of 50 percent, the breakeven passenger number is decreased to only 34. The maximum load should be specified on the chart.

4.25   Revenue and Cost Presentations

Revenue and cost are two important indicators, and their relationship defines the magnitude of profit potential (if any). Figure 4.45 shows the following:

Revenue potential based on the available seats on the airplane (RA)

Revenue based on the number of passengers carried (RP)

Cost based on the operation of the aircraft (CA)

Revenue potential could be expressed in cents per available passenger-mile. Revenue is based on cents per revenue passenger-mile (yield). Cost is based on cents per available passenger-mile. Figure 4.45 indicates the relationship of these three factors.

Comparing revenue and cost figures for different airlines, the airline average operating revenue and cost range at short distances tends to be higher, whereas it is lower for long distances. This situation was discussed earlier in this chapter under "Airplane Fuel Burn Characteristics" (Sec. 4.12). When the number of passengers carried is equivalent to the number of available seats on the airplane, then $RP_1 = RA_1$ (this is the case when load factor is 100 percent).

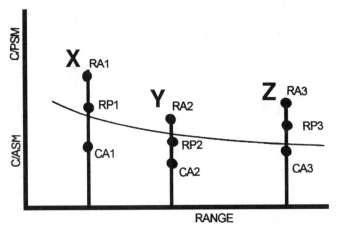

Figure 4.45   Revenue and cost versus range.

Note: The distance between RP_1 and CA_1 is related to profit, and the distance between RA_1 and CA_1 indicates profit potential.

Figure 4.46 approaches this revenue-cost relationship from a different point of view. The horizontal scale represents revenue, and the vertical scale is cost; both are in cents per available passenger-mile or revenue passenger-mile or, for cargo, in cents per available ton-mile and yield in revenue ton-miles. A 45-degree line starting from the intersection of the horizontal and vertical scales should be labeled "Revenue = cost." Points on the chart represent revenue and cost. If these points are between the horizontal scale and the 45-degree line (Revenue = cost), the operation is profitable. Beyond this "Revenue = cost" line, the operation is showing a loss.

The following comment is presented for each aircraft shown on this chart: For aircraft *X*, profit is defined between the point RP_1 and the intersection of the revenue (horizontal) line and the line labeled "Revenue = cost." Furthermore, the airplane has a good potential for revenue growth between points RP_1 and RA_1. Aircraft *Y* is an airplane operating at a loss with its present market share. RP_2 is between the "Revenue = cost" line and the vertical cost line. Its profit potential is very limited, and only a small growth can be anticipated with limited profit. Aircraft *Z* has a profitable operation, but the growth of its profit potential is limited.

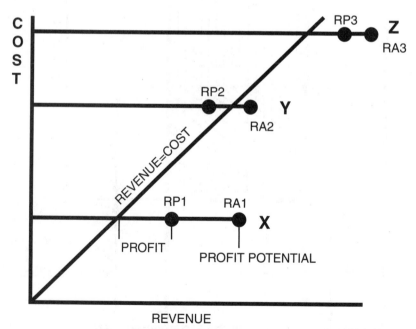

REVENUE

Figure 4.46 Revenue and cost presentation.

Taking another example for the purpose of explaining these relationships, Table 4.10 presents data for airplane types 1, 2, and 3. Airplane 1 is the first choice because it generates the highest profit and has a good profit potential. Not only is Airplane 2 less profitable than Airplane 1, its maximum profit potential is very limited. Airplane 3 has losses, but by increasing its revenue or market share, it could be made profitable.

Taking the figures from Table 4.10 for Airplane 1, a bar chart explains this relationship in Fig. 4.47.

TABLE 4.10   Profit, Revenue Cost Comparison

Items	Airplane 1	Airplane 2	Airplane 3
Maximum revenue potential	28	29	32
Revenue	18	27	26
Cost	10	21	28
Profit*	8	6	−2
Maximum profit potential	18	8	4

*Profit + indirect operating cost before taxes.

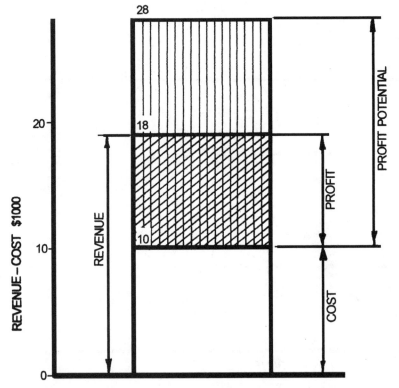

Figure 4.47   Revenue, profit, profit potential, and cost relationship.

4.26   Airplane Performance Information

Block speed, block time, and block fuel are presented in a graphic format in Figs. 4.48 through 4.50. They are developed from an airplane operations manual by calculating speed, time, and fuel for the range of the airplane.

Speed, time, and fuel affect operating cost. The fastest airplane can carry more payloads over a network for a given time period. This is reflected in the block time versus range curve. More time (slower airplane) for the same trip distance generates higher operating costs because most of the cost elements are based on block time (e.g., pilot's remuneration). Similarly, block fuel for a given trip should be the lowest.

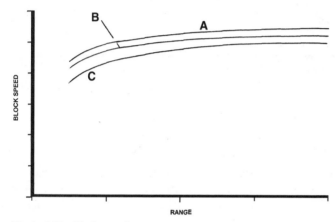

Figure 4.48   Block speed versus range.

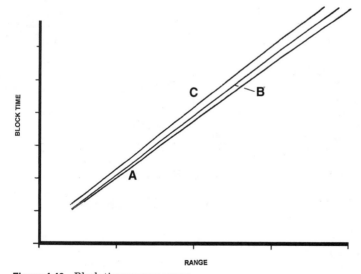

Figure 4.49   Block time versus range.

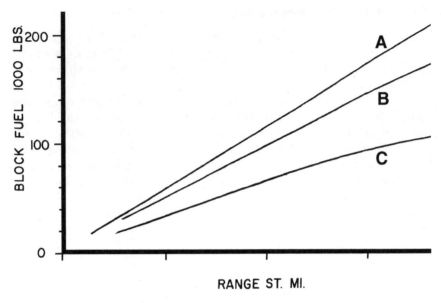

RANGE ST. MI.

Figure 4.50 Block fuel versus range.

4.27 Airplane Weight Ratios

Sometimes two or more aircraft weights and their relationship to the number of passengers or the payload are evaluated. The next few tables provide a comparison between a DC-8-73 and a DC-10-30 airplane (Table 4.11).

The passenger and operating weight relationship is presented in Table 4.12.

For the DC-8-73 airplane, it takes 666.6 lb of operating weight empty for one passenger, and for the DC-10-30 airplane, it takes 810.7 lb.

TABLE 4.11 Basic Weight Statements

Aircraft type	DC-8-73	DC10-30
Max. takeoff weight, lb	355,000	555,000
Max. landing weight, lb	275,000	403,000
Max. zero-fuel weight, lb	261,000	368,000

TABLE 4.12 Operating Weight Empty and Number of Passenger Relationship

Aircraft type	DC-8-73	DC-10-30
Operating weight empty, lb	172,000	263,500
Number of passengers	258	325
Operating weight empty/no. of pass.	666.6	810.7

There is about a 26 percent difference between the number of passengers carried and a 21 percent difference between the operating weight empty per passenger. The DC-8-73 airplane has a better relationship considering the operating weight empty and its number of passengers.

Cargo operations. The basic data for the DC-8-73 and DC-10-30 are listed in Table 4.13. Table 4.14 lists operating weight empty and structural load figures.

For the DC-8-73 aircraft, each pound of load corresponds to 1.35 lb operating weight empty. For the DC-10-30, this weight is 1.74 lb.

The structural payload is 74 percent of the operating weight empty on the first airplane and only 57 percent of the second one. Tare weight–payload relationship is shown in Table 4.15.

TABLE 4.13 Weight Statements

Aircraft type	DC-8-73	DC-10-30
Operating weight empty, lb	150,000	234,000
Structural payload, lb	111,000	134,000
Pallets, upper level	18	30
Lower level	—	8
Pallet size	108 × 96 in	108 × 96 in
Pallets weight at 250 lb each	4,500	9,500
Net structural payload, lb	106,500	124,500
Volume, ft^3	9,873	18,996

TABLE 4.14 Op. wt. Empty and Payload Relationship

Aircraft type	DC-8-73	DC-10-30
Op. wt. empty/struc. payload	150,000/111,000 = 1.35	234,000/134,000 = 1.74
Payload/op. wt. Empty, %	74	57

TABLE 4.15 Load, Tare Weight Relationship

Aircraft type	DC-8-73	DC-10-30
Structural payload, lb	111,000	134,000
Tare weight, lb	4,500	9,500
Net payload, lb	106,500	124,500
Tare wt./net payload	4.22	7.63

TABLE 4.16 Volume Comparison

Aircraft type	DC-8-73	DC-10-30
Total volume, ft^3	9,873	18,996
Net payload, lb	106,500	124,500
Density, lb/ft^3	10.78	6.5

For the DC-8-73 airplane, the tare weight takes only 4.22 percent of the net payload, and on the DC-10-30, this is 7.6 percent. Density figures are listed in Table 4.16.

Low density is well suited for smaller package operations. This type of comparison adds another dimension to the evaluation process. Low operating weights are always preferred due to users fees that are based on weights.

4.28   Airplane Fuel Burn Comparison

A few fuel comparisons are presented in this chapter. Fuel cost is a significant part of operating cost, so it is important to scrutinize every aspect of fuel consumption.

Trip fuel comparison—Passenger operations. Two airplanes were selected for comparing their fuel consumption. The DC-8-73 airplane is a reengined version of the DC-8-63 airplane whose JT3D-7 engines were replaced with CFM-56-1 engines in order to comply with the required noise limitation defined by noise regulations. This CFM engine has better fuel consumption, but it is slightly heavier, and the extra weight is noticeable in cargo configuration. Table 4.17 lists the data.

The DC-8-73 aircraft (CFM-56) has about 17 percent better fuel consumption than the DC-8-63 (JT3D-7). This manifests in the fuel per passenger and fuel per trip distance indices. Table 4.18 is prepared for cargo operations.

Note: Trip distance is 1750 nautical miles. The DC-8-73 carries 1300 lb less maximum payload due to its heavier engine (CFM-56) weights.

TABLE 4.17   Distance Fuel Relationship

Aircraft type	DC-8-63	DC-8-73
Trip distance, nautical miles	1500	1500
Trip fuel, lb	46,000	39,000
Number of passengers	258	258
Lb fuel/no. of passengers	178	151
Lb fuel/trip distance	30.6	26

TABLE 4.18   Load and Trip Fuel Relationship

Aircraft type	DC-8-63	DC-8-73
Trip fuel, lb	59,500	50,500
Payload, lb	104,000	102,700
Lb of fuel/payload	.5721	.4917
Payload/fuel	1.74	2.03

Reviewing these tables, the following conclusion can be reached: The DC-8-73 airplane requires less than one-half pound (0.4917) of fuel to carry 1 lb of payload versus the DC-8-63 airplane, which needs 0.5721 lb of fuel for the same 1-lb payload at a trip distance of 1750 nautical miles.

By reversing this relationship, the payload-fuel ratio for the DC-8-63 is 1.74 and for the DC-8-73 is 2.03. This indicates that the DC-8-73 can carry 2.03 lb of payload on 1 lb of fuel, whereas the DC-8-63 can carry only 1.74 lb of payload with the same amount of fuel. This is about a 17 percent improvement over the DC-8-63 engines (JT3D-7). As a result of the reengining, the airplane offers a more economical operation.

Fuel burn in pounds divided by trip distance is presented in Fig. 4.51. Aircraft type B burns 12 percent more fuel than aircraft A, which is considered as the base to compare other airplanes. Aircraft C burns 16 percent more than the airplane considered for the base.

For a ton moved 1 mile, a comparison could be made (Fig. 4.52). For the amount of fuel burned in pounds, calculating the figures for airplane A, which is the base, and airplanes B and C, the following is established: Airplane B requires 0.12 lb and airplane C requires 0.7 lb more fuel to move 1 ton for 1 mile than aircraft A. The best aircraft is type A for this range.

Fuel per seat comparison—Passenger operations. Figure 4.53 was prepared to compare fuel per seat versus block fuel. The best-performing airplane has the lowest fuel per seat and the lowest block fuel per trip.

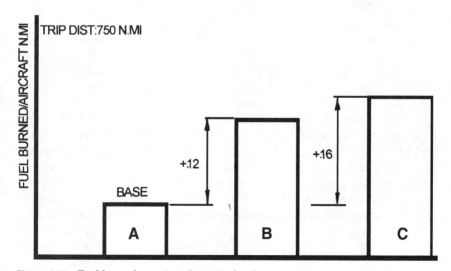

Figure 4.51 Fuel burned per aircraft nautical mile comparison.

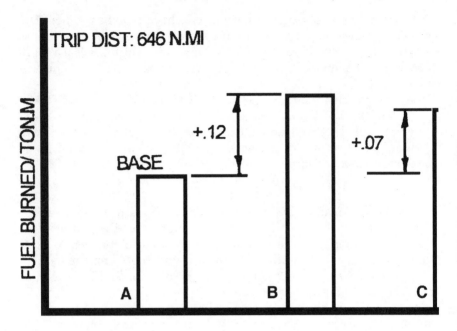

Figure 4.52   Fuel burned per ton-mile comparison.

The comparison was prepared for the DC-8-60 and the DC-8-70 series airplanes.

The figure clearly shows the better performance of the 70 series airplanes at 2000 nautical miles. They are powered with CFM-56 engines and have an improved fuel consumption. As was mentioned earlier, moving along on passenger guidelines toward the zero point shows a better-performing airplane. For example, the DC-8-63 carrying 258 passengers has a fuel-passenger ratio of 135 lb, and for the DC-8-73 airplane, the fuel-passenger ratio is 116 lb. This is a 16 percent improvement. These figures are valid for a trip distance of 2000 nautical miles.

A similar chart for cargo operations is shown in Fig. 4.54. The fuel-payload ratio is plotted versus block fuel. The slanted lines represent weight lines, and in this case, cargo weights on the chart are indicated in thousands of pounds. The slight changes in payload are caused by the change in operating weight empty due to the heavier CFM-56 engines (four engines). The figure clearly reflects the advantages of the reengining (CFM-56) as far as fuel consumption is concerned.

Productivity, or ton-mile per gallon, is another measure of fuel efficiency for a comparison between two or more airplanes. Figure

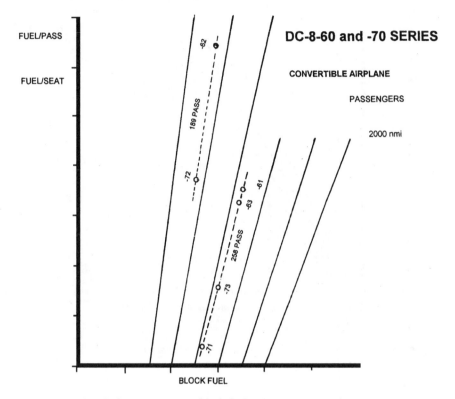

Figure 4.53 Trip fuel per seat versus block fuel.

4.55 shows the ton-mile per gallon of fuel values for various types of airplanes. The airplane with the higher ton-mile per gallon value is preferred.

Comparing fuel efficiency for an 875-nautical-mile trip length, the two trip fuel figures in Fig. 4.56 show some 7 percent difference in that airplane A burns more. This 7 percent does not say anything about the airplane's performance; consideration of how many passengers are carried with this amount of trip fuel furnishes a good basis for comparison of the two airplanes. In this case, airplane A has a higher fuel pounds per passenger seat value, and airplane B has a 27 percent lower fuel pounds per seat value. This means that airplane B burns less fuel for each passenger, and this is a better performance for airplane B.

The efficiency of the fuel consumption for a given airplane can be shown in the next chart by comparing the number of passengers that

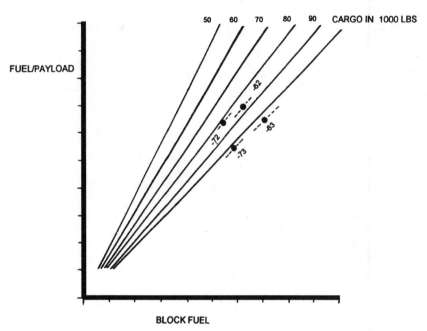

Figure 4.54   DC-8-60 and DC-8-70 series trip fuel per payload versus block fuel.

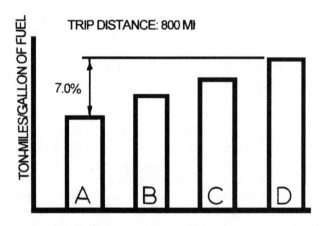

Figure 4.55   Ton-mile per gallon of trip fuel.

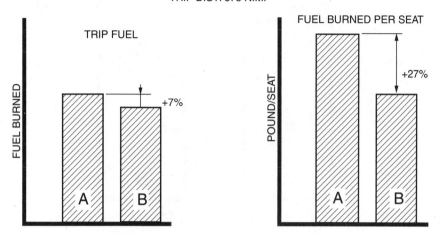

Figure 4.56   Trip fuel burn and trip fuel per seat value comparison.

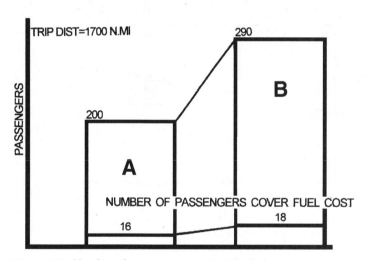

Figure 4.57   Number of passengers covering trip fuel cost.

cover fuel cost with the total number of passengers. Airplane A requires 16 passengers to cover fuel cost, as opposed to airplane B, which requires 18 passengers and carries a total of 290 passengers, whereas airplane A carries 200 passengers over a distance of 1700 nautical miles (Fig. 4.57).

Considering the relationship between breakeven (fuel cost) number of passengers and total number of passengers, aircraft B is better by 1.8 percent, as shown in Fig. 4.58. The trip distance is 1200 nautical miles.

Another measure of fuel efficiency is presented in Fig. 4.59. It is expressed in pounds of payload per pound of fuel burned over a wide

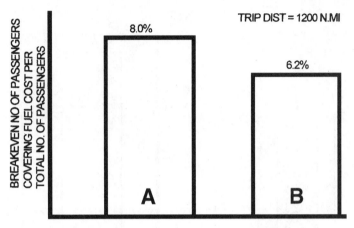

Figure 4.58 Breakeven number of passengers covering fuel cost per total number of passengers.

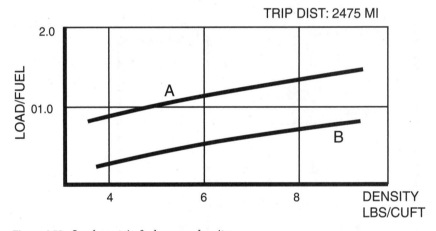

Figure 4.59 Load per trip fuel versus density.

range of densities. Let's say that aircraft A can deliver 1 lb of cargo per 1 lb of fuel at a density of 5.3 lb/ft^3. Aircraft B at the same density value delivers only 0.6 lb of cargo for each pound of fuel.

Another approach is to compare fuel per seat or per payload and express the differences in percentages, as shown in Fig. 4.60. Here, one airplane fuel per load relationship is the base, and any other airplane's figure can be expressed in percentage related to the base airplane. In similar fashion, block fuel, trip cost, and direct operating cost in percentages can be compared and presented in bar-chart form.

Another method was developed comparing relative fuel burn per seat versus relative block fuel. Both are expressed in percentages. The

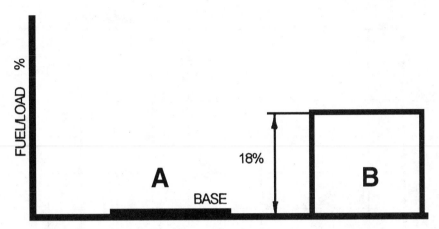

Figure 4.60   Trip fuel per load comparison.

airplane to be evaluated is placed into the origin of this coordinate system. Any other airplane figures are placed in one of the four quadrants of this presentation. For example, the airplane that is placed in the second quadrant has a fuel burn per seat percentage that is higher than that of airplane Z at the origin. However, its relative block fuel (expressed in percent) is lower. Any airplane placed in the first quadrant is a better airplane than airplane Z (Fig. 4.61).

Figure 4.62 shows the number of passengers versus DOC in cents per available seat-miles, and there is a comparison between airplanes

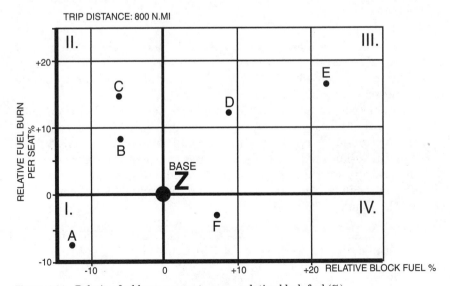

Figure 4.61   Relative fuel burn per seat versus relative block fuel (%).

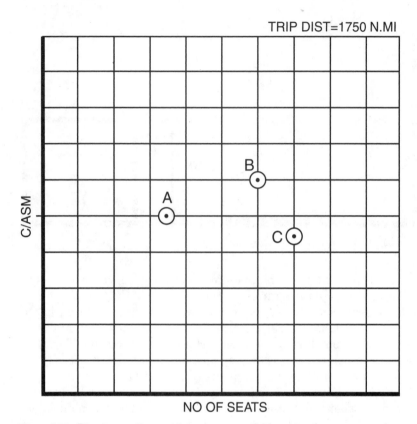

TRIP DIST=1750 N.MI

NO OF SEATS

Figure 4.62   Direct operating cost in cents per available seat-miles versus number of passengers.

A, B, and C. The preferred airplane is C because it carries the highest number of passengers with the lowest cents per seat-mile cost. For cargo operations, the horizontal scale should be payload in pounds, and the vertical scale should be in cents per ton-mile.

Another type of chart has DOC in dollars per nautical mile on horizontal scale and DOC in cents per available seat-miles or cents per available ton-miles on the vertical scale. The chart is prepared for a 650-nautical-mile trip. The slanted lines are load lines expressed in numbers of passengers (or payload in pounds) (Fig. 4.63).

Trip cost can be calculated in two different ways taking the numbers from the figure for aircraft A.

1. For this aircraft, the trip cost is $12.26 per nautical mile, and when multiplied by a trip distance of 650 nautical miles, it furnishes a trip cost of $7969.

TRIP DIST; 650 N.MI

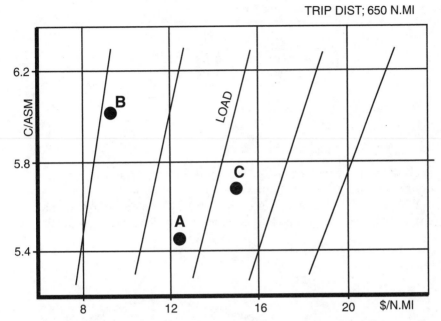

Figure 4.63 Direct operating cost in cents per available seat-miles versus dollars per mile.

2. For this airplane, cost is 5.45 cents per available seat nautical mile. Trip distance is 650 nautical miles. The number of passengers carried is 225. Thus

$$0.0545 \times 650 \times 225 = \$7969$$

The results are shown in Table 4.19.

Table 4.20 ranks the three airplanes by their profits. In order to see the procedure, the following assumption is considered: passenger yield as 12 cents per passenger revenue miles.

Note: Revenue = f(yield, no. of passengers, distance).

TABLE 4.19   Cost Data Comparison for Three Types of Aircraft

Airplane type	A	B	C
Cents/ASNM	5.45	6.05	5.68
No. of passengers	225	153	270
Trip cost, $	7969	6016	9964
Dollars/trip	12.26	9.256	15.33
Trip cost/no. of passengers, $	35.41	39.32	36.90

TABLE 4.20   Ranks by Profit Three Aircraft

Airplane type	A	B	C
Revenue $	17,550	11,934	21,060
Cost $	7,969	6,016	9,964
Profit* $	9,581	5,918	11,096
Rank	2	3	1

*Profit + overhead before taxes.

4.29   Airplane Productivity

A productivity curve could be prepared for any type of airplane. It is defined as ton-miles per hour versus range (Fig. 4.64).

With an increase in speed (see Fig. 4.48) and constant payload, productivity increases with range to the maximum payload range point. Beyond

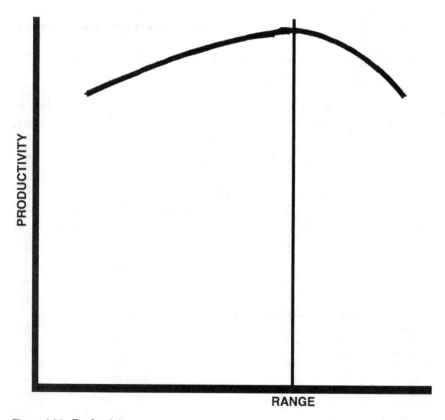

Figure 4.64   Productivity versus range.

this point, due to decreasing payload, the productivity also decreases. Speed increases, but it cannot balance the loss due to decreased payload. These charts now enable the user to identify the range of best productivity indicated on the DOC cents per ton-mile curve. It has two components: speed and tonnage. To have a second look at these components, Table 4.21 shows the data.

A smaller airplane has two disadvantages versus a larger airplane as far as productivity is concerned:

1. Smaller airplanes carry less payload.

2. Smaller airplanes fly at slower speeds.

There are some additional considerations that have to be discussed here. Productivity consists of tonnage, distance, and time or tonnage and speed (speed = distance/time).

Tonnage. A smaller airplane has a lower trip cost. Let's say a B737 requires $3000 for a certain trip and a B747 requires, for the same trip, about $6600. The B747 may carry three to four times more passengers, and for this reason, its unit cost expressed in cents per passenger-mile is significantly less than the B737. In general, the larger the aircraft, the less is its unit cost and the higher is its productivity.

Time. There is another factor having a certain effect on cost, and this is time. More revenue is generated when an airplane is flying faster for a given time period. Airplanes do not contribute to profit while sitting on the ground.

Let's assume that the flight originates from an airport with average congestion. Taxi-out time at the departure airport and taxi-in time at the destination are recorded as a total of 20 or 30 minutes. The effect of a set of short trips lowers the daily utilization, and this, in turn, increases the operating cost. Thus frequent short trips cause the airplane to spend more time on the ground with more taxi fuel and taxi time. Consequently, the yearly utilization of an airplane greatly affects its profit picture.

For longer trip lengths, the taxi time–trip time relationship is more favorable, and the daily utilization is improved. The same is applicable

TABLE 4.21   Productivity Factors

At 31,000 ft	B737/DC-9	757/DC-8	DC-10/L-1011	B747
Mach no.	0.76	0.80	0.82/0.83	0.84
Speed in knots	446	469	481/487	493
Speed in mi/h	514	540	554/561	568

to taxi fuel and trip fuel. In conclusion, on a yearly basis, longer trips have fewer departures, less taxi-in and taxi-out time, and less taxi fuel; therefore, longer trips contribute to lower units costs.

Speed. Smaller airplanes fly slower and generally have shorter trips, so their operational Mach number is lower. Larger airplanes fly longer trips at a higher Mach number. As shown on the block speed versus range chart (Fig. 4.48), speed increases with range. For a period of let's say 1 year, the airplane could cover more miles and generate more revenues.

Airplane size effect. A larger airplane can fly faster, its average trip length is longer, and it can carry more load expressed in tons. Furthermore, there are additional advantages due to size and speed:

- More flying time
- Less taxi time and taxi fuel
- More distance covered
- More tonnage carried
- Less maintenance work required (cycle per flight)
- Less passenger processing cost
- Less station cost
- Less aircraft servicing cost
- Less airport landing and other fees
- Higher and better utilization

When trip distance is doubled, trip fuel, trip time, and trip cost are *not* proportionately twice as much.

To see these elements on the opposite end of the spectrum, consider the operations of Qantas and Southwest Airlines. Charging relatively higher ticket prices for shorter trips, Southwest is able to overcome the disadvantages caused by the short ranges of its operations.

In conclusion, airplane productivity is a function of size and speed. Airlines have to optimize these two elements based on their route structure in order to achieve higher revenues.

There are two ways to evaluate productivity, namely, applying equal utilization or equal miles. The vertical scale of Fig. 4.65 is expressed in revenue ton-mile per year in millions. In this comparison, the larger airplane is better by 173 percent where equal utilization is concerned and by 167 percent where equal miles are the basis for comparison. For passenger operations, revenue passenger-mile per time unit should be applied.

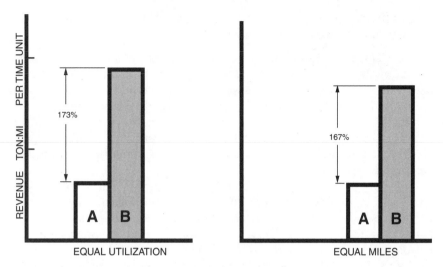

Figure 4.65   Equal utilization and equal miles comparison.

4.30   Airplane Profit Potential

A graphic presentation is available to indicate profit potential for a given airplane over its range. The first step is to draw a payload versus range curve. The load could be cargo, passengers, or both (Fig. 4.66). Once the payload versus range curve is established, the direct

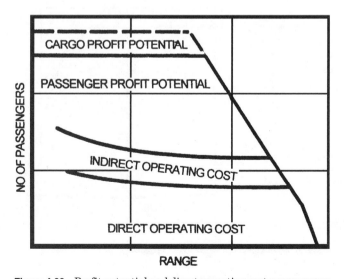

Figure 4.66   Profit potential and direct operating cost versus range.

and indirect operating costs are placed on the chart, expressed in number of passengers (or in payload for cargo operation).

Note: The maximum number of passengers is 250. The airplane can carry in its lower lobe an additional cargo of 6400 lb. Converting this weight into numbers of passengers, this would be 32 passengers (6400/200), where 200 lb is the average weight of passengers plus baggage.

By subtracting direct and indirect costs expressed in number of passengers, the profit potential is obtained by the remaining number of passengers (Fig. 4.67).

This chart is prepared for a passenger operation. The airplane carries a full load of passengers (275) up to 3500 nautical miles. As the range increases beyond this, the passenger load decreases and is traded for fuel to gain range. Knowing the cost of operation, the breakeven passenger load is defined. The difference between the full load and the breakeven load is the profit potential. For example, at 500 nautical miles, the profit potential is the difference between the full load of 275 passengers and the breakeven load of 155 passengers. Thus 120 passengers are generating the profit. At 3500 nautical miles, the breakeven load is 77 passengers, so 198 passengers are generating the profit. Beyond 3500 nautical miles, the profit potential declines because the airplane cannot carry the full load of passengers.

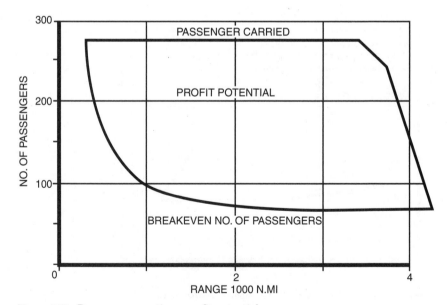

Figure 4.67 Passenger operations: profit potential versus range.

For a given trip, the same evaluation is presented here. The trip distance is 2450 nautical miles, and an enroute wind is considered. Although the direct and indirect operating costs are twice as much for the larger airplane, the profit potential is *not* twice more. This is presented in Fig. 4.68.

For a type A airplane, 37 passengers cover direct costs, while 35 passengers cover indirect operating costs. The total number of passengers is 147. Thus 75 passengers are generating profit. The breakeven load factor is 48.9 percent. For a type B airplane, the breakeven load factor is 55.3 percent. The operating cost (DOC + IOC) for the type B airplane is about twice that of the type A airplane, and the profit potential should be at least twice as much. For type A, the profit-generating passenger number is 75, and for type B, this is only 108 passengers.

The amount of cargo in the lower deck of passenger airplanes contributes to the profit potential. Figure 4.69 presents the magnitude of this contribution as a function of range. The cargo is presented in equivalent number of passengers. A scale of load factor can be assigned to this cargo.

This airplane carries a maximum of 369 passengers, and the cargo contributes an equivalent of 66 additional passengers. At 210 lb per passengers plus baggage, this is a total of 13,860 lb of cargo. A load-factor scale can be placed on this chart. Carrying this additional cargo of 13,860 lb, the maximum range of the airplane is 4550 nautical miles. Carrying only a 50 percent load factor, or 6930 lb, the maximum range is 4700 nautical miles. Carrying no additional load, only the maximum number of passengers (369), the range of the airplane is 4900 nautical miles.

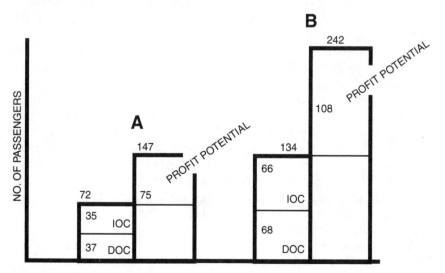

Figure 4.68   Profit and cost comparison.

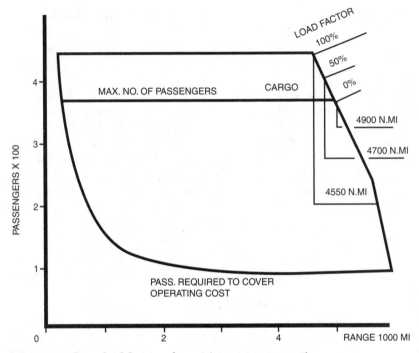

Figure 4.69   Cargo load factor and range in passenger operations.

The number of profit-generating passengers versus range is presented in Fig. 4.70. The highest profit range is located where the cents per passenger-mile is the lowest. This area is close to the maximum payload range point. This graph could be shown with a second (vertical) scale for the actual profit potential in dollars (DOC only).

For two airplanes, the overlapping area (Fig. 4.71) is where both airplanes are profitable. Based on the airline route network, the proper airplane with the highest profit potential is preferred.

For cargo operation, a breakeven payload versus range chart could be developed. The breakeven load is presented on the graph in order to see the profit potential as a function of range. Here, three lines are presented on the curve due to tare-weight influences. Pallet weights are less than container weights, so the airplane can carry more payload in palletized configuration (Fig. 4.72). Considering no tare weight indicates the maximum profit potential. The weight difference between a palletized and a containerized configuration could be some 20,000 lb that do not generate profit.

A graph similar to the one in Fig. 4.71, presented for cargo airplanes, is shown in Fig. 4.73. The common area of profitability is shaded.

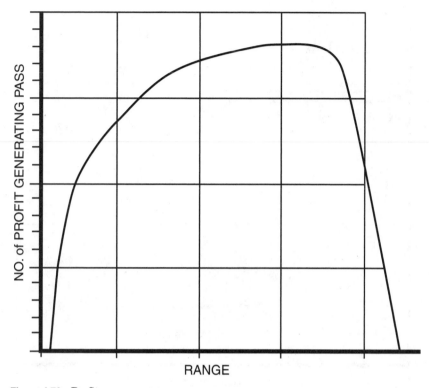

Figure 4.70 Profit versus range.

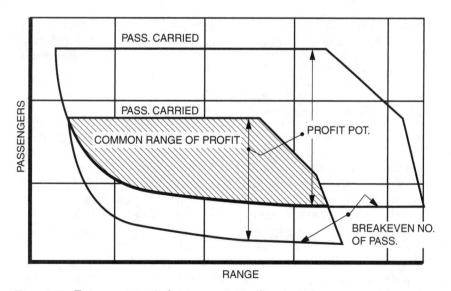

Figure 4.71 Two passenger airplanes: common profit potential area versus range.

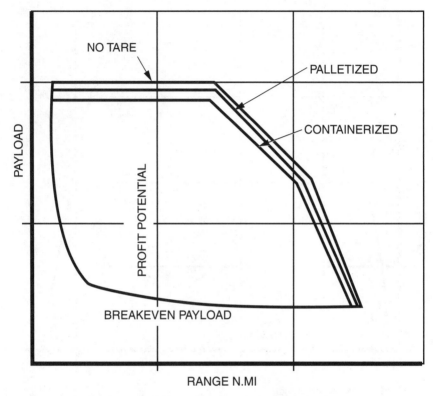

Figure 4.72   Profit potential with various tare weights versus range.

Note: Once trip cost and ticket price and/or freight revenues are known, the breakeven load (passenger and/or freight) is defined based on DOC only. Any load beyond the breakeven load is contributing to profit and IOC.

4.31   Economic Evaluation of Selected Trips

Figure 4.74 shows payload versus cost in cents per ton-mile for the JFK to London nonstop trip (standard day, enroute wind). An enroute tailwind of 21 knots is considered (85 percent probability of not being exceeded). Until the load is less than 80,000 lb, one single type A aircraft is an economical aircraft for this trip. Introducing a one-stop operation with a slight increase in cents per ton-mile (two landing fees, station costs, and additional block times and block fuel contribute to a higher operating cost), the payload-carrying capability could be extended up to 90,000 lb. Operating two type A airplanes is economical until the daily load is below 160,000 lb. Once the available daily payload is above this load, a type Z aircraft proves to be more econom-

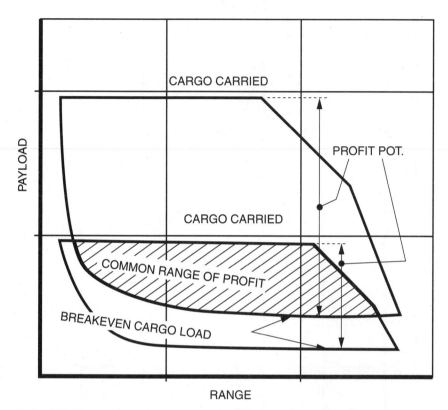

Figure 4.73 Two cargo airplanes: common profit potential area versus range.

ical. Introducing a one-stop operation, the daily available load can be extended for the two type A aircraft up to 180,000 lb, but for 160,000 lb and more, the type Z aircraft is more feasible. There is also another consideration: Having two type A aircraft available, the flexibility of scheduling two airplanes has certain advantages. On the other hand, cost figures would be higher because they would be doubled; for example, landing fees would cost twice as much as usual.

The marketing forecast is an important tool in selecting the proper airplane. The study for the charts shown here was conducted for JFK to LHR trip with an enroute tailwind. It is recommended to develop these kinds of charts for some primary trips in both directions. The payload here represents the daily available load.

Cost and revenue are shown in Fig. 4.75 for an eastbound flight from JFK over the Atlantic. This trip consists of two segments, namely, from New York to London and then to Frankfurt, and this is a one-stop operation. Two type Y aircraft are shown in this study and compared with a larger type of airplane named type X.

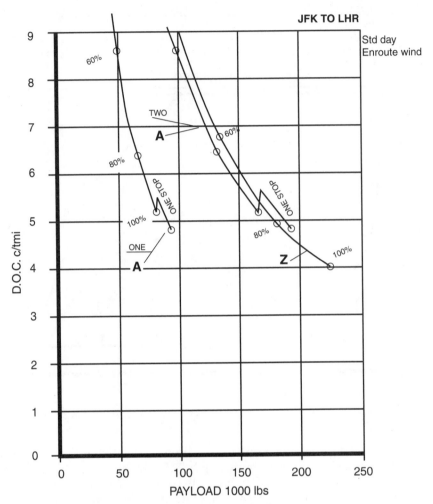

Figure 4.74 Cost in cents per ton-mile versus payload.

There are two scales on this chart. The first scale (horizontal) indicates the available load per day for this operation. Assuming 312 trips per year, the second (horizontal) scale is the yearly potential. The vertical scale indicates the dollar figures for revenue and cost for daily and yearly values at the time of preparation of this example. The revenue was based on 30 cents per pound for these trips. For one type Y airplane, the breakeven load is about 65,000 lb, or about 75 percent load factor. This is the point where the cost (horizontal) and the revenue (slanted) intersect. Below this intersection the chart shows operating losses, whereas profit is indicated above it (based on DOC only, before taxes). The shaded sections on this chart show the profitable areas for both aircraft.

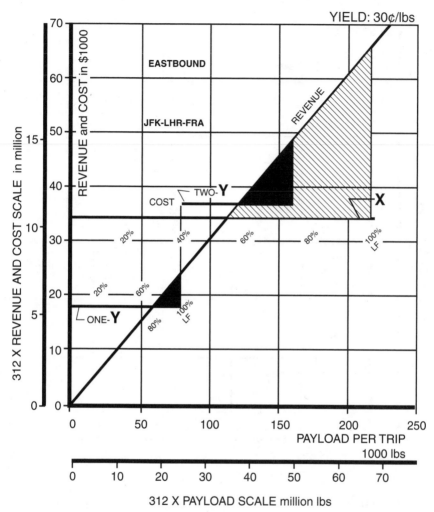

Figure 4.75   Revenue and cost versus payload.

With two type Y airplanes, the breakeven load is 130,000 lb. Airplane type X is profitable from 120,000 lb and up. For this reason, a type X airplane is preferred when the daily load is above 120,000 lb. The shaded area is profitable.

Another way to show profit or loss is the presentation in Fig. 4.76 for the same eastbound trip: New York to London to Frankfurt. The vertical scale shows the profit. The breakeven line is where the profit is zero. Above this line, the profit is positive, and below this line, losses are presented. The upper horizontal scale shows the load capability for a given trip. The lower horizontal scale shows it for yearly operations. Here too, the shaded areas indicate the areas of profit for this operation.

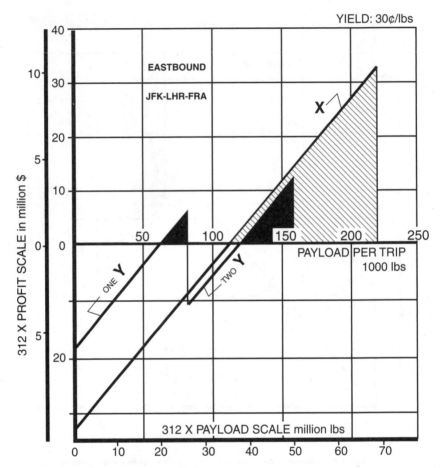

Figure 4.76 Profit/loss versus payload.

This chart indicates the difference between revenue and cost. The vertical scale on this chart is the profit scale. Reviewing the results on this chart, the conclusions shown in Table 4.22 can be drawn.

This table shows the yearly profit and losses as a function of load carried on these aircraft. It shows the losses when insufficient load is carried, the breakeven load, and the airplane maximum load-carrying

TABLE 4.22   Profit-loss Presentation for Two Airplane Types

One Y, Yearly load		Two Y, Yearly load		One X, Yearly load	
Loss	Profit	Loss	Profit	Loss	Profit
0–18	18–26	0–36	36–52	0–35	35–65

In millions of pounds (British).

capability. Profit is generated between the breakeven and the maximum load.

The next step is to proceed with marketing analysis. In selecting the proper aircraft, several other marketing factors should be considered. One of these would be the seasonal fluctuation. Perhaps one type Y airplane would be enough for certain markets and for seasonal peaks, and short-duration leasing, a type X airplane would be the solution. Just the reverse may be true by purchasing one type X and leasing one or two type Y airplanes for the peak season.

Profitability can be presented as a function of the profit potential seats per breakeven number of seat relationship. For cargo operations, profit potential, cargo in pounds per breakeven cargo in pounds, is used. The higher this value, the better is the profit, as shown in Fig. 4.77.

Note: Passenger operation: maximum seats is 394, breakeven seats is 189, and profit potential is 205 passengers. Profit potential per breakeven load is 205/189 = 1.08, or 8 percent.

Figures 4.78 and 4.79 are a good visual presentation of the relationship between profit potential and breakeven values. In Fig. 4.79, the total load is on the vertical scale, and the breakeven and profit-generating loads are presented in a different manner.

In Fig. 4.80, the payload is a function of profit per ton-mile for a certain trip length. Entering with net payload (no tare weight included) and moving up to the airplane line, the profit/ton-mile or loss per ton-mile is shown. (Tons can be converted to number of passengers.) Three types of aircraft are shown in this graph. In order not to clutter the figure, only aircraft type B shows its payload capabilities at various load factors.

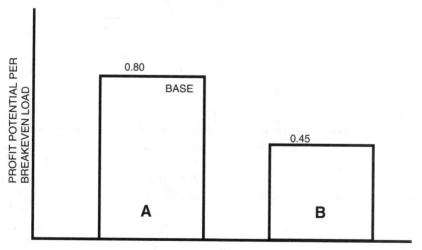

Figure 4.77   Profit potential per breakeven load.

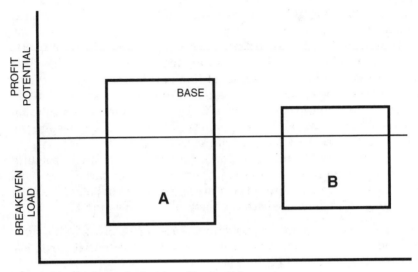

Figure 4.78   Breakeven load and profit potential.

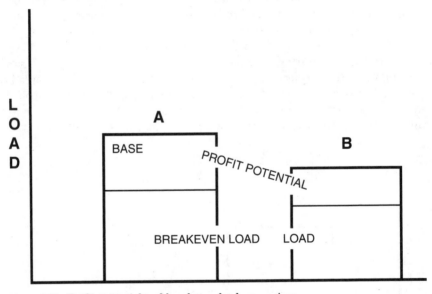

Figure 4.79   Profit potential and breakeven load comparison.

Figure 4.80 shows the minimum load the airplane has to carry to be profitable for this selected city-pair. Examining a longer or shorter trip length, the profit picture can change in any direction.

For a type A airplane, a minimum of 42,000 lb is the breakeven load; for type B, it is 74,000 lb, and for type C, it is at least 110,000 lb at a range of 2000 nautical miles.

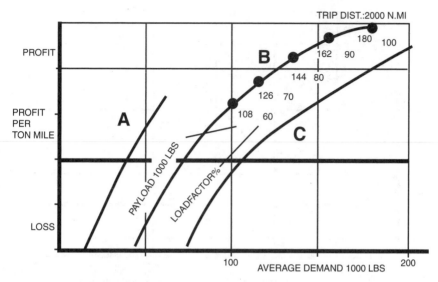

TRIP DIST.:2000 N.MI

Figure 4.80   Profit comparison for various types of airplanes.

Figure 4.81 shows for a selected airplane type the annual revenue and profits for passenger operations. For up to about 65 passengers, there is no profit, and the annual revenue is about $5 million. This is the breakeven point where revenue equals cost. Carrying—for the sake of argument—250 passengers, the annual revenue is about $16.4 million, and the profit a little above $1.2 million. This chart is prepared for a certain trip length. It is assumed that the takeoff weight is not limited, and an enroute wind is applied. This chart shows the breakeven number of passengers and the associated annual revenue and profit. Furthermore, for any number of potential passengers, the annual revenue and profit are easily extracted.

Another approach is a comparison between various airplanes concerning the amount invested per seat. This is presented in Fig. 4.82. Three airplanes are compared; type A airplane investment per seat value is 17 percent higher than the airplane selected as the base airplane, whereas type B is 21 percent higher. The comparison is presented in bar-chart form.

A comparison of relative operating cost is presented in Fig. 4.83, where the horizontal scale is the relative cost per trip in percentage and the vertical scale is relative passenger- or ton-mile cost in percentage. The slanted lines could represent numbers of passengers or tons. Positioning the selected airplane to the zero point on both scales, any other airplane can be positioned on this chart related to the selected airplane. They will fall in four quadrants, as shown in Table 4.23.

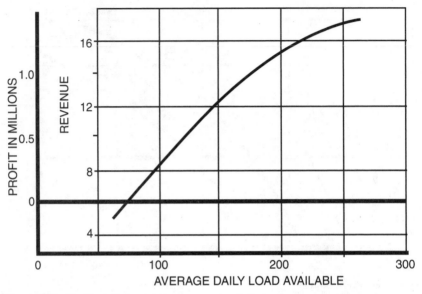

Figure 4.81 Profit and revenue versus average daily load.

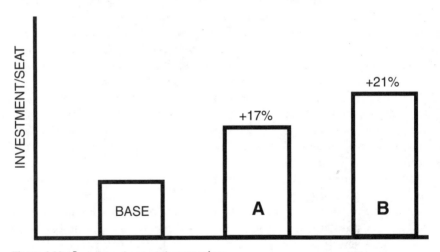

Figure 4.82 Investment per seat comparison.

The next step is to position all the other airplanes into these quadrants. This presentation shows their relationship, but above all, the airplane in the fourth quadrant indicates that its trip cost and ton-mile cost are lower than the selected airplane in the zero position or at the reference point.

TABLE 4.23   Quadrants Designation

Position	Relative cost per trip, %	Relative ton-mile cost, %
First quadrant	Negative	Positive
Second quadrant	Positive	Positive
Third quadrant	Positive	Negative
Fourth quadrant	Negative	Negative

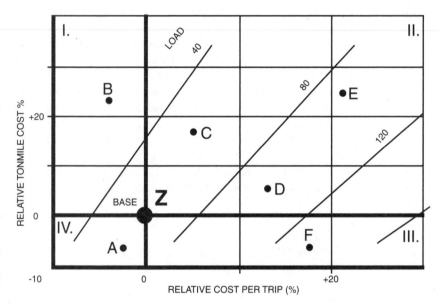

Figure 4.83   Relative ton-mile cost versus relative trip cost.

Similar charts can be prepared where relative indicators could be developed for evaluation purposes.

A graphic presentation is shown in Fig. 4.84 for the profitability of two airplanes. This bar chart indicates the relationship between various cost elements. The condition should be spelled out and listed on the chart or on an attached sheet.

This chart indicates the relationship between DOC, IOC, profit potential, and cargo-carrying capability, if any, for both airplanes, and it could be compared with other type(s) of airplane(s) in the same manner. Substituting loads for passengers and applying yield per cargo operations, a similar chart could be developed for freight operations.

For a type A airplane, 46 passengers generate revenue that covers direct operating cost and 52 passengers cover indirect operating cost. This aircraft is able to carry a total of 160 passengers. The profit potential is based on 160 − 98 = 62 passengers. This airplane carries a certain amount of load on the lower level, and this is converted into the equivalent number of passengers, which extends the profit potential range of this airplane. This load is presented for various load factors and is indicated on the chart.

Aircraft type B does not carry any additional cargo load. The conclusion of this chart is that the type B airplane has about one-third less profit potential than type A. Only 134 − 93 = 41 passengers generate profit, and no cargo is carried. Its operational cost is very close to that of aircraft type A. Based on the figures presented, type A is the preferred choice, assuming all other factors are equal.

Profit potential can be presented for the economic life of the airplane by developing the cumulative profit potential chart (Fig. 4.85). The yearly profit potential is presented with an escalation rate included. It is based on aircraft performance, yearly mileage, and naturally, an escalation rate. This chart compares two or more aircraft that are in service for the same time period (4 years); one generates $4.08 million in profit, and the other generates only $1.80 million. Furthermore, assuming the same amount of profit, it takes one aircraft 5.8 years and the other 8 years to generate the same profit (Fig. 4.85). A similar presentation could be developed for direct and indirect operating costs.

4.32   Time Effect on Profit

In certain cases, time has an eroding effect on profit. In order to demonstrate this, a small table is prepared for a few selected aircraft.

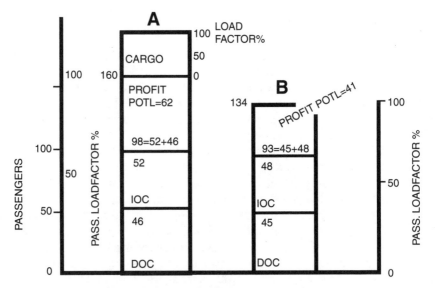

Figure 4.84   Profit potential: direct and indirect cost comparison.

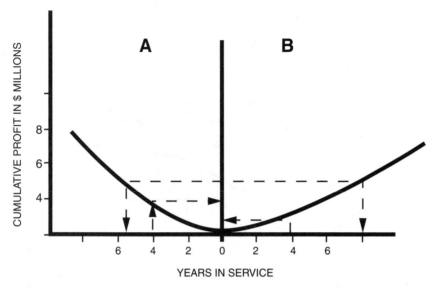

Figure 4.85   Cumulative profit comparison.

In this table, the direct operating cost is shown side by side for the years 1980 and 2000. Increments are in percentages (Table 4.24).

The direct operating cost figures are industry averages for the indicated time periods. To illustrate this point, Fig. 4.86 is presented. Without going into an elaborate discussion of buying power, constant or current dollars, or inflation, this graph theoretically could indicate the following:

1. The cost of operation as a function of time is escalating at a certain rate (Table 4.24).

2. The operational revenue is increasing due to the airline's increased market share over the years. This increase could be limited only by the airplane capacity, and at that point in time, management should consider changes because without change, profit cannot grow anymore.

3. Airplane maximum capacity is constant.

TABLE 4.24   Historical DOC Data

Aircraft	DC-9-30		DC-10-30		B727-200	
Year	1980	2000	1980	2000	1980	2000
DOC in dollars per block-hour	1400	2000	4200	5000	1970	2500
Percentage	42		19		26	

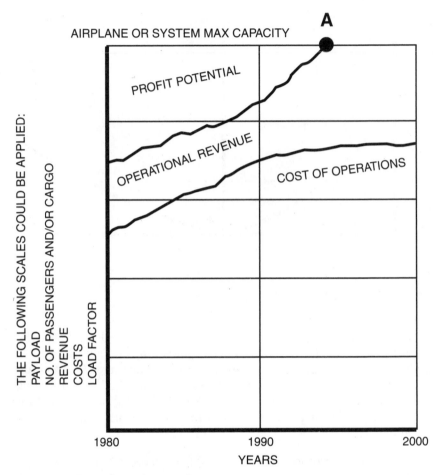

Figure 4.86   Time effects of various factors.

4. Beyond point A, the airplane cannot carry more passengers, and the cost is increasing slightly. Thus profit is decreasing in relationship with the cost increase.

It can be stated that the airplane has the same maximum seat capacity in 2000 as it had in 1980. The profit in face of increased operating cost is decreasing. There are three ways to increase profit:

- Increase the number of seats, if possible.
- Increase revenue by raising the ticket price (by nature, these increases have a limited effect on profit).
- Purchase a new airplane with more seat capacity.

Note: The vertical scale could be load factor, where breakeven load factor would be the indicator of cost, operational load factor the actual passengers carried, and airplane capacity 100 percent load factor. Furthermore, this scale can be expressed in tons, passengers, or dollars. If the operational revenue hits the maximum, shown as *A* in the graph, revenue cannot be increased. This defines the airplane's useful life for this particular schedule, market, and airplane combination. From this point on, profit growth is reduced to zero.

Note: The ground rules for economic studies may vary from airline to airline and from time to time. The fuel price leaped from 10 cents per gallon to close to $1 per gallon. In 1996, the price was around 50 to 58 cents per gallon. Inflation rate also escalates cost elements. Utilization varies with airlines. Depreciation schedules change with tax laws. The studies presented here were conducted at different times with different ground rules, cost elements, depreciation schedules, etc. The graphs shown are for the mere purpose of studying the methodology of how to establish the right procedure for the right evaluation. At the time of an evaluation, the prevailing values, cost elements, assumptions, taxes, and depreciation schedules are applicable.

In conclusion, it can be stated that any economic comparison must set its own ground rules. These rules always should be spelled out before any aircraft evaluation takes place. Because of changing conditions (tax laws, fuel prices, depreciation, etc.), each report and chart should be clearly labeled with the date of preparation.

5

Cargo and Terminal Operations: Total-Cost Concept

5.1 Air Cargo Market and Cargo Operations

The cargo industry has a long past that started way back in 1907. A 19-year-old man named James Casey initiated merchandise home delivery from a department store to its customers in Seattle. His company, called American Messenger Service, later became United Parcel Service.

In 1918, the U.S. Congress assigned the Army to carry mail on the East Coast, and 1 year later, Boeing launched its B-1 mail carrier airplane. In 1925, the United States Postal Service established nationwide mail service. In 1926, Varney Airlines received by bid the mail-carrying services on the West Coast. During the same year, Robertson Aircraft was flying mail between Chicago and St. Louis. Charles Lindbergh became its chief pilot. This company later became American Airlines. In 1933, Douglas Aircraft manufactured the twin-engined DC-3. It had 21 seats and, by the early 1940s, carried the bulk of domestic passenger service. In 1938, the Civil Aeronautics Board (CAB) was established to regulate the passenger and cargo industry. In 1942, American Airlines started its cargo service across the nation with DC-3 airplanes.

During World War II, the Air Transport Command and the Naval Transport Service exposed servicemen to handle and ship military cargo by air. At the end of the war, the veterans came home with the idea of establishing commercial air cargo services. One of these veterans was General Claire Lee Chennault, who founded Flying Tiger Line in California. It operated for about 40 years in the Pacific area. There were several other airlines operated by veterans, but neither the demands nor the airplanes could sustain a profitable operation. Competition from traditional airlines was very tough, and for this reason, these fledgling companies could not survive.

Flying Tiger Line on the West Coast and Seaboard World Airlines on the East Coast were the two largest and most successful cargo airlines after the war.

The Berlin Airlift took place in 1948 for a period of 15 months. This was a good way to demonstrate to the world the significance of air cargo service. Many useful lessons were learned on the importance of logistics.

In 1958, Douglas and Boeing Aircraft Manufacturers ushered in the jet age; DC-8 freighters and the Boeing 707 aircraft were introduced. Their contribution to the cargo industry was twofold, namely, speed and volume. In order to move goods efficiently in and out of the airplane, pallets, igloos, and containers were standardized. To handle the increased volume and save time, computerized cargo terminals were established.

In 1968, the wide-body Boeing 747 aircraft was introduced to the industry. Its cargo-carrying capability of 250,000 lb was about 2½ times more than the DC-8's volume. Other, smaller wide-body airplane types were built, the Lockheed 1011 and the DC-10, and later its derivative the MD-11. In 1972, F. W. Smith established Federal Express to carry checks between banks and the U.S. Federal Reserve. Also in 1972, Seaboard World Airlines received its first B747-200. It planned to carry sea containers, but later, for economic reasons, it ceased to use them. A year later, Flying Tiger Line introduced its first B747 airplane.

In 1978, the Airline Deregulation Act was signed, and the following year, Flying Tiger bought Seaboard World Airlines. According to its plans, it would create an efficient Atlantic and Pacific operation. Nine years later, Federal Express bought Flying Tiger for its routes, fleet, and customer base.

In 1993, Boeing launched its B747-400 airplanes, which had a higher payload than the previous 747 series aircraft.

There are a few well-known companies handling and carrying cargo. Some are listed below:

Atlas Air

Burlington Northern Air Freight

Danzas

DHL

Emery

Kitty Hawk

Polar Air Cargo

SwissGlobalCargo

TNT

This century will see the development of new and improved services in air cargo. The present paperwork will be taken over by an efficient computer network. Improvements in cargo handling on the ground are also on the list. The preceding is a very short description of air cargo history. There are a few books available on this subject that deal with the matter in greater depth.

All economic forecasts draw an optimistic picture of the future of the air freight business. Freight will grow about twice as fast as passenger traffic, and this will be about 7 to 7.25 percent.

Total freight carried in 1997 was about 26 million tons and about 103 billions in ton-kilometers. Tables 5.1 and 5.2 show the increase in freight development over recent years. Many factors affecting air freight growth have positive values, such as the low unemployment rate and factory production. All in all, an optimistic picture is emerging as far as the future of air freight is concerned.

There are a few clouds on the horizon that hopefully will not affect the future. Can the superstructure carry the additional load? If nothing else, clogged airports cause delays in growing proportions. The top 25 airports accounted for some 11 million aircraft movements, and this was an increase of 1.3 percent over the previous year (1997). Chicago had some 850,000 aircraft movements, and this is the world's busiest airport. For the same year, manufacturers received orders for 1309 airplanes, and next year orders amount to some 1460 airplanes. These new airplanes need modern terminal buildings, efficient airports, modern cargo-handling equipment, and a smoothly running traffic control system.

The United States and South America will be a slow-growth area, whereas Europe and South Africa have good potential for an increase of 6 to 7.5 percent in cargo traffic. Europe and Asia will grow at a somewhat higher rate. Similarly, the Pacific market is also expected to expand. Over the next 15 years, a steady growth can be expected in international air cargo traffic despite inevitable ups and downs. The domestic cargo traffic growth is predicted to be half the international growth. This prediction changed when problems in Asia affected the world economy. The expected figures now are lower; however, air transportation by its nature still has a bright future. The United States developed the North American Free Trade Agreement (NAFTA) to expand its economic growth on this continent. The unique character of air cargo contributes to this growth.

Table 5.1 presents the world's freight volume figures based on historical and forecasted freight ton-kilometers.

The relationship between unit revenue and unit cost (Table 5.2) indicates a trend that looks very promising for the future because the revenue shows a higher rate of growth than the unit cost.

TABLE 5.1   World freight volume in ton-kilometer

Year	1985	1990	1995	2000	2005
Freight ton-kilometer	40	54	83	120	160E

Note: Freight ton-kilometer in billions.

TABLE 5.2   Unit revenue and unit cost history

Year	1990	1992	1994	1996	1998
Unit revenue	10.01	10.20	10.90	11.75	12.75
Unit cost	9.80	10.15	10.48	10.85	11.57

TABLE 5.3   Yearly fuel cost variations

Year	1990	1992	1994	1996	1998	2000
Fuel cost	78.50	63.60	54.10	66.40	51.50	65.60

Fuel cost has more effect on operating cost because freight airplanes are heavier than passenger airplanes and consume more fuel for the same trip length. It is interesting to see fuel price variations in the last 10 years (Table 5.3).

The tragic events of the September 11, 2001 attack on the United States will have unpredictable consequences on the air cargo business.

5.2   Future of Air Cargo

Intense globalization—knowing no international borders—will increase trade between nations in the future. Cargo traffic is predicted to have a growth increase in selected market areas.

Air cargo can be classified into two distinct groups: time- and value-sensitive groups.

Time-sensitive group

- Perishables, such as flowers, fruits, and vegetables
- Live animals
- Bakery products
- Various other food products
- Obsolescent items, such as apparel, footwear, etc.
- Emergency items, such as drugs, machinery parts, etc.

Speed of air shipment enables producer to

Decrease perishability

Handle emergencies

Gain market advantages

Profit by distribution economics

Value-sensitive group

▪ Medicines

▪ Electronic components

▪ Photographic equipment

▪ Chemicals

▪ Machine parts

▪ Fragile goods

Certain items are seasonal, and the demand is unpredictable. Inventories generally are kept low; hence speedy replenishing and just-in-time (JIT) supply are sometimes required.

Shipping by air has other advantages too; it does not need heavy packaging to reduce damage, the goods are not exposed to long travel times (such as by ships) where they are not watched around the clock, and the exposure to theft or deterioration is much less. Insurance and packaging costs therefore are lower than with other types of transportation. By using air transportation, inventory and warehousing costs and sometimes ground transportation costs are reduced significantly. A large inventory requires greater expenses, and it could become obsolete within a short time period. Once it is obsolete, the goods have to be handled anew, causing additional labor and other costs. The total cost should be analyzed, not just the air freight transportation cost only. The cost of capital is less; similarly, insurance and other expenses are reduced. Certain warehouses could be eliminated altogether. For example, rent at JFK Airport is about $14 per square foot with an additional maintenance and insurance cost. This illustrates clearly that warehousing costs are not negligible.

Manufacturers are working now on a global scale. For example, material is shipped from one continent, where it is mined, to another, where parts are made. The parts are then shipped to a third continent to be assembled, and from there to the warehouse (if any) or to the market. Logistic management now has an important and larger role in the planning process than in the past. This is what we call *multinational manufacturing.* It is well known that Texas Instruments ships chips to Singapore for manufacturing and then returns the product to the United States. Air cargo plays an integral part in this process.

In the future, air cargo has the potential to grow by concentrating on carrying perishables, emergency items, inventory-replenishment items, and any other commodities for which air shipment is more favorable. Furthermore, production growth in high-value goods requires faster movement. Factors contributing to the growth of air cargo are more industrialization in developing nations, higher interest rates resulting in lower inventories, and more efficient use of funds for higher profits.

5.3 Cargo Market

The magnitude of the transportation budget is about one-quarter of our gross national product. Freight transportation bills run about $150 to $170 billion per year. However, the share of freight transportation related to mail and passenger transportation is relatively constant. Only 0.3 to 0.6 percent by weight and 5 to 35 percent by value of our international trade goes by air.

A small percentage of cargo is moved only by air over the Atlantic market. Since it is a fraction of the total traffic, it has a potential in the future for further increases. Certain items will never be carried by air, such as wheat, metal ingots, and coal. The value in dollars per pound decides whether the transportation mode will be airplane or ship.

A new concept has developed: the so-called total-cost concept. Air shipping by itself is more expensive, but considering the total cost of shipping and the time savings, it may be competitive with other modes of transportation. Introduction of the total-cost concept changes the prevailing idea about the high cost of air transportation. Based on this concept, all cost elements such as warehousing, heavy crating, pilferage, shipping damage, insurance, total transportation time, and the cost of money have to be considered. Money invested in cargo is tied down for a longer period of time because of slow shipping speeds. Large inventories prevent money from being used in other areas. This justifies replenishing inventories on short notice, which is available by air only.

Air shipping at first glance costs more than other means of transportation, but looking at all the cost elements, in the final analysis, air transportation results in savings in certain instances. The express category of shipment, consisting of small packages, letters, or documents, has shown a high growth rate since the mid-1970s. One example is Federal Express, which developed the small package business with its express delivery system and door-to-door service. Its service is reliable and punctual. Besides Federal Express, United Parcel Service is another integrated operator with its own airplanes and trucks. Some freight forwarders collect and consolidate small shipments into larger ones and negotiate volume discounts with the airlines; others, like Emery, have their own fleet of airplanes.

Segmenting the cargo market contributes to effective marketing. A few examples are presented below:

By direction

By commodity

By shipment weight

By length of haul

By price elasticity

By direct versus indirect shippers

By scheduled versus charter requirements

By size of pieces

By cost of service

By profitability of service

Air cargo markets are characterized by their diversity:

By density

By need for pickup and/or delivery

By degree of shipper leverage on carrier

By need for special handling

By need for special carriage

By departure/arrival time requirements

By transit time requirements

By claims experience

By bulk versus unitized

There are various additional presentations in segmenting:

By true origin-destination

By airport origination-destination of shipment

By flight origin-destination

By customs requirements

By unit load device type

Considering aircraft type:

By aircraft configuration requirement

By aircraft size requirements

By belly/combi versus all cargo

By shipment frequency

Some other items are listed below:

By yield per pound or ton

By flag carrier

By general commodity versus specific commodity rating

By individual specific commodity rates

By commodity perishability

Cargo market information is collected, selected, and evaluated; it is the basis for further forecasts. Reliable forecasts come from reliable data. Here are some sources for cargo database development.

United Nations' trade data. Virtually all countries report the value of their imports and exports to the United Nations. There is a report based on a five-digit commodity breakdown, value, weight, and so on.

U.S. import/export statistics. U.S. imports and exports are quite detailed and include information on air shipment and individual customs districts.

- Statistics are kept between 43 customs districts and other countries
- Five-digit commodity breakdown
- Value and weight
- Air and vessel modes
- Special trade data service providing market share and other data

For the future, major opportunities will exist, and some airlines with good management will benefit. The airline recognizing and acting on its customers' (ever-changing) needs will be the most profitable.

Future opportunities

- Increased understanding of the value of time
- Rising real gross world product
- Wider distribution of the world's wealth
- Easier market entry and exit
- Industrial miniaturization

Future problems

- Real cost increases
- Airport/airway constraints
- Stronger competition
- Stricter environmental requirements as far as noise, smoke, and traffic are concerned
- Closer public scrutiny
- Capital
- Management inattention to problems

5.4 Cargo Characteristics

Air cargo, by definition, consists of freight, mail, and express items. Table 5.4 shows the differences between cargo and passenger loads.

Our exports are not in balance with imports; in fact, U.S. imports exceed exports. As an example, air cargo directionality is shown between the United States and the Far East in Fig. 5.1.

5.5 Cargo Seasonality

Passenger traffic volume generally is less between January and March and between October and the middle of December. It increases between April and September and between Thanksgiving and Christmas. Cargo could be low in the summer and high during the winter. In the case of winter garments, the highest volume is experienced between the middle of August and the middle of November. This illustrates the seasonality of air cargo. A comparison in Fig. 5.2 shows the seasonality of passengers and garments.

TABLE 5.4 Cargo characteristics

	Cargo	Passenger	Cargo problems
Move	Economic reason	Personal reasons	
Direction	One direction	Round trip	Backhaul
Weight	Heavy per volume	Light per volume	Variable/shapes/weight
Trip	Predesigned	Individual choice	Each movement controlled
Timing	Late-evening hours	Daytime	Curfew/noise
Intermodal	Yes	No	
Routing	Not important	Important	
Activities	Brokers, packers, and trucks	None	

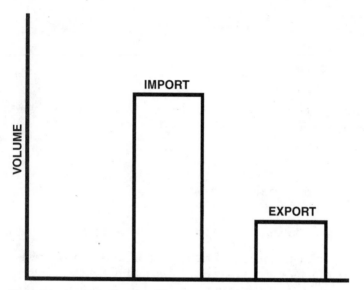

Figure 5.1   Air cargo directionality.

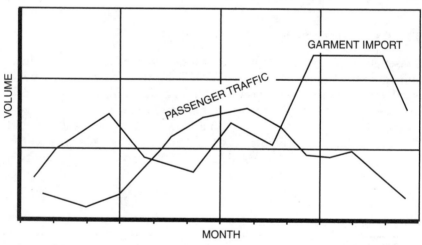

MONTH

Figure 5.2   Passenger and textile wear seasonality.

5.6   Cargo Airplanes

The best known dedicated freighter is the B747F airplane. Earlier, Canadair built its swing-tail CL-44, Lockheed the Hercules L100 series (where the floor is close to the ground for easy loading), and the Armstrong Whitworth Argosy Airplanes. The cargo-carrying capability of these freighters and the excess capacity of passenger airplanes (in the lower hold) are available. Special LD containers were developed where enough belly space is available or pallets (with lower heights) also can be used to hold cargo in the belly of a passenger plane. The B747F could carry 9 pallets up to 64 inches high in its lower compartment. Several airplane types could carry a full load of passengers on the upper deck as well as a sizable load below deck.

The problem is that passengers have to depart on time in a time period that may not be convenient to cargo shippers. Nevertheless, this cargo is the frosting on the cake for such carriers by generating additional revenues.

Convertible airplanes are able to move between their options, which could be passengers in the summer and cargo in the winter or cargo on weekdays and passengers on weekends. They are valuable because of their convertibility. There is a penalty to be paid for the use of these types of aircraft, however. The convertible aircraft is heavier (operating weight empty) than the pure freighter. Some airlines use "combi" airplanes, where both passengers and cargo are carried on the main deck due to a lack of enough passengers. Sometimes this means that the combination of available passengers and cargo together is able to generate profit, while alone, each would result in operating losses. With the exception of the B747 aircraft, all previous airplanes were designed as passenger airplanes with the ability to also carry cargo. The large volume of the B747 freighter will allow it to handle cargo demands well into the future.

The largest cargo airplane in the world is the Russian Antonov 225 airplane, with a payload of 250 tons.

5.7   Short Historical Background of Cargo Loading and Equipment

Basically, there are two methods of loading cargo onto an airplane. The first method is loading by hand, which was the only method at the beginning of the airplane industry. Besides passenger transportation, air mail was the first source of cargo revenue. The mail, packed in sacks, was placed on the seats and tied down. Even today, numerous small aircraft throughout the world carry small packages in this fashion. Small packages are accumulated, transported to the airplane by cars and/or trucks, and hand loaded. This type of loading is very ineffective because there is no way to position the cargo efficiently. Once the packages are inside the airplane, there is no time to reload them for better positioning. It is quite difficult to achieve good volume utilization with hand loading.

The second method is based on the need to load large-sized cargos quickly. It is a rather costly proposition to keep a large aircraft on the ground unnecessarily. In order to save time, space, weight, and money, the use of a flat panel, called a *pallet* or an *igloo,* was introduced, and then later, *containers* were introduced.

5.8 Development and Description of Airborne Pallets, Igloos, and Containers

Air cargo dates back to the time of World War I. Since then, airplanes have grown in size and carry increased amounts of cargo. Bulk loading became more cumbersome, and a mechanized pallet system evolved. In order to achieve an efficient loading system, standardization had to be introduced inside the airplane, as well as in the ground-support equipment. Standardization was an absolute necessity. Many organizations such as Air Transport Association (ATA), International Standards Organization (ISO), and Society of Automotive Engineers (SAE) developed an efficient loading system with standardized equipment. This enabled the interchange of pallets and containers on different types of airplanes. The military developed its own system. Here we deal with commercial cargo-handling equipment only. A few pallet and containers data are selected for presentation.

5.8.1 Pallets

The ATA netted pallet is shown in Fig. 5.3. Weights and dimensions are listed for a few pallet types in Table 5.5.

The relationship between tare weight and maximum loading capacity is 2.29 to 3.07 percent for the M4 pallet. The M5 pallet is a wider pallet, and its tare weight is 4.36 percent of the maximum loading capacity.

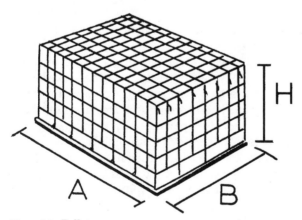

Figure 5.3 Pallet.

TABLE 5.5   Pallet specifications

Type	Dimensions (A × B) (in)	Height (H) (in)	Pallet + net weight (lb)	Volume (ft³)	Weight limit (lb)
M2	238.5 × 96	96	890 + 77*	1183	25,000
M4	125 × 88	86	250 + 35*	490	10,200
M5	125 × 96	118	610 + 45	705	15,000
A	108 × 88	87	200 + 50*	345	9,000

*Aluminum with balsa core.

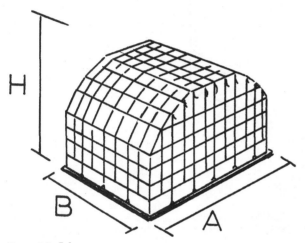

Figure 5.4   Igloo.

5.8.2   Igloos

Sometimes packages are slippery (mail in nylon bags) and nets cannot hold them securely. To use the cubic volume of the pallet and reduce the time for stacking efficiently, a compartment can be placed on the pallet, called *igloo,* that fits the inside of the airplane. In addition, igloos can be preloaded by customers and sealed to prevent theft. The ATA has developed specifications for igloos. They can be structural or nonstructural.

A structural igloo is a rigid shell with an aircraft pallet, and it restrains cargo without the use of nets. A nonstructural igloo is a bottomless rigid shell in combination with an aircraft pallet and net (Fig. 5.4).

Igloos are used on a variety of airplanes. Their tare weights are between 3.39 and 6.1 percent of their maximum carrying capability. A smaller-sized igloo, Type A4, has a tare weight–maximum load ratio between 5.87 and 7.1 percent for structural and 3.97 and 5.51 percent for nonstructural versions (Table 5.6).

TABLE 5.6   Igloo specifications

Type	Dimensions $(A \times B)$ (in)	Height (H) (in)	Tare weight (lb)	Volume (ft³)	Weight limit (lb)
A1, A2, A3	125 × 88	87	452–810	390–460	8000–13,000
A4	108 × 88	87	390–570	300–380	8000–10,000

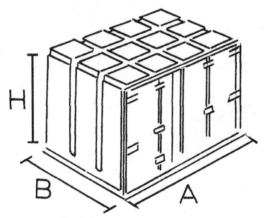

Figure 5.5   Container.

TABLE 5.7   Main deck container specifications

Type	Dimensions $(A \times B)$ (in)	Height (H) (in)	Tare weight (lb)	Volume (ft³)	Weight limit (lb)
M1	125 × 96	96	660–1150	570–610	15,000
M2	238.5 × 96	96	2100	1170	25,000
M3	125 × 88	96	930	560	15,000

5.8.3   Containers

Figure 5.5 presents an M1 container. Its tare weight–maximum weight ratio is between 6.82 and 7.6 percent for structural and 4.4 and 5.6 percent for nonstructural versions. The tare weight percentage for M2 structural containers is between 8.32 and 8.46 percent of its 25,000-lb load limit.

The M3 container is smaller than the M1 container (88 in). Its tare weight is 6.1 percent of its maximum load (Table 5.7).

5.8.4   Lower-lobe containers

For the lower lobe, a set of containers was standardized. A few are shown below. The LD-3 container is shown in Fig. 5.6. Its tare weight takes 7.0 to 10.5 percent of its maximum carrying capability. The LD-2 is smaller and is used on B767 airplanes. The tare weight is about 7.0 to 9.8 percent of total load (Table 5.8).

The LD-6 container (Fig. 5.7) has a tare weight percentage between 5.5 and 6.9 percent. It takes the whole width of the belly of the airplane (Table 5.9).

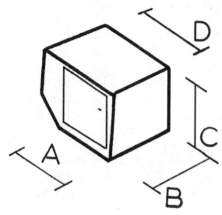

Figure 5.6   LD container.

TABLE 5.8   LD-1, LD-2 and LD-3 lower lobe container specifications

Type	A	B	C	D	Tare weight (lb)	Volume (ft³)	Weight limit (lb)
LD-1	61.5	60.4	64	92	210–370	160–173	3500
LD-2	47	60.4	64	61.5	190–260	120	2700
LD-3	61.5	60.4	64	79	246–370	149–158	2700

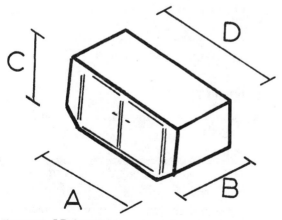

Figure 5.7   LD-6 container.

TABLE 5.9   LD-6 and LD-8 lower lobe container specifications

Type	A	B	C	D	Tare weight (lb)	Volume (ft³)	Weight limit (lb)
LD-6	125	60.4	64	160	385–485	316	7000
LD-8	96	60.4	64	125	390–540	245	5400

A wide variety of containers is available, so when selecting a container, the tare weight ratio to the maximum load capability should be considered. Generally speaking, for medium- to large-sized commercial jet airplanes, it is important to choose the right container type with the lowest tare weight.

Controlled-environment containers were developed in order to carry commodities requiring special treatment. Special containers have been developed for carrying cattle, horses, sheep, wine, flowers, etc. Some have refrigeration systems or use dry ice, and others are heated as needed. Some can carry automobiles.

Intermodal containers can be loaded on airplanes, transported by ship, or lifted to "ISO" chassis as part of surface transportation. They are built to specification to satisfy all the necessary requirements.

5.9   Tare Weight Effect on Payload versus Range Curve

In an earlier chapter when presenting a payload versus range chart for cargo operations, it was mentioned that the tare weight is considered zero. This shows the airplane's theoretical load capacity. From this baseline, applying the proper tare weight figures, according to airline policy, airplane, and type of load, the actual potential of the airplane can be obtained.

Table 5.10 has been prepared for a DC-10 airplane for palletized and containerized cargo. This should provide some idea of the actual tare weight for this airplane.

In a mixed configuration where pallets and containers are intermingled, the expected tare weight could be between 8800 and 26,691 lb.

5.10   Terminal Loading Equipment

Such equipment can be classified into two groups: passenger and cargo.

5.10.1   Passenger loading equipment

For passenger loading, various solutions are available. The most frequently used types are as follows:

1. The terminal buildings are designed in such a way that a powered, flexible corridor called a *jetway* extends to the airplane door level.

TABLE 5.10   DC-10 Tare weight presentation in palletized and containerized configuration

DC-10	Palletized	Containerized
Upper deck, number of	22	17
Weight (each)	300 lb	1099 lb
Volume (each)	494 ft^3	987 ft^3
Size	88 × 125 in	88 × 125 in
Type	**A1 netted**	**Contoured**
Upper deck tare weight	6600 lb	18683 lb
Upper deck volume	10864 ft^3	13026 ft^3
Lower deck, number of	5/5	26
Weight (each)	300/140 lb	308 lb
Volume (each)	379/265 ft^3	154 ft^3
Size	88 × 125/61 × 125 in	Container/LD3
Type	**A1 netted**	**LD-3**
Lower deck tare weight	2200 lb	8008 lb
Lower deck volume	3720 ft^3	4514 ft^3
Bulk volume	510 ft^3	510 ft^3
Total tare weight	8800 lb	26691 lb
Total airplane volume	**14598 ft^3**	**21293 ft^3**

The passengers simply enter from the terminal building into this corridor and walk into the airplane. There is no problem in boarding handicapped persons, and no weather affects this type of loading. Once the airplane is ready to depart, this flexible corridor is withdrawn (it moves on wheels) from the airplane door to the terminal building and is ready to serve the next airplane.

2. Rolling staircases are moved to the airplane. Passengers can walk out from the terminal building and climb up these staircases to the airplane entrance door. Sometimes these staircases are mounted on trucks and parked next to the airplane in such a way that the upper end of the staircase entrance platform matches the airplane door level. This type of loading has two disadvantages:

   a. It is exposed to weather conditions.

   b. It is difficult to cater to handicapped people.

3. The passengers board a bus at the terminal building, and the bus is driven to the airplane on the ramp. Here, the passengers are discharged and enter the airplane via a staircase. Certain bus types have the capability to line up the chassis to the terminal exit level; the bus moves to the airplane and elevates its chassis to the airplane door level. Weather conditions do not affect this type of loading. Since passengers can follow instructions and move independently, loading them is simple and requires less expense in terms of loading equipment than cargo.

A short description of baggage handling is presented here. Baggage is taken to the airplane and loaded into the belly. This can be done by a mobile belt conveyor or any other loader available and suitable for this job (Fig. 5.8).

Removal of passenger baggage is the reverse operation. The baggage is moved from the airplane to the sorting area, removed from the loading unit, and, by means of a conveyor belt, is moved out to the passenger waiting area. It arrives in the passenger baggage area and is transferred to the carousel (Fig. 5.9). Sometimes it comes out on a conveyor belt. There are many ways to bring the baggage to the passenger area, and the method used depends on such factors as the size of space, traffic and available resources.

5.10.2   Cargo loading and equipment

Cargo does not follow instructions and does not move under its own power; movement of cargo has to be planned to the minutest details. For these reasons, cargo equipment is diversified, and various pieces have been developed for specific uses or for a given type of cargo. There is a wide variety of cargo loading equipment, and some is discussed below.

Loading pallets, igloos, and containers.   Pallets are platforms where cargo is placed and secured with nets. They can be loaded well before the airplane arrives, and enough time is available to arrange the packages on the pallet for better volume utilization. At the destination airport, packages or loads are either delivered directly from the pallet at the truck dock or delivered with the pallet to the customer. This type of handling is called *unit load device* (ULD) handling.

Figure 5.8 Mobil belt conveyor loader.

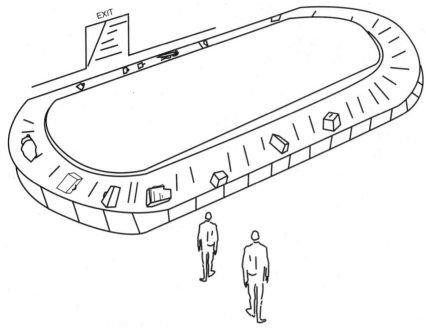

Figure 5.9   Passenger baggage carousel.

To build a pallet, a template is used that accommodates the contours of the airplane, as shown on Fig. 5.10. In this way, the pallet will fit the inside of the airplane properly. Across the pallet, a net secures the packages.

Pallets on Douglas and Boeing airplanes are 88 × 125 in. On the B747, there is a provision to handle 96 × 125 in pallets for maximum cube utilization.

For flexible pallets, a pallet supporter is necessary. This is a rigid frame with rollers on the top surface. The pallets can move in one direction only (Fig. 5.11) and have locks on the top to keep them from rolling off during towing.

A pallet on a pallet supporter (it has no wheels) is moved to the airplane by a forklift, which elevates the pallet up to the airplane door. Another type of pallet supporter has wheels and can be towed to the aircraft. A small tug attached to the side of the pallet dolly ensures easy maneuvering.

LD-3 container dollies are used mostly in the airline industry. In order to carry LD-2 containers, an additional LD-2 dolly stop is installed. A pallet transporter has inverted casters on the top and wheels on the bottom. It requires locks so that the pallet cannot roll off the top. Casters having a diameter of 2 to 3 in are mounted in an inverted

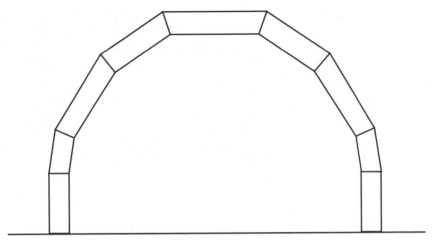

Figure 5.10 Template for pallet buildup.

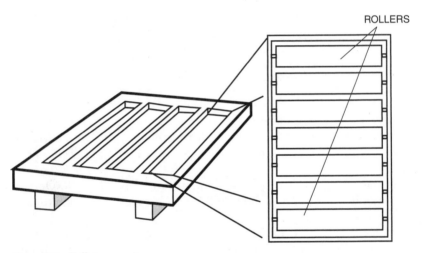

Figure 5.11 Pallet supporter.

position, free to swivel. The load, when placed on these casters, can be shifted in any direction because the casters follow the motion of the unit. After the pallet is secured, the transporter can be towed or pushed to the aircraft. Some are even self-propelled.

Forklifts. A wide range of lifting equipment is available. The forklift is one of them (Fig. 5.12). Forklifts are rated by their lifting capability. When lifting an 88-in-wide pallet, the center of gravity is assumed to be in the middle of the pallet. This means that a force could exert a momentum on an arm of 44 in away from the forklift and could overturn it. In order to prevent this, a 20,000-lb forklift is used to handle a maximum load of 10,000 lb only.

For the B707 and/or DC-8 aircraft, a special loader was developed. A ladder-type device up to the airplane's main floor lifts the pallet or container. The unit itself can be folded and stored on a pallet and flown with the airplane. Its advantage is that it can be used at airports without lifting equipment. Its disadvantage is that it takes a pallet position (Fig. 5.13.)

Figure 5.12   Forklift.

Figure 5.13   Loader.

When a pallet enters the airplane, it rests on a ball transfer plate (Fig. 5.14). The pallet has to be rotated 90 degrees in airplanes when loading thru cargo doors on the side. This is the purpose of the ball transfer plate, which allows the pallet to be moved forward or aft.

5.11   Cargo Systems Elements

Rollers are attached to the airplane's floor to facilitate moving pallets and/or containers. Douglas developed a shuttle system that is electrically powered, completely self-contained, and controlled by one person. The B747 has a mechanized drive that automatically moves pallets or containers in the aircraft. Thus two people are able to handle loading. The restraining systems are also automatic, except for the end restraints, which are set manually.

Cargo system hardware includes roller assemblies, casters, sill rollers, and drive wheels, etc. A roller conveyor assembly is shown in Fig. 5.15. An omnicaster roller assembly is shown in Fig. 5.16.

For safety purposes, once the pallets, igloos, and containers are loaded inside the airplane, proper tie-downs have to be applied. Forward restraint is the most important device affecting safety. There are three basic methods of achieving tie-down in the fore and aft directions:

1. *Individually fastening the load to suitable tie-down points in the floor.* This method is used not only on older aircraft but also on

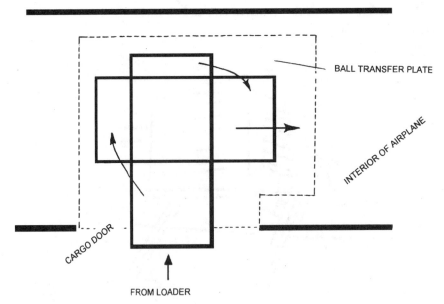

Figure 5.14 Side-door loading ball transfer plate.

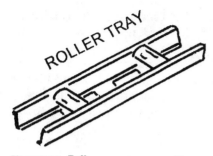

Figure 5.15 Roller conveyor assembly.

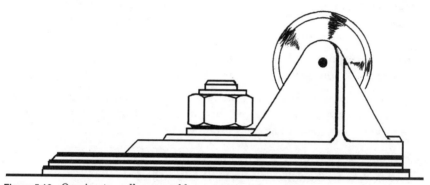

Figure 5.16 Omnicaster roller assembly.

larger, bulky, and/or dense loads that cannot be placed on pallets or fit into containers and yet must be tied down properly.

2. *Net used on pallet.*   The load from the net must then be taken into the aircraft basic structure by the pallet itself or by fastening the net directly to the aircraft basic structure.

3. This method is the most current and is used widely. First, a net is fastened to the pallet, and this net absorbs about 3*g* forward loads (during normal taxiing or running into soft ground); second, in a crash situation, the individual constraint offered by the net to the pallet would be assumed to fail. Meanwhile, the load would be restrained by a large net across the front of the cargo compartment. This net deflects while absorbing the load and stops it. Sometimes a cross-wise net is used and tied to the aircraft structure between each pallet.

Besides the forward barrier net, 9*g* individual pallet retention systems are available. The basic overthrow net is fastened to the forward and rear edges of the pallet in such a way that it will put the load directly into the floor fittings. In this way, the overthrow net serves a dual purpose because it takes out vertical loads and retains the load under conditions of emergency deceleration. The net shifts forward in such a way that the outline of the net would assume a fore and aft component.

In the past, there has been a lack of standardization of tie-down methods. The result has been that a pallet could not be readily transshipped from one type of aircraft to another one. This additional handling was an obvious source of delay and expense that had to be eliminated. Today, this problem effectively has been solved.

For emergency landing conditions, the FAR 25.561 regulation describes the barrier net installation. This should satisfy $9g$ forward, $2g$ upward, and $\frac{1}{2}g$ sideways requirements (Fig. 5.17). Forward barrier net installation should stop the load before it reaches the flight crew in the cockpit.

5.12   Development of the Container System: Intermodal Transportation

Besides pallets and igloos, a container system also was developed. This system has several advantages for air carriers, as well as shippers. The advantages for air carriers include:

- Preloading
- Lower cost for handling cargo in the terminal

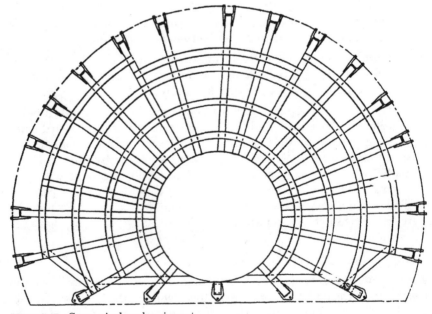

Figure 5.17   Cargo airplane barrier net.

- Reduced loading time on/off the airplane
- Safety (avoidance of theft or damage to cargo)

For shippers, the advantages include

- Preloading
- Lower cost of preparation because of a faster procedure
- Increased weight capability
- Better volume use
- Decreased damage and theft
- Faster transit time

Some disadvantages include

- Large investment needed
- Weight penalty
- Repair cost
- Back haul

The sizes of containers are generally standard: $8 \times 8 \times 10$ ft. The length can vary by as much as 20, 30, or 40 ft. Other significant factors in container usage are the relationship between tare weight and weight carried and tare-weight-to-volume ratio (density). Two types of containers are examined and presented in Table 5.11. For an M1 container, see Fig. 5.5.

The loss due to tare weight is 7.2 percent with M1 and 8.36 percent with M2 containers. This tare weight requires additional fuel to carry the extra weight. As a rough estimate, for 10 M1 containers, the tare weight is about 10,000 lb. The tare weight for 10 pallets is about 3000 lb. Thus there is a difference of 7000 lb in carrying containers up to a distance of 3000 nautical miles on a large cargo airplane, and an additional 1750 to 2100 lb (261 to 313 gal) of fuel would be needed for such a trip. Expressed in dollars at 60 cents/gal, the additional cost per trip

TABLE 5.11   Containers weight-volume comparison.

		M1	M2
Internal volume	ft^3	588	1178
Tare weight	lb	1080	2090
Size	ft	$8 \times 8 \times 10$	$8 \times 8 \times 20$
Weight limit	lb	15000	25000
Tare wt/wt. limit	%	7.2	8.36
Tare/volume	lb/ft^3	1.8	1.7

would amount to $156 to $187. Assuming 300 trips per year, the total cost would be $46,800 to $56,100. When a decision has to be made between pallets and containers, this additional cost also should be taken into consideration.

The large scale of containerization prompted cargo airlines to search for any reasonable way to cut costs in their operations. The idea of leasing or renting instead of owning cargo containers was the next logical step. According to industry officials, approximately one-half million air cargo containers are owned by airlines. They cost between $4000 and $8000 each. An airline has to have sufficient containers to fill its airplanes. As a rough estimate, a few sets of containers could cost around $200,000 per airplane.

Having a leasing company own the containers would be advantageous to the airlines because leasing the containers would cost less than owning them. Such an arrangement does not require tying up large amounts of capital, and it has additional benefits as well. It would completely standardize the containers, and at the same time, some tare weight improvement could be achieved. This would show up as a substantial fuel and/or cost savings for each airline. Naturally, some other problems have to be ironed out, such as developing a smoothly running distribution system in such a way that no cargo should be stranded due to a lack of containers. Procedures have to be established for repairing damaged containers and covering repair cost. Possibly, airlines could establish their own pooling company and work out a well-functioning system. This is a worthwhile goal because savings for each airline could be substantial. The rental for a container would range from $3 to $5 per day.

Before going deeper into this subject, a short description and definition of *intermodalism* is in order. Intermodalism consists of two factors:

1. Containerization of the freight by the shipper in a special receptacle designed to interface with the carrier(s) vehicle(s)

2. Unit transfer of the containerized freight between two or more transportation modes

The term *intermodalism* did not come into common usage until the development of unit transfer of over-the-road truck trailers among road and rail modes and sea containers. Simple interline transfer of non-containerized freight between modes is not intermodalism in the strict sense of the term because containerization by the shipper is a prerequisite on intermodalism. Transfer of airline igloos between aircraft and over-the-road flatbed trucks comes close, but it is not intermodalism in the strict sense either because the igloo is not designed to interface with the truck; indeed, it is a very inefficient receptacle for long-haul over-the-road carriage. It has to be unloaded from the trailer and reloaded onto an aircraft. Similarly, when the cargo is transferred to truck trailer, it has to be removed from the igloo and reloaded into the trailer.

Intermodalism permits the shipper to load freight at its own facility into a specially designed truck trailer. The same trailer travels over the road to the airport of origin and flies as a unit to the airport of destination, then traveling over the road once again to the consignee's warehouse truck platform.

With the introduction of larger airplanes, the increased number of pallets and containers required a more sophisticated loading method. A new generation of cargo-handling equipment has been developed to target the following improvements:

- Faster lifting and loading devices for the larger number of pallets and containers.

- Proper ground equipment for 10-, 20-, and 40-ft containers handled singly or 40-ft containers handled in tandem.

- Handling equipment that can be adapted to higher airplane sills, such as the B747 with a 16-ft sill height, compared with the B707 or DC-8 with 11- and 10-ft sill heights, respectively.

- Requirement of direct transfer of containers to a standard over-the-road chassis.

- No longer does the shipper have to make the shipment fit the aircraft. Instead, the container is built to fit the shipper's need and still fits the aircraft.

- Containers are sealed and locked at the shipper facility. They will be unlocked at the consignee facility. Thus the opportunity for pilferage or tampering with the goods is eliminated.

- The needs of the handling equipment and facilities are identical to those used in loading surface-only trailers.

- Containers are brought directly to the staging area. Rehandling of the shipment is not required. In minutes, the container chassis is uncoupled, and the tractor is free to depart or pick up another intermodal unit for delivery to a consignee. Delays are minimized, and security is improved.

To speed up the transfer of containers, intermodal containers were developed. An $8 \times 8 \times 20$ ft intermodal container is shown in Fig. 5.18. It is placed on an intermodal adapter. The adapter is necessary when a lightweight intermodal container needs structural support. With a heavy structural base, no adapter is necessary. Handling of an air/land intermodal container is shown in Fig. 5.19.

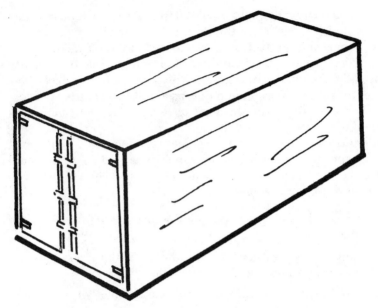

Figure 5.18   Intermodal air container.

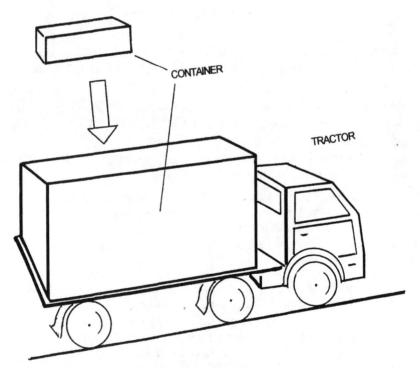

Figure 5.19   Air/land intermodal container handling.

The intermodal adapter can handle two 10-ft containers or one 20-ft container. As soon as it is loaded on the trailer, it is ready to go to its destination. In Fig. 5.20, the intermodal container arrives at the ramp, is quickly disconnected, and then is transported by powered ramp equipment to the airplane.

To move this trailer efficiently and position it parallel to the aircraft loading unit, it was necessary to develop a power transporter underneath the chassis. In this way it was possible to position the trailer accurately. The resulting design was called the *super yard horse* or just the *yard horse* and is shown in Fig. 5.21.

This short description highlights certain advantages of intermodalism:

1. Shippers are already geared to use these containers. Built to standard, they meet all the strength and safety specifications for over-the-road trailers.

2. The rectangular interior of the container eliminates the loss of capacity associated with the curved contours of igloos, which are shaped to fit the interiors of airplanes. No longer does the shipper have to make the shipment fit the aircraft; instead, the container is built to fit the shipper's needs.

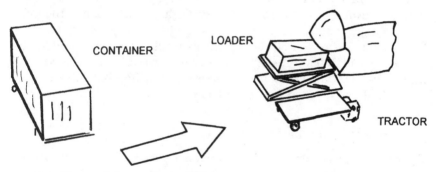

Figure 5.20 Intermodal container handling.

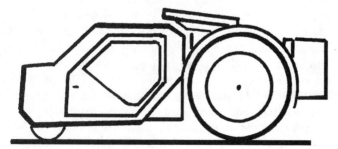

Figure 5.21 Super yard horse.

3. The interiors of these containers are lined with standard fittings, providing the capability for tie-downs and restraints, double decking, garment hanging, and a variety of other configurations.

4. With these containers, the time-consuming rehandling of cargo is eliminated. There is no need to place the cargo into and remove it from special aircraft-type containers. Costly driver and tractor delays are eliminated, and there is no waiting time.

5. These containers meet recognized worldwide customs requirements, which shortens waiting time at airport facilities.

6. Containers are sealed and locked at the shipper's location. They remain under seal until the consignee unlocks them at its own facility. This eliminates pilferage and/or tampering with the goods during transit.

7. The rates for this transportation mode are simplified and economical.

8. This approach provides the shipper with all the advantages of door-to-door service.

The economics of sea and land transportation support 40-ft containers, and only a few airplanes can handle them. Intermodal containers are heavier than standard containers and require additional fuel for transport. For example, on a trans-Atlantic flight, an additional 1000 lb of cargo carried requires an extra 240 to 280 lb of fuel per trip. Assuming 320 trips per year and 60 cents/gal, this would generate an additional cost of around $6000 to $8000 per year per aircraft. In the early days, intermodal containerization looked very promising, but later the economics did not work out as hoped for.

5.13 Heavy Lifting Devices

The newer types of loading equipment have greater weight-lifting capabilities and higher speed. The B747 nose opening is 100 in. At least 2 in is needed for the rollers at floor level, so this leaves 98 in, which is only 2 in greater than the 96-in vertical dimension of an 8 × 8 ft container. This requires a loader with high accuracy. The floor of the loader should be parallel with the airplane floor; if not, the container would rub against the airplane door, causing extensive damage during the loading process. Single-stage mobile scissors loaders are available in a wide range of load capabilities. They can easily handle 10- and 20-ft containers or pallets (Fig. 5.22).

A four-post loader has the same capabilities as a scissors loader. Both can handle special oversized cargo. A maximum elevated height of 18 ft is sufficient. A ground-transfer height of 20 in is standard (Fig. 5.23).

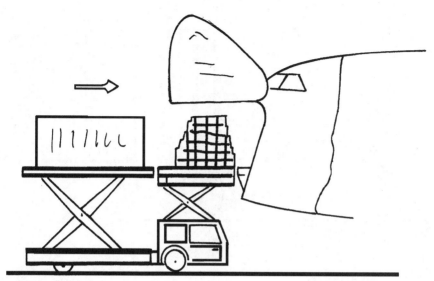

Figure 5.22 Scissors loader.

Figure 5.23 Four-poster loader.

Many systems and equipment are available for loading airplanes. Only a few will be discussed here. In Fig. 5.24, a container is lifted from a truck using top-corner ISO fittings. Another system uses rollers. Loading begins at the warehouse, goes through the roller system straight to the airplane nose, and then moves onto the main deck of the airplane.

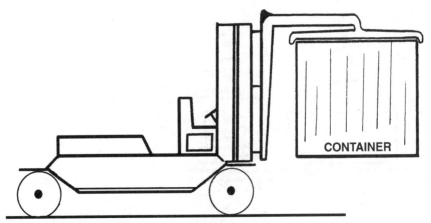

Figure 5.24 Truck lifting device.

5.14 Odd-Sized Loads

Loads travel on pallets and in containers. Special and/or odd-sized loads offer new opportunities in air transportation. Transportation of long pipes, power stations, odd-sized objects, automobiles, and jet engines frequently is done by air. Large jet engines are divided into core module, fan module, parts package, and gantry and can be shipped on three pallets.

5.15 Loading Efficiency

Loading efficiency is based on the relationship between nominal and actual load. During the buildup of pallets, it is very difficult to use all available cubic space because of the variety of package sizes and shapes. There will be empty spaces in various locations on the pallet. This means the total available space of a type A netted pallet having an internal volume of, let's say, 311 ft^3 cannot be fully used. Loading efficiency figures are based on actual measurements.

Loading efficiency is presented in Table 5.12. To illustrate the use of these numbers for a midsized aircraft, an example is given in Table 5.13. The conclusion that can be drawn from these figures is that while the theoretical volume is 311 ft^3, the effective volume is 248 ft^3 (0.80 × 311). In some cases this also can limit the maximum load based on a certain density, thus increasing cost and reducing profit.

5.16 Cargo Door Loadability

Loading cargo through the side cargo door has certain limitations. It is not enough to push the load through the door, but the depth of the load could be limiting due to the airplane's geometry. For this reason, charts

TABLE 5.12   Loading efficiency

	Pallet	Container	Bulk
Midsized aircraft	0.80	0.80	0.70
Large aircraft	0.83	0.84	0.70

TABLE 5.13   Pallet specification

Weight limitation	8000 lb
Tare weight of pallet	210 lb
Weight of nets	30 lb
Internal volume	311 ft^3

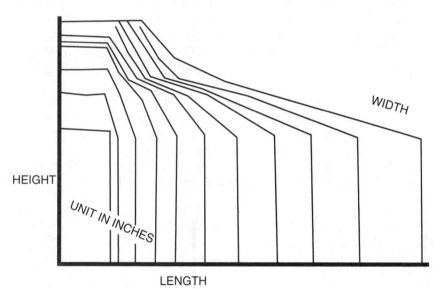

LENGTH

Figure 5.25   Typical airplane side-door loading chart.

are prepared indicating the sizes of boxes that can go into the belly of an airplane. One chart is shown in Fig. 5.25. Any load that is within the boundaries shown on the chart can be loaded into the airplane without any difficulties. Each airplane has its own unique loading chart.

5.17   Loading Time

Each airline has to decide what kind of loading system should be installed. Two factors must be considered: money available and operational time requirement. The time required is based on the operational schedule. If the schedule permits a longer ground time—without influencing the profit picture—operation with a forklift is satisfactory.

For forklift operations, the main deck can be loaded and unloaded with a caster bed using a 25,000-lb forklift. A 20,000-lb forklift with a caster bed can handle the lower-deck cargo. Tugs and pallet dollies can be used between the aircraft and the cargo terminal. The estimated turn-around time of 1½ hours is very good considering the low investment in cargo loading equipment. The time required is shown in Fig. 5.26. With this kind of loading arrangement assuming 68 tons and 2 hours and 28 minutes of loading time, the loading efficiency is about 27 tons per hour.

A more expensive solution is the use of mobile loaders. For main-deck loading, the airplane requires a conventional mobile loader with a height capability of 17 ft. The lower deck can be serviced by the mobile loaders currently used on all wide-body jets. The cargo handling times will vary for the type of loader used. The times shown in Fig. 5.27 are representative of a typical loader and transporter operation. Moving 68 tons over a 2-hour time period would indicate a 34 ton/h loading capability.

The use of a loading dock for main-deck cargo handling provides the lowest cargo handling time. The lower cargo decks are serviced by a mobile loader/transporter. A fixed loading dock probably would represent the largest investment in cargo-handling equipment, as shown in Fig. 5.28. Assuming the same 68 tons and a moving time of 1½ hours would indicate a loading capability of about 45 ton/h.

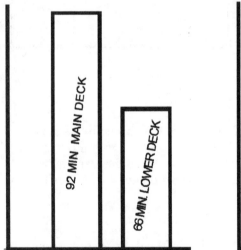

Figure 5.26   Forklift loading time.

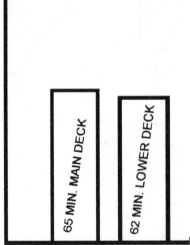

Figure 5.27   Mobile loader loading time.

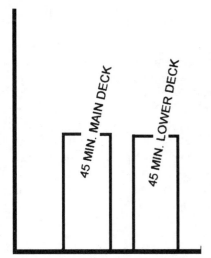

Figure 5.28   Loading time of a permanent loading dock.

5.18 Weighing Airplanes

The air freight business transports goods, and the transportation cost is based on weight. The weight of the product is furnished by the customer. There are no businesses in the world where a company takes the customer's word for the quantity involved except the air freight business. This may cause the airlines to receive less money for their services than they should.

For safe operation, the captain of an airplane needs to know the weight of the load and the center of gravity of the airplane. Weighing individual packages or pallets before loading is cumbersome, time-consuming, and error-prone.

Two basic weight and balance systems are used:

1. Electronic strain gauges can be built into the axles of the aircraft. These gauges total the weight on all landing gear assemblies, and then the center of gravity is computed. During loading, any time the center of gravity approaches its limit, the system shuts off power to avoid tipping of the airplane.

2. Hydraulic pressure is measured on airplane struts, totaling weight and computing center of gravity.

Both systems are reliable and used by many airlines.

5.19 New Trend In Cargo Handling

A new trend is observed in cargo operations. Cargo operators strive for faster loading times to reduce ground time. The first major innovation was the development of very close roller trays, every 3 in instead of the traditional 10 in. These rollers use a very low friction plastic bearing material. The second innovation was the introduction of continuous vertical guides coated with a teflon-impregnated substance that permits pallets or containers to slide along without catching. A new foot-activated locking device was introduced. Now, all locks and restraints can be operated by foot, so no bending is required by the handling personnel. Applying movable side guides, one person can adjust the different widths of pallets. For example, an 88 × 125 in pallet in one position and a 96 × 125 in pallet in another position can be loaded into the airplane without difficulty.

Weight savings were accomplished by removing power loaders. This saves about a few hundred pounds in weight.

5.20 Loading Doors

Different aircraft manufacturers design different cargo loading doors. To mention a few interesting aircraft layouts, the CL-44 aircraft has a swing-tail design. In many years of operation, it was observed that this mechanism hardly caused any problems. The B747 has a nose cargo door in the front of the airplane, permitting the loading of long objects. Such shipments would be impossible to load through a side door. The "Super-Guppy" with a swing door opening to the left side is shown in Fig. 5.29.

A different approach is shown in Fig. 5.30 for the Lockheed L-100. There is an inward door opening under the high tail in order to provide for straight-in loading. It has a low-level floor at truck-bed level.

Figure 5.29 Super-Guppy.

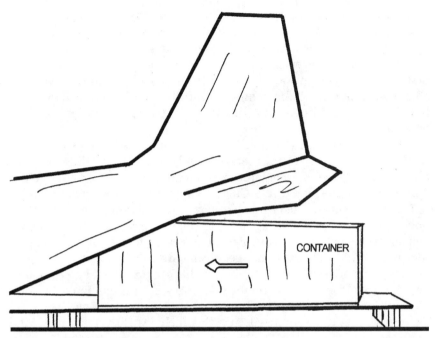

Figure 5.30   Loading under the high tail.

5.21   Oversized Airplane Capability

A new type of cargo market is emerging, namely, the oversized/over-weight cargo market. The special design of such aircraft opened up a new trend, making air transportation possible for oversized cargo. One of these aircraft is the Russian Antonov AN124-100 airplane. The cargo area is 21 ft wide, 14½ ft high, and 120 ft long. A fold-down hydraulic ramp at both ends of the airplane permits on/off movements. At the nose of the airplane, a large flip nose door is available. The aircraft has a unique ability to "kneel" on its front undercarriage. Two overhead cranes fit the airplane internally. A larger, six-engine Antonov AN225 aircraft has a 250-ton payload capacity, as opposed to the AN124 with 150 tons.

5.22   Freight-Handling Terminals

For cargo operations, the infrastructure is very important. Air cargo spends a relatively short time in the air (trip time). Problems occur when cargo is on the ground. The shipment has to be moved, processed, and delivered to its destination. The two most labor-intensive periods are at the departure and destination locations. Various cargo terminals were built in the past, but the flow of freight in these buildings is relatively similar.

On arrival at the cargo terminal from the shipper's ground transportation, a shipment is unloaded from the trucks, processed, and then moved to the holding area. The next step is to sort and assign the shipment to a passenger or cargo airplane—whichever the case may be—and move it to the waiting area. Once the assigned airplane arrives, the shipment is moved to the ramp and into the airplane. For storage of import and export items, interline transfer, bonded items, etc., special areas are available.

The cargo arriving in an airplane at a terminal building is processed in the check-in area. From there it could go to the interline transfer area or to a bonded storage area. Shipments arriving from outside the United States are placed in a Customs area, where they are subjected to Customs clearance before being moved to a holding area. From the holding area, shipments are moved to the delivery area and then onto delivery vehicles. Each airline designs its cargo terminals according to its specific needs and customer requirements.

In the case of outbound flow, airway bills are prepared, and packages are checked, labeled, and weighed. Export declarations must be completed. This is checked by Customs. Samples are taken, if necessary. All paperwork has to be finalized, and this includes Customs and airline documents.

In the case of inbound flow, at arrival, cargo documents and cargo are checked, sorted, and sent to the proper locations. The cargo is checked against the airway bills. If the cargo is imported, Customs is also involved. Cargo is now sorted and, in case of perishables, sent to cold storage or any other special area, if required. Once the paperwork is completed and Customs releases the freight from overseas origin, it is transported to the consignee.

This may sound very simple, but in reality, the process is a little more complex.

The next step in the development of cargo handling is paperless documentation of cargo flow. To speed the process up and make it less costly, standardization is highly overdue.

This was a short description of air freight movement at terminal buildings. Two important factors affect the cost of air freight transportation:

1. *Labor-intensive process.* At departure and arrival stations, cost savings can be achieved.

2. *Efficient documentation.* It should be simple and effective, striving to achieve a paperless flow of data by optimizing computer use and standardizing formats.

Sometimes, too much time is spent clearing cargo through Customs. The cargo is held for too long because of language problems and lack of standardized documentation. These are areas for improvement in the near future. Curfews could contribute to late deliveries of air freight. Such improvements are the next goal of the industry.

A pallet or container storage are shown in Fig. 5.31. The pallets are stored in bins. A trolley on its own tracks moves up and down and transports the pallets to and from the aircraft.

5.23 Airplane Conversion (Passenger to/from Cargo)

To achieve maximum profit potential for an airplane, airlines have the capability to convert their airplanes from passenger to cargo configuration, and vice versa. This can be carried out per trip, day, month, season, or any time at all.

QC cargo systems. The QC (quick change) system was introduced first on the B727 aircraft. It reduces conversion time to less than 15 minutes with a five-person crew. Unlike other conversion systems, this procedure does not change the cargo hardware. The seats are mounted on pallets, two rows to pallet with the carpet on. These pallets are then attached to the floor with eight latches per pallet. The carpeted seat-pallets fit snugly against each other and are free of gaps. The passengers

Figure 5.31 Freight storage.

are seated about 1¼ in higher than normal, but this is not noticeable. On the B727-200 aircraft, the weight difference for the standard-wing airplane between passenger and the all-cargo configuration is about 2700 lb (passenger configuration is heavier). Table 5.14 lists the conversion items. A few items from the passenger configuration and a few items from the cargo configuration are presented below. Conversion from the all-passengers to all-cargo configuration requires the total removal of seats, carpet segments, and hat racks. It also requires installation of the cargo system hardware and barrier cargo net.

C systems. This is a flexible system for cargo operations. The seats are removed, and cargo hardware and the barrier net are installed with a fume-type bulkhead between passengers and cargo. The area can be changed at each stop, having the maximum flexibility for carrying cargo. A complete change from all-passenger to all-cargo configuration could take about 60 minutes with a six-person crew.

5.24   Animal Passengers

The transportation of large animals such as horses, cows, and wild animals (Fig. 5.32) originated in Europe in the early 1920s. It was always news when racehorses or rare animals were transported by air. The introduction of large jet liners opened a growing market for carrying animals on airplanes. They were walked up into the airplane and into their stalls. DC-8 and B707 airplanes were the first aircraft to carry a maximum cattle load of 65 head. The greater capacity of the B747 freighter permitted a larger number of animals to be carried. Cows were kept in metal-frame pens and loaded as containerized cargo, instead of walking them to the cargo doors up ramps. In 1978, Flying Tigers initiated the first all-cattle flights from Seattle to Asia. Disposable wooden pens replaced the metal pens; in this way, the tare weight was reduced for the return flight.

As more and more animals flew by airplane, methods were developed by manufacturers for handling animals during their transportation.

TABLE 5.14   Conversion Items

Install passenger kit	Remove cargo kit
Carpets	Sill roller assembly
Seats	Lock and roller trays
Lavatories	Side guides
Galleys	Caster trays
Drop ceiling	Center guides
Partitions	Power drive units
Life rafts	End stops and locks
Overhead stowage	Loader controllers

Figure 5.32   Animal transportation.

Animal transportation consists of three stages: preflight, the flight, and postflight handling. Animals are sensitive to many environmental effects:

- Sudden changes in temperature and humidity
- Possibly vibration
- Excessive and/or loud noises or noises in the frequency audible only to animals
- Pollution, air quality, etc.

The airplane's lower compartment is constructed to satisfy FAA requirements to carry animals. Animals generally are categorized by size due to their weight/metabolism relationship. Furthermore, the shipper furnishes the temperature limits, indicating the expected temperature range. For example, bears belong to the large animal group. The temperature should be in the range of 35 to 85°F. During travel, the animal is in a crate or cage. The cages have to be clean, escape-proof, and properly sized. It is important that pre- and postflight animal waiting areas be clean, well-ventilated, free from drafts, in the proper temperature range, at least 5 to 6 in above the floor, and at least 4 to 5 in on all sides away from other containers. In addition, the waiting areas should not be exposed to heat sources such as sunlight or radiators. Similarly, the proper temperature range should be observed during ground loading and loading into the airplane. It is advisable during the summer months to precool the airplane and load animals as quickly as possible, close the doors, and be ready to taxi out, take off, and have all the air-conditioning units operating as soon as possible. A good practice is to carry large animals on the main floor and have temperature probes at certain intervals on the ceiling to monitor the temperature from the cockpit. If a problem develops, corrective action can be taken before harming the livestock.

Animal transportation is on the increase because it is fast and has the least exposure to stressful environments. Animals arrive in a much better condition than by slower transportation modes. An arrangement for carrying horses in stalls is shown in Fig. 5.33.

Emery Air Freight Corporation recently transported four bottle-nosed dolphins and four beluga whales from San Antonio, Texas, to Florida and then to Ohio Sea World.

5.25   Cargo Distribution Costs

Airplane manufacturers started to conduct research in the distribution cost domain. There were large differences in cost between surface and air cargo transportation. In order to open up a larger market for

Figure 5.33   Animal transportation in stalls.

air transportation, it had to be shown that air transportation had advantages over other means of transport. As a consequence of these studies, the "total distribution transportation cost" concept was developed. Every element was scrutinized, and in the final analysis, it was shown that air transportation in a certain high-value area could be more economical than surface transportation. One of the major advantages of air transportation is explained by the speedy replenishment of inventories. As a consequence, no money was unnecessarily tied down in slow-moving stocks, thus freeing money to be used for other investments. The basic objective of the total distribution cost concept was to minimize distribution cost for a required level of service. This can be achieved if transport is managed and controlled as an integrated system.

The method of obtaining a comparison for both transportation modes is based on the product's value (Fig. 5.34). Generating transportation for surface and air as a function of product value and plotting the points on a graph leads to the following conclusion: Low product value generally is associated with low transportation cost. Similarly, the opposite is also true, namely, high product value is associated with high transportation cost. At a certain product value, both transportation costs are equal. This is the potential threshold point. Above this threshold value, air transportation is the more economical choice.

Analyzing air transportation cost elements, the following conclusion can be drawn:

1. Air transportation line-haul cost is the most expensive phase of operation.

2. Many of the air transportation cost elements are less costly.

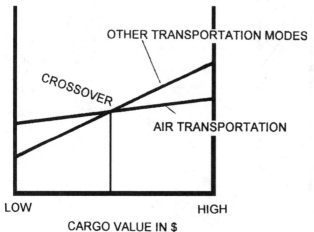

Figure 5.34   Distribution cost evaluation.

For example, packaging is a small expense in air transportation because placing a product on a pallet or inside a container requires minimal or no packaging cost. Another savings is in the shorter transit time, which could significantly reduce inventory costs to the point that warehouse expenses may be reduced or eliminated completely.

Considering these factors, and by adding up all the cost elements, air transportation could beat the cost of other transportation modes. This could represent savings to customers and bring additional profit to the airlines (Fig. 5.35).

5.26   Shipping Time and Shipment Cost Comparison

The shipper's interests are in shipment cost and shipping time. There is an inverse relationship between these two factors. When schedules are tight, air shipment is always the solution. When time is not pressing, a slower transportation mode could be used.

In making a decision about mode of transportation, various other factors also should be considered. The highest transport cost is by air and

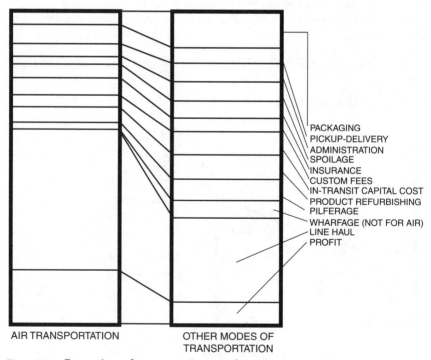

AIR TRANSPORTATION   OTHER MODES OF
                            TRANSPORTATION

Figure 5.35 Comparison of transportation cost elements.

the lowest is sea transportation. Any other mixed (sea-land, air-land, etc.) mode of transportation costs somewhere between the other two modes.

Sea transportation is used generally for bulk and low-dollar-value commodities. In sea transportation, the following additional costs are considered:

- Transportation cost, ground
- Crating/packaging cost and time
- Warehousing cost
- Administration cost
- Insurance cost
- Interest cost

Let's assume that the value of a certain commodity is $500,000 and transportation time is 35 days. At 8 percent yearly interest related to the 35 days, there would be an additional cost of $3836. Moreover, there is a possibility that during this long transit time, theft, pilferage, or damage may occur.

Air transportation is used for high-dollar-value commodities and for perishables such as fruit and flowers or seasonal items such as garments. The following cost items are considered and compared against sea transportation costs:

- Transportation cost, air
- Crating/packaging cost and time, none*
- Warehousing cost, none*
- Administration cost, less
- Interest cost, none*

For sea transportation, ground transportation consists of the trips from the factory to the seaport and at the destination seaport to the customer's warehouse. For air transportation, it is a different picture; the goods are moved from the factory to the airport and at the destination airport are moved to the customer's warehouse. Ground transportation costs generally are less for air shipments. Moreover, due to short transportation times, there is hardly a chance for theft or damage.

Figure 5.36 presents transportation cost versus transportation time. The shorter the transportation time, the higher is the transportation

---

*It indicates no cost at all or a very small amount only.

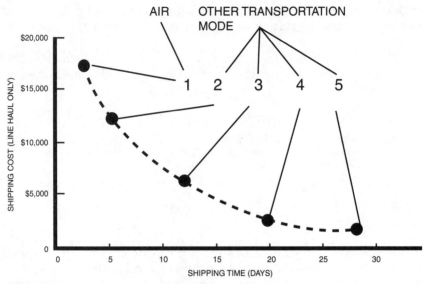

Figure 5.36   Transportation time versus transportation cost.

cost. The following conditions are shown in this figure for a Tokyo-Amsterdam trip:

1. All air transportation   air[*]
2. All air split charter via San Francisco   air/sea
3. Air/sea link:
   Tokyo–San Francisco   sea
   San Francisco–Amsterdam   air
4. Sea/land/sea:
   Tokyo–San Francisco   air
   San Francisco–Boston   rail
   Boston–Amsterdam   sea
5. All sea transportation   sea[†]

There is no clear-cut answer as to what type of transportation should be used. In each individual case, all factors have to be considered to make the best decision.

---

[*]High transportation cost.
[†]Lowest transportation cost.

5.27 Sample Analysis Comparing Ocean versus Air Shipping: Case History

This is an actual case for a transoceanic shipment. The study was conducted by a now-defunct cargo airline's sales department.

5.27.1 Basis for the analysis

1. Decrease in transportation time
2. Elimination of ocean packing
3. Elimination of unpacking at destination
4. Simplification of documentation
5. Reduction in documentation cost
6. Reduction in accessory charges, labor, and insurance premiums wherever applicable
7. Thorough reduction in time in transit, partial or complete elimination of back orders
8. Increase in return on investment
9. Better customer satisfaction

5.27.2 General observations

From a discussion with Alfa Company's representative, it was learned that the factory usually has a backlog of orders. The contract has a penalty clause for late delivery; however, this clause apparently has never been enforced. It is generally felt that the long delivery time has a negative effect on maintaining a good relationship with customers. Air transportation would expedite delivery and help to satisfy the consignees in Country X in Europe.

The basic transportation data discussed in this study pertain only to *one single shipment*. A single-unit cost-factor module will be established and will enable Alfa Company to determine the difference between ocean and air freight for this particular commodity on a yearly basis.

The ocean packing is considered rather heavy for the commodity shipped. In addition, packing charges are substantial, amounting to approximately $2.08/ft^3$.

5.27.3 Method applied

■ Basic data were compiled from the invoices of one particular shipment.

- Aside from studying the effect of airlifting this particular shipment in lieu of ocean shipping, a single-unit cost-factor module was constructed for perusal by Alfa Company.

- The most advantageous way of air shipping was established based on the presently effective commodity rates.

- Container shipping did not prove to be advantageous because of the density per cubic foot of this freight.

Taking the elimination of ocean packing into consideration, it can be calculated that the total volume for air would be approximately 1140 ft^3. At a total weight of 8723 lb. A total of 2.54 pallets is required with a density of 7.7 lb/ft^3. A net usable space per pallet of approximately 412 ft^3 (450 ft^3 − 8.5% for stowage loss in pallet buildup) is considered (Table 5.15).

5.27.4   Application of basic data

- The net shipping weight per case has been adopted as the air shipping weight, thus reflecting the weight decrease through the elimination of ocean packing.

- The air freight quotations are based on the lowest applicable cargo airline's regular published rates presently effective from New York to Country X in Europe.

The following surface transportation cost is considered:

Sea—From factory to seaport

From seaport to consignee

Air—From factory to airport

From airport to consignee

For calculation of the cost of capital investment for goods in transit, a conservative factor of 6 percent yearly interest is being used.

- Transit time for ocean, door to door, is 35 days.

- Transit time for air, door to door, is 6 days.

TABLE 5.15   Ocean and air transported weights

Weight	Ocean, lb	Air, lb
Gross weight	11,594	—
Net weight	8,723	8,723
Tare weight	2,871	—

5.27.5 Door-to-door transportation cost: Summary

All the expenses are tabulated side by side for each kind of transportation element. The savings shown in Table 5.16 may not be significant, but in the case of 100 shipments, it is quite a substantial amount that cannot and should not be ignored.

5.27.6 Cost of capital investment while goods in transit

	Ocean	Air
35 days by ocean at 6% per annum		
$\dfrac{35 \times 6 \times 45{,}958}{360 \times 100}$	$268.00	
6 days by air at 6% per annum		
$\dfrac{6 \times 6 \times 45{,}958}{360 \times 100}$		$46.00
Total cost of capital investment	$268.00	$46.00
Total cost of transportation	$51,146.00	$48,580.00
Difference in favor of air		$2,566.00

5.28 Sample Analysis for Reduced-Inventory Cost Calculation

This set of calculations presents the advantages of reduced inventory and the advantages of air shipment.

The following information is available:

Value of present inventory	$900,000
Inventory carrying costs at 27%	$243,000
Inventory reduction, %	75

TABLE 5.16   Transportation cost comparison

Items	Present method	Air freight
Invoice value	$45,958.00	$45,958.00
Packing cost	2,763.00	0
Shipping documents preparation	17.00	5.00
Transport to port of departure	486.00	118.00
Transatlantic transport cost	1,093.00	2,355.00
Insurance premium	251.00	98.00
Delivery to "YYY"	310.00	0.00
Total direct transit cost	$50,878.00	$48,534.00

Inventory carrying costs consist of

- Cost of money
- Deterioration
- Warehouse storage
- Obsolescence
- Handling costs
- Administration
- Insurance
- Miscellaneous items

A detailed analysis would indicate the amount of these costs, expressed in percentages. For this example, a 27 percent figure was selected as an average value.

The value of the inventory	$ 900,000
Reduction of inventory	75%
Reduced inventory	$ 225,000
Inventory removed	$ 675,000
Removed inventory carrying cost at 27%	$ 182,250

Value of present inventory	$ 900,000
Yearly turnover rate	3
Yearly value of inventory	$2,700,000
Value per pound	$ 4.70
Total pounds	574,468
Difference in surface and air transport rate	23.2 cents
Additional air transport cost	$ 133,276
Miscellaneous additional air transport costs	$ 12,178

Removed inventory carrying cost at 27%	$ 182,250
Less total additional air transportation cost	$ 145,454
Net savings	$ 36,795

The duration of air transportation is around 2 to 3 days. Surface transportation time is between 36 and 46 days. Considering a 6 percent interest rate, it is easy to calculate the loss of money due to the long surface transportation time. An additional benefit could be the elimination, if possible, of a warehouse or a lower warehouse floor area that could reduce rent.

Chapter

6

Economics of Charter Operation

6.1 Introduction

The question has arisen from many sources as to why scheduled airline fares are so much higher than those of charter airlines. In order to answer this question, a side-by-side comparison should be made of direct and indirect operating costs for both types of airlines.

6.2 Direct Operating Cost Comparison

6.2.1 Flight operations

The difference, if there is any, is in the remuneration of the crew. It may amount to an insignificant sum of money and does not represent important savings. In Europe, for example, there is no difference at all between the salaries of charter flight and scheduled flight crews.

6.2.2 Fuel cost

This depends on location and on the negotiating power of the airline. Sometimes charter airlines cannot receive a discounted price on their return flights because of the small amount of fuel purchased, and therefore, they have to pay a higher fuel price.

6.2.3 Depreciation

Aircraft utilization by a charter airline generally is higher than that of scheduled airlines; as a consequence, the amount of depreciation (expressed in dollars per block-hour) may decrease. A charter airline can fly practically around the clock. Its passengers can fly late at night or in the very early morning hours, whereas scheduled airlines are bound to certain peak hours. Furthermore, charter airlines are able to benefit from the use of their airplanes even during the lowest seasons by leasing them to other operators. Basically, there are no significant differences between scheduled and charter airline direct operating costs.

6.3 Indirect Operating Cost

Charter airlines are able to lower their airport charges by flying at certain times of the day or the week when they can take advantage of lower fees or an alternative airport in the vicinity of their departure or arrival area. For example, instead of using JFK in New York, EWR is available as an alternative, or instead of LHR in London, Gatwick Airport may be used, where the airport fees are considerably lower.

Further savings can be realized for aircraft servicing and passenger handling by subcontracting these services. This seems to be more expensive, but it does not require the charter operator to keep a large number of employees on payroll. There is no station manager, ground equipment, maintenance, or spares, and the charter operator keeps the cabin crew to the legal minimum and only for the season.

Food for passengers is less elaborate and less costly. No connecting flights are involved. Thus there is no responsibility on the part of the charter airline to provide passengers with lodging if they miss their connecting flights.

Passenger ticketing is simple and requires minimum personnel at a minimum cost. There are no extra expenses for selling tickets, paying sales personnel, or giving commissions to travel agents. In other words, there is no superstructure. Charter airlines also know ahead of time their expected cash flow because charter passengers make their ticket payments in advance.

6.4 Seating Capacity

Another advantage for charter airlines is the airplane seating capacity (Fig. 6.1). By minimizing the pitch (distance) between two rows of seats, a dense seating capacity is established. In addition, the number of seats abreast could be increased to the maximum. There is no first class or business class. For example, on the DC-10, the galley is positioned in the belly, and as a result, the DC-10 has more seats on the main deck. A scheduled airline cannot afford to have the same seating density because it would lose passengers. On a larger airplane, this seat-density difference could mean as much as 60 to 120 passengers.

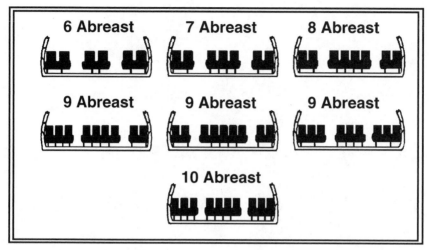

Figure 6.1   Seating configurations.

6.5 Overhead

Charter airlines do not have to maintain a large department for forecasting and planning because their cost structure is not very complicated. All in all, their overhead expenses are lower than those of scheduled carriers.

6.6 Load Factor

This is another area where charter airlines are ahead of scheduled airlines. While a scheduled airline experiences a load factor of 58 to 65 percent, charter airlines have a load factor of at least 80 to 85 percent. This is one of the significant factors in a charter operation.

This short description has highlighted the major differences between charter and scheduled airline expenses. It has been shown how the charter airlines' advantages bring down their operating costs and enable them to accept lower revenues and lower ticket prices, thereby luring more passengers away from scheduled airlines.

7

Fuel Conservation and Its Economic Impact

7.1 Flight Department Fuel Conservation

Until 1973, the cost of fuel was about 10 to 12 cents/gal, just a small percentage of the direct operating cost (DOC). This cost has changed drastically since 1973 and is currently a sizable part of the DOC. This drastic change prompted the air carriers to embark on a fuel conservation program. The manufacturers, through their in-house publications, placed fuel conservation on the top of their list and disseminated valuable information and recommendations on fuel savings.

Two departments play an important role in fuel savings, namely, the flight department and the maintenance department. Fuel conservation is synonymous with cost savings. It is advisable to know

1. The method of saving fuel
2. The result in terms of dollars

7.1.1 Flight operations

Airlines have conducted performance monitoring since the early days of their operations. Two methods evolved and are widely used: the statistical approach and the cruise-performance audit.

The statistical approach compares the actual and planned data for each sector.

1. Weights
   a. Takeoff

b. Landing
c. Zero fuel
d. Passengers
e. Freight

2. Fuel consumption
   a. Trip fuel
   b. Reserve fuel
   c. Etc.

Uplift is compared for actual and planned weights and operational data per trip, per station, per season, etc. By evaluating the figures obtained, a trend of increased fuel burn could be apparent during a longer period.

The cruise-performance audit is applicable for a shorter time. Its accuracy is about 1 percent in specific range. Airlines with long-range operations take advantage of this method, taking numerous readings during a flight. For airlines operating on short ranges, the statistical approach would serve better.

Airlines routinely record and analyze cruise data obtained during a fleet performance audit by plotting these data versus time. The result can point out problems, and airlines can take corrective steps to prevent further deterioration of performance. A typical evaluation is shown in Fig. 7.1.

Airlines use both methods with variable sophistication, based on the resources available. These activities are considered as fuel savings when higher fuel consumption is detected and corrective steps are taken. After the 1973 fuel crunch, each activity related to fuel consumption was studied for the sole purpose of fuel or cost savings. To see the significance of fuel savings, two airplanes were selected, a small and a large one:

Airplane	B737	B747
Average flight time	0.8 h	5.5 h
Average distance	280 nautical miles	2600 nautical miles
Average fuel burn, gal/h	850	3700
Average fuel burn, lb/h	5500	25,000
Annual fuel, gal/h	1.9 million	13.1 million
1 percent fuel savings	19,000 gal	131,000 gal
At 50 cents/gal	$9500	$65,500

The fuel savings of one airplane per year is shown above. Considering a fleet of few hundred airplanes, yearly fuel savings is a large amount.

There are many ways to save fuel in flight operations. We will consider here only procedures that are safe and can be integrated into air-

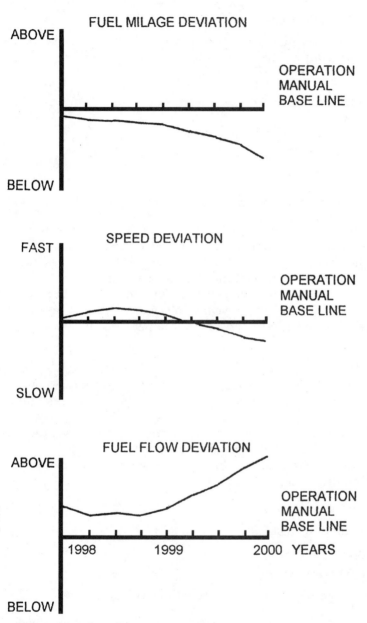

Figure 7.1   Aircraft performance monitoring.

line operations without any cumbersome process and without any loss of payload and revenue. Some of the various methods will be described briefly below.

7.1.2   Takeoff procedure effect on fuel burn

Studies were conducted on fuel burn savings related to takeoff procedures. Two procedures were considered:

1. Takeoff
   a. Cleanup
   b. Cutback
   c. Climb, maximum climb thrust
2. Takeoff
   a. Cutback
   b. Cleanup
   c. Climb, maximum climb thrust

A fuel savings of 5 to 100 gal exists between the two procedures in favor of procedure 2. This amount is a function of airplane weight, flap retraction height and speed, and different takeoff flap settings, noise abatement, and reduced takeoff thrust requirements.

Using an average 50-gal savings per departure for a medium-sized airplane and 320 departures per year at 50 cents/gal, a fuel savings of an average of $8000 per year can be realized per airplane.

7.1.3   Climb speed

The climb schedule for fuel economy is slower for certain types of airplanes. For others, a slow speed schedule for lighter weights and a higher speed schedule for heavier weights is more effective. To arrive at the best solution, airplane climb performance should be studied, analyzed, and evaluated for optimal climb schedule.

Here is a quick description of this analysis. Figure 7.2 shows a typical climb profile.

1. Brake release to 250 kias to 1500 ft

2. 250 kias climb to 10,000 ft

3. Accelerate to climb speed at 10,000 ft

4. From 10,000 ft to top of climb

5. Accelerate to cruise speed and proceed to a common point

Note: 250 kias is instrument-indicated airspeed in knots corrected for instrument error only.

Using variable climb weights and speeds to a common point would define the optimal fuel savings.

Following the optimal climb speed schedule, a fuel savings of approximately $2000 to $5000 per year per aircraft could be realized based on

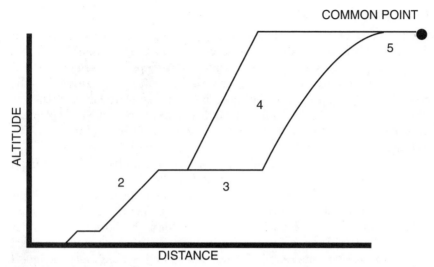

Figure 7.2 Climb schedule.

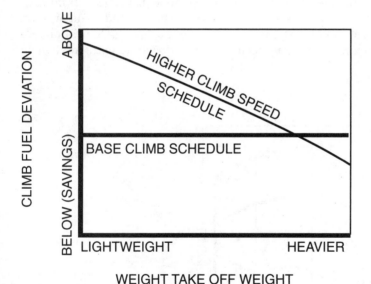

Figure 7.3 Optimal climb speed schedule.

the type and size of the airplane. Figure 7.3 shows the fuel savings for a medium-sized airplane. For a baseline, a slower climb speed schedule should be selected. Applying a higher climb speed schedule, a comparison would indicate the fuel saving for the takeoff weight regime.

7.1.4 Cruise

For the optimal range condition, the long-range cruise was accepted for improved stability. This is just 1 percent below the (optimal) maximum range speed. Good stability means that after an upset caused by any turbulence without any corrective action, the aircraft rapidly returns to the stabilized cruise speed. Slower speed has less excess thrust to return to the stable range. One percent less fuel consumption below the optimum is sacrificed in exchange for better stabilization of the airplane.

Thus the speed will be 10 to 20 knots higher without any appreciable loss of fuel economy, and the airplane will be in a stable flight regime. A typical specific range chart is shown in Fig. 7.4.

The line labeled "Max. Range" is the location of the best fuel economy; however, 99 percent of this maximum range, as mentioned before, is called *long-range cruise*. On one end of this chart, the buffet boundary is limiting; on the other end is the placard speed. This type of cruise chart is used to calculate the required cruise speed and related fuel mileage. It furnishes fuel flow in pounds per hour for any operating conditions for a given altitude.

At certain altitudes and weights, there is a point where fuel mileage is optimal. Cruise level should be close to this altitude and weight. However, this is not always the case because of operating limitations. To keep the airplane on target speed, frequent thrust adjustments would

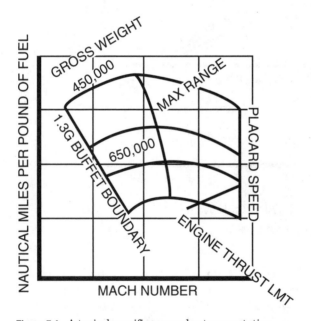

Figure 7.4 A typical specific range chart presentation.

be required. For operational reasons (instead of flying the airplane with a constantly changing speed at long-range cruise), constant Mach numbers were accepted based on the type of airplane. This resulted in satisfactory fuel economy and satisfactory airplane stability.

7.1.5 Mach number's effect on trip fuel and trip time

Change in Mach number would affect cruise time and cruise fuel. A faster Mach number shortens trip time and increases trip fuel consumption. Slower speed results in lower fuel flow and longer trip time. As shown in Fig. 7.5, a 3000-mile trip for this type of airplane flying at Mach 0.82 instead of Mach 0.80 would be about 10 minutes shorter and would burn 3.8 percent more fuel.

Here, another question should be addressed very briefly. Some people believe that at lower Mach numbers the fuel flow should be higher due to the higher angle of attack. It is well known that airplane drag consists of parasite, induced, and compressibility drags. All three drag elements have to be considered (not only one of them) when fuel flow is examined. As airplane speed decreases, fuel flow (lb/h) and drag (all three elements) also decrease.

Fuel flow varies with speed, and their relationship furnishes fuel mileage. Speed divided by fuel flow gives fuel mileage, and vice versa. A dimension analysis would be the following:

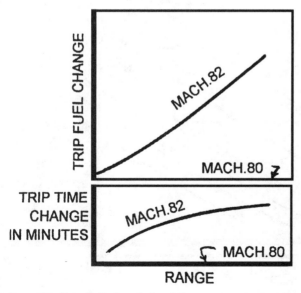

Figure 7.5 Speed effect on fuel and time.

$$(\text{Miles/hour})/(\text{lb/hour}) = \text{miles/lb}$$

The maximum range point on the fuel mileage curve is easily spotted (see Fig. 7.4).

7.1.6 Cruise altitude

The best fuel economy is defined as the best mileage out of every pound of fuel at any given condition. Flying at the optimum altitude furnishes the very best fuel economy. Theoretically, we should fly in a climb-cruise schedule, but this is not practical for operational reasons. Figure 7.6 is a graphic presentation of this practical solution, where we fly in climb-cruise configuration based on traffic conditions.

The optimum could be achieved in such way that the airplane should fly initially 1000 or 2000 ft above the optimal altitude (the optimal altitude increases about 1000 ft/h). As the flight progresses, after about an hour, the airplane is flying at optimal altitude. Continuing the flight, it will now fly below the best altitude. Flying above and below the best altitude on a given sector is the best way to solve this problem. This step-climb approach can be repeated (traffic permitting) until the airplane approaches the final destination. It is not a perfect solution, but it is the best available for this condition. Furthermore, it shows the penalty in percentage when not flying the optimal altitude.

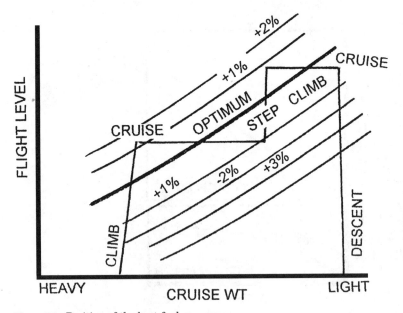

Figure 7.6 Position of the best fuel economy.

Figure 7.7 shows the various cruise configurations. It shows the fuel mileage increase or decrease as a function of high Mach number, minimum-cost Mach number, or long-range cruise. This chart shows the optimal altitude for these three conditions.

7.1.7   High Mach number

This is used when time is important; for example, to reach a destination on time for connecting flights or when crew time is critical. Fuel consumption in this case is high.

7.1.8   Minimum-cost Mach number

Airlines have established minimum-cost speed. Cost figures were calculated for time and fuel as a function of speed. Combining these two cost elements, a U-shaped curve was developed. For easier understanding, the relationship is shown graphically in Fig. 7.8. In this figure, the fuel cost is plotted versus Mach number. Due to increased drag, the faster the airplane flies, the more fuel is used up. This is not a linear relationship.

Figure 7.9 shows time cost (e.g., crew and maintenance) versus Mach number. This is a linear relationship.

To obtain the minimum cost, which is a U-shaped curve, time and fuel cost elements should be added up at the proper Mach number (Fig. 7.10). The lowest point of this curve is the location of the minimum cost and its associated Mach number.

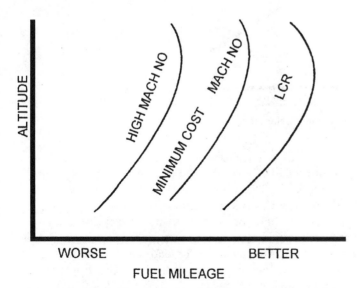

Figure 7.7   Locations of optimal altitude for various airplane speeds.

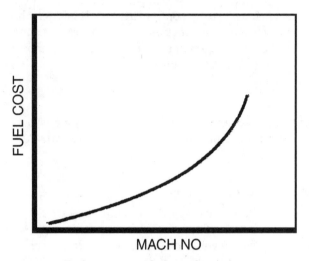

Figure 7.8   Fuel cost versus Mach number.

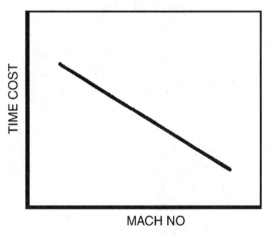

Figure 7.9   Time cost versus Mach number.

For midsized airplanes such as the B707, DC-8 series, B757, and B767, the minimum-cost Mach number is around Mach 0.80; for the DC-10 series and Lockheed 1011, Mach 0.82 or 0.83 is generally selected, whereas the B747 usually has a Mach number of 0.84.

A brief explanation of the concept of cost index is presented here. Boeing introduced the method. Cost index is a number that is put into an airplane's onboard computer to calculate, for example, minimum cost for a given trip. Cost index consists of two factors, namely, cost of time and cost of fuel. Their relationship defines cost (see Figs. 7.8, 7.9, and 7.10). The lowest point on Fig. 7.10 is the lowest-cost figure, and the associated

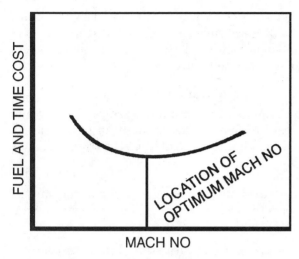

Figure 7.10   Cost versus Mach number.

Mach number is the most economical cruise speed. This approach allows each operator to build its own cost index, so it is easy to adjust and present a quick answer. The cost savings can be significant if these numbers are used in conjunction with an onboard computer.

A quick explanation follows for how to determine the best speed economical for a given trip (Fig. 7.11).

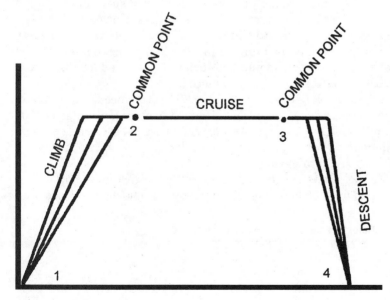

Figure 7.11   Trip optimization.

Climb. Economical speed is the speed where, for a set of climb speeds, the lowest cost is obtained. Here, time cost and fuel cost are applied. The range of the calculation is confined between points 1 and 2, where point 2 is a common point at cruising altitude.

Cruise. Economical speed is the speed where, for a set of cruising speeds expressed in Mach numbers, the minimum cruise time cost and minimum cruise fuel cost are found. The variation in cruise speeds affects time cost and fuel cost. For this calculation, the cruise region is between points 2 and 3.

Descent. This is calculated in the same way as climb, for various descent speeds, and selects the best-cost speed. The range for this calculation is between points 3 and 4.

Note: The operating manual presents various fixed climb and descent speed schedules. The optimum of these fixed climb and descent schedules should be then selected.

7.1.9 Long-range cruise

A payload versus range curve is shown in Fig. 7.12. The cruise speeds are long-range cruise and Mach 0.86. The speed and fuel consumption relationship is presented in Table 7.1. The effect of different cruise speeds is also shown.

An extreme example would illustrate the effect of speed on range. An airplane can carry a maximum of 376 passengers. However, there is a requirement for a 6000-nautical-mile trip. The airplane has to carry 50,000 lb of load, or 250 passengers (at 200 lb for each passenger and baggage). Cruising speed should be Mach 0.86 (selected for this example). For this trip, zero wind is considered.

Figure 7.12 indicates that at Mach 0.86 cruise speed, the range of maximum payload is 5300 nautical miles. The maximum range is 6000 nautical miles at zero payload. At long-range cruise, the range at maximum payload is 5600 nautical miles. The maximum range is 6650 nautical miles at zero payload.

By selecting a long-range cruise, the 6000-nautical-mile trip can be accomplished with 50,000 lb or 250 passengers. The dashed lines indi-

TABLE 7.1 Time, fuel and range versus cruise speed

Items	Long-range cruise	Mach 0.86
Trip time	Longer	Shorter
Trip fuel	Less	More
Range	Longer	Shorter

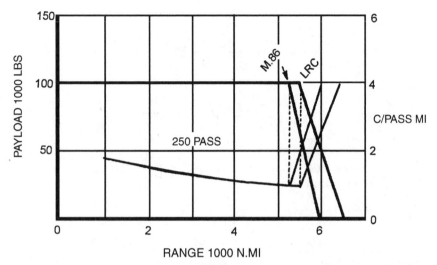

Figure 7.12   Payload versus range considering mach 0.86 and long-range cruise.

cate the relationship between the maximum payload range and minimum cost.

Direct operating cost in cents per passenger-mile is presented on the
same chart as a function of range. From this chart, the following conclusions can be drawn:

1. Cruising at mach 0.86, the range of the payload is shorter by about
   400 miles than when cruising at long-range cruise (LRC).

2. The direct operating cost, expressed in cents per passenger-mile,
   increases significantly beyond the range of maximum payload.

7.1.10   Nonoptimal cruise altitude penalties

Flying below optimum has a higher fuel penalty. Flying above optimum
is preferred, but this should not exceed a certain altitude based on the
type of airplane because of reduced stability and maneuver margin and
higher fuel penalties. The following tables reflect the penalties. The
conditions are

Aircraft	Wide body (three engines)
Cruise speed	Mach 0.82
Cruise altitude	35,000 ft
Yearly utilization	2400 h
Cruise weight	420,000 lb

Fuel penalties for flying at different cruise speeds:

	LRC	Constant Mach no.
2,000 ft above optimum	1%	2%
At optimum	0	0
2,000 ft below	1%	2%
4,000 ft below	3%	3%
8,000 ft below	7%	10%
12,000 ft below	14%	20%

Fuel penalties for flying different altitudes for a given mach number for a year's operations:

	Penalty	
	Altitude	Cost ($)/aircraft/year
Optimum	2000 ft	48,000
Optimum	0	0
Optimum	2000 ft	17,000
Optimum	5000 ft	99,000
Optimum	7000 ft	108,000

7.1.11   Optimum cruise procedure

There are different optima for different airlines. Four categories can be established for these optima

1. Carry the highest load with the least cost

2. Carry the highest load

3. Operate with the least cost

4. Operate with the fastest speed or shortest trip time due to competition and/or crew scheduling problems

Generally, airlines try to achieve item 1. However, given the exigencies of the real world, each schedule consists of a mixture of these goals.

7.1.12   Two- versus three-pack operation

Studies have indicated that operating one air-condition pack less in flight could save about 0.3 to 0.8 percent fuel flow. The decision to shut down one pack should be based on passenger load and the time when food is served. This would amount to $4000 to $10,000 per year depending on the airplane's fuel consumption.

7.1.13 Descent

The maximum fuel economy during descent can be achieved by idle thrust whenever possible and using a slow-speed descent. Descent tables are published in operating manuals.

Slow-speed descent generates the following fuel savings (320 trips per year):

Equipment	Per trip, lb	Fuel/yr, lb
B707	250	80,000
B727	200	64,000
B737	80	25,600
B747	300	96,000

The range of fuel savings potential is between 80 and 300 lb per trip. Assuming 320 trips per year and 60 cents/gal, the range of the potential fuel savings could amount to $2200 to $8600 per aircraft per year depending on the size and type of the airplane.

For premature descent, the following penalties are presented in pounds per minute:

B707	80 lb/min
B727	60 lb/min
B737	40 lb/min
B747	200 lb/min

Assuming 100 minutes per year when premature descent is occurring, $350 to $2000 would be the yearly penalty.

A two-segment descent is steeper than standard. At 500 ft, the flight path joins the standard descent profile. Noise is reduced in this way over a noise-sensitive community. However, wake vortex and safety are other considerations.

7.1.14 Landing

Numerous aircraft have two or more flap settings in landing configuration. For the DC-10, landing with 35 degrees of flap versus 50 degrees of flap would result in about 150 lb of fuel savings per trip. The lower landing flap setting could lead to more wear and tear on the brakes due to higher landing speeds. On the other hand, the structural life of the flaps is extended. For this type of airplane, assuming 320 trips per year at 60 cents/gal, the fuel savings potential would be around $3300 to $4400.

7.1.15 Holding

To minimize fuel, holding maneuvers should be conducted at the highest possible altitude in clean configuration at 1.5 V_s minimum maneuvering speed. A 1.0 percent increase in fuel consumption could be experienced compared with the slats-extended configuration. Table 7.2 shows the increased holding fuel flow related to clean configuration (for a midsized airplane).

TABLE 7.2 Holding fuel

Items	Increased fuel flow, lb/h
Gear extended	7000–11,000
Slats extended and gear up	1250–1450

Note: Since holding is not done on a daily basis, it would be misleading to calculate fuel savings for a yearly operation. Holding is done mostly in winter time when bad weather causes delays or in the summer when on holidays a large number of airplanes arrive at about the same time to a congested airport. For each trip, the increased fuel burn would generate an additional cost based on the quoted figures.

7.1.16 Flight planning

In order to achieve fuel conservation, flight plans have to be accurate. When a performance audit discovers that a flight plan specifies more fuel than the airplane actually burns, it should be corrected as soon as possible because carrying additional fuel leads to unnecessary fuel consumption and expenses.

Another important factor in accurate flight planning is the correct airplane weight (operating weight empty), since applying an incorrect weight may result in an incorrect uplift.

Some airlines tailor their flight plans to tail number for high-fuel-burning airplanes, and in this way they do not penalize the rest of the fleet. Based on the airplane, flight plans should be flexible and should provide (1) minimum time, (2) minimum cost, (3) minimum fuel, or (4) any other requirement needed by the operation. Saving 1 minute in a flight plan could save around $20,000 to $30,000 per year in fuel consumption.

Another goal is to reduce reserve and contingency fuel. An alternative airport should be selected carefully. Redispatch should be used whenever possible, and the shortest routing should be searched for, just to mention a few items where fuel savings are possible.

7.1.17 Ground rollout and taxi

Every phase of operations is scrutinized for potential fuel savings, and so is taxiing the airplane. The following should be considered:

1. In certain cases of one engine shutting-down, the breakaway thrust is larger on the operating engines. There is practically no fuel savings in this phase of operations.

2. For a typical three-engine airplane such as the B727-200, average taxi-out fuel figures are listed below with the alternate choices for this operation.

Three-engine operation and no auxiliary power unit (APU)	1200 lb
Two-engine operation with APU	850 lb
Two-engine operation with no APU	770 lb
Two-engine operation, no APU, and X-bleed	1000 lb

Note: For a two-engine airplane, any one of the engines can be shut down when APU is operating. For a four-engine airplane, the N_2 or N_3 engine may be shut down. For high-bypass engines, a period of 5 minutes should be observed after landing before shutting down engines. An analysis should be conducted to define the best savings method for each airplane and engine combination.

Certain operational procedures are associated with engine shutdown during taxiing. Each airline has to make its own decision on what approach should be used. It should consider additional workload, maintenance, simplicity in the procedures, safety, and the economics of this operation.

Running two versus three engines would result in additional fuel savings of about $8400 per year. Taxiing accounts for a large amount of fuel burn at airports. For a JFK/SFO/LAX/ATL size airport, this can be an estimated 30 to 40 million gal/year for taxi-out for takeoff and an estimated 12 to 18 million gal/year for taxi-in.

7.1.18 Ground operations

The following items are examples of fuel saving procedures:

■ Engine starts should not be initiated until passenger boarding is completed.

■ ATC departure delay should result in engine shutdown.

■ The closest runway should be used, if possible.

■ Taxi fuel may be reduced by intersection takeoff, if possible. This requires the availability of airport analysis marked with a note, e.g., "Intersection takeoff on runway 12L from taxiway Z."

■ Runway turnoff should be initiated after landing as soon as possible to minimize taxi distance to terminal.

■ The use of APUs should be kept to a minimum. Actually, APUs should be used for starting the engine only. For air conditioning, engine bleed is more economical than APU bleed. Taxiing with one engine shut down could save about 50 to 120 lb/min in fuel. Assuming 1 minute of fuel savings and 320 departures per year, this would amount to $1200 to $3000. Adding the savings in APU costs could add about 5 to 25 lb/min or up to $600 per year.

For the APU versus other power sources, Table 7.3 presents a comparison. This is for a narrow-body airplane: one APU with a typical overnight load (minimum electrical load and interior lights). Fuel cost is considered 50 cents/gal, and electric power costs 0.05 cents/kWh. In order to compare fuel consumption between towing and taxiing, Table 7.4 is presented.

Conditions used for this table:

■ All engines running

■ Fuel flows in lb/h

■ Fuel cost 50 cents/gal

■ Fuel density 6.7 lb/gal

Tow-bar-less tractor operation is a fuel saver. It saves about 70 percent fuel and is a one-person operation versus two-person operation

TABLE 7.3   Approximate fuel comparison to obtain ground electrical power

Requirement	APU (1)	Diesel/GPU*	C. E. power†
Energy form	26–44 gal/h	2.0–5.0 gal/h	11–40 kWh
Cost	$12–$22/h	$1.0–$2.50/h	$0.55–$2.00/h

*GPU = Ground power unit.
†C. E. power = commercial electrical power.

TABLE 7.4   Fuel and cost omparison for towing or taxing

Aircraft	Engine fuel flow	Engine $ cost	APU fuel flow*	APU $ cost†	Tractor fuel flow	Tractor $ cost	$ Total APU † TR
B747	7200	537	900	67	60	4.50	71.50
DC-10	4500	335	500	37	60	4.50	41.50
B707	4200	313	300	22	60	4.50	26.50
B727	3180	237	280	20	60	4.50	24.50

*Fuel flow in lb/h.
†Cost of towing airplane with APU running.

with the conventional tow-bar tractor. This tow-bar-less tractor is equipped with a special cradle, and its operator docks with the aircraft nose wheel.

7.1.19 Center of gravity

Fuel savings can be obtained by shifting the center of gravity (CG) aft. By loading the airplane in this way, the magnitude of savings would be about 0.3 to 0.5 percent in fuel. For example, moving 5000 lb of cargo from forward to aft would yield a fuel mileage improvement of 0.5 percent. A yearly estimate is not possible because fuel savings do not occur regularly.

7.1.20 One-stop versus nonstop flights

On long trips when the airplane is close to its range limit (e.g., 6000 miles), introduction of an enroute fuel stop could reduce trip fuel, depending on the airplane and engine combination (aircraft performance should be analyzed for these conditions). This fuel savings may result in an opportunity to carry more payload, thus producing more revenue. However, two landing fees, aircraft handling, trip times, and other costs, if any, also should be considered.

7.1.21 Fuel tankering

Sometimes it is worthwhile to tanker fuel due to the variation in fuel prices. Caution should be exercised, however, because carrying fuel costs fuel too. The following items play a part in the decision to tanker fuel:

- Price difference
- Aircraft type
- Trip distance
- Enroute wind and temperature
- Cruise speed
- Flight level
- Payload
- No fuel available at destination

When lack of fuel is the reason for tankering, trucking fuel or exchanging allocation with another airline so as to have fuel at the station should be considered. Figure 7.13 presents a chart related to fuel tankering. By tankering, some operators could save 1½ to 3 percent of their fuel costs.

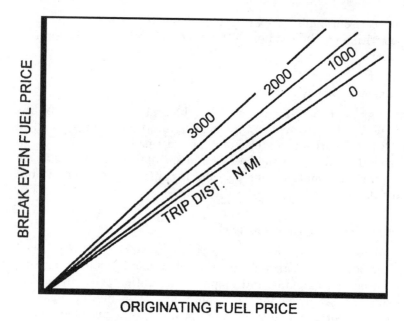

ORIGINATING FUEL PRICE

Figure 7.13   Fuel tankering.

7.1.22   Fuel dumping

Fuel dumping could occur when shortly after takeoff an emergency situation arises and the airplane has to return and land. In order to land at the maximum landing weight, the excess fuel is dumped (environmental authorities are not in favor of fuel dumping). The other alternative is to have an overweight landing.

A typical dump quantity in airline operation for a B727 would be around 15,000 lb, and for a B747 around 100,000 lb. The airline industry could dump around 28 to 35 million pounds per year, which is a huge waste of money or fuel.

The decision to dump fuel or execute an overweight landing is based on airline policy. The Federal Aviation Authority (FAA) has certain regulations related to overweight landing. In addition, manufacturers issue advice, procedures, and prescribed inspections after an overweight landing.

7.1.23   Fuel penalty from carrying excess weight and/or fuel

Aircraft weight is one of the many other parameters that has an effect on fuel consumption. The higher the landing weight, the higher is the amount of fuel consumed. Carrying excess weight results in fuel penalty.

Fuel penalty is a function of the range flown. For example, at the 4500-nautical-mile range, a wide body 3 engine airplane will burn ½ lb additional fuel for each additional pound of weight carried. For each type of airplane, a similar chart could be developed (Fig. 7.14).

There is a strong motivation to reduce the amount of excess weight carried. It is important to keep the "Empty operating weight" of the aircraft to a minimum, remove excess operational items, and above all, keep reserve fuel from being unreasonably high. The weight sensitivity of a wide-body aircraft is indicated in Table 7.5. For example, adding 10,000 lb of extra fuel to the fuel computed in the flight plan, e.g., on a trans-Atlantic flight (about 3200 nautical miles), would require 2800 lb of fuel just to carry this extra fuel. The airplane would arrive at the destination with 7200 lb of fuel only (from the original 10,000 lb of fuel). This would amount to a loss of $209/trip (at 50 cents/gal). When projected, a year's worth of penalties would amount to a loss of $62,700.

7.1.24 Reserve fuel

For domestic and international operations, the Federal Air Regulation (FAR) specifies the required reserve fuel (Table 7.6). Some air carriers

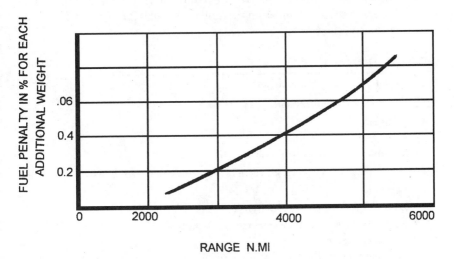

RANGE N.MI

Figure 7.14 Fuel penalty for carrying extra weight.

TABLE 7.5 Landing weight change versus block fuel change

Payload	Landing weight	Block fuel
70,000	360,000	178,650
50,000	340,000	167,817

add extra fuel beyond the legal requirement (see fuel penalty above). Sometimes, a discretionary amount of extra fuel is added based on the captain's decision, which may or may not be warranted.

This extra fuel could have been saved. Another source of savings would be the selection of a closer alternate airport (if available). In case no alternate is selected, the requirement described in FAR 121.643 should be followed:

> However, no alternate airport is required if
>
> 1. For at least 2 hours before and 2 hours after the estimated time of arrival, the ceiling at the airport to which the flight is dispatched is forecast to be at least 1000 ft...and
> 2. The visibility at the airport is forecast to be at least 3 miles.

The deletion of an alternate could reduce the landing weight to about 3500 lb for the B737 and around 10,000 lb for the B747 airplanes, and naturally, it would reduce the trip fuel by 1.2 to 3.5 percent.

7.1.25 Domestic reserve fuel

Within the continental United States, the minimum requirement is to be able to fly the airplane from the destination to the designated alternate airport and to have enough fuel on board for a normal cruise

TABLE 7.6 Reserve fuel definitions

Reserve Fuel: FAR 121 Requirements
121.639: U.S. domestic air carriers
All operations:
■ Fuel to alternate
■ 45 minutes at normal cruise consumption (LRC)
121.645: U.S. flag, supplemental air carriers and commercial operations
Jet operations:
Any operation outside the 48 contiguous states
■ Fuel to fly additional 10 percent of total flight time
■ Fuel to alternate
■ 30 minutes at holding speed at 1500 feet above alternate 121.647: Factors for computing fuel required
Consideration shall be given to:
■ Wind and weather conditions forecast
■ Anticipated traffic delays
■ One instrument approach and possible missed approach at destination
■ Any other conditions that may delay landing

consumption of 45 minutes according to FAA regulations (ICAO fuel reserve at alternate is 30 minutes). Figure 7.15 presents a graphic layout for these conditions.

Note: A typical quantity of reserve fuel assuming a 200-nautical-mile alternate for a DC-9 is about 6500 lb and for a B747 about 38,000 lb.

7.1.26   International reserve fuel

FAA international reserve policy consists of two parts:

1. After reaching the destination, to fly for an additional 10 percent of the total flight time at normal cruise speed (ICAO policy is 5 percent of trip fuel).

2. Airplane should have enough fuel to fly to the designated alternate and, when there, to hold for 30 minutes at 1500 ft above airport elevation, descend, and land.

A graphic presentation is shown here to illustrate the requirements (Fig. 7.16). The reason for describing in more detail the reserve fuel calculation is the significance in fuel savings of applying a proper reclearance procedure that may increase the load on the airplane.

Note: A typical reserve fuel quantity assuming a 5.5-hour trip and an alternate of 200 nautical miles for a DC-8-73 is about 15,000 lb and for a DC-10-30 about 21,600 lb.

7.1.27   Redispatch or reclearance procedure and its revenue potential

A good knowledge of the international reserve fuel regulation is necessary before considering studying redispatch procedures. Quoting from FAR 121.631: "However, the dispatch or flight release may be amended

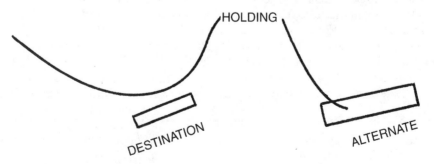

Figure 7.15 Domestic reserve fuel.

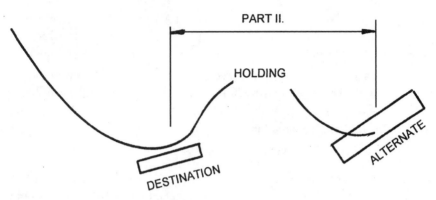

Figure 7.16 International reserve fuel.

enroute to include any alternate airport that is within the fuel range of the aircraft as specified in Paragraphs 121.639 through 121.647."

The international reserve fuel regulation includes a certain amount of contingency fuel. This contingency fuel is a function of trip length or trip time. At the arrival at the destination airport, a 300-nautical-mile trip has less contingency fuel requirement than a 5000-nautical-mile trip. The reasons behind this rule are as follows: This extra fuel should cover navigational errors, enroute weather, and/or other miscellaneous reasons contributing to the increase in trip fuel beyond the predicted fuel burn. Forecast error would be just one of these reasons.

Due to improved weather forecasting and computerized and more sophisticated navigational techniques, the probability of using the contingency fuel significantly has decreased. Once the airplane approaches the destination airport (from a long trip), it has a large amount of contingency fuel left on board. By introducing enroute dispatch, this large amount of reserve fuel can be reduced to increase payload or range. The fuel to alternate and holding fuel are independent of trip length. Contin-gency fuel depends on trip distance and time, and this is the area where redispatch could be applied. Reclearance for ZRH-JFK trip is shown in Fig. 7.17.

As an example, a trans-Atlantic flight is chosen from ZRH to final destination at JFK. Selecting Halifax for initial destination, two flight plans are prepared. The top of descent point at the initial destination is the starting point for analyzing the reserve fuel distribution. At this redispatch point, if sufficient fuel is on board, the flight is recleared to the final destination. The initial flight plan called for Zurich to Halifax, with Bangor as the alternate airport. Applying redispatch at abeam Halifax, the airplane is recleared to JFK and Washington, D.C., as alternate. The contingency fuel from Halifax to JFK is significantly less than that from Zurich to Halifax and alternate fuel to Washington, D.C. This fuel difference is used to fly the airplane from Halifax to JFK, with the

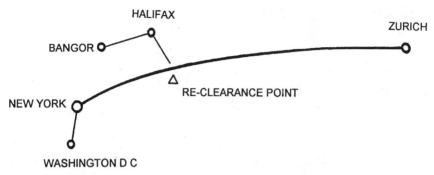

Figure 7.17   Reclearance procedure.

payload destined to Halifax. This payload is naturally higher than the flight plan showing for the ZRH to JFK trip. With redispatch, the large amount of contingency fuel was used to permit the airplane to continue the flight to its final destination. Higher payload and less contingency fuel contribute to a more economical operation. The benefit of redispatch is the increased payload on flights where takeoff weights are limited. For the same trip length, this increase in payload is significant and could amount to a savings of approximately 80 percent of the contingency fuel.

The fuel required for 10 percent of the flight time from Zurich to Halifax is about 12,600 lb (for a three-engine, wide-body airplane). By applying for reclearance—from Halifax to New York—the fuel required is now 2,000 lb, producing a savings of 10,000 lb or additional payload.

Enroute dispatch is used commonly to increase payload. It would be a waste not to apply it whenever legal and possible.

A few remarks about redispatch: In case of insufficient fuel at the redispatch point, the flight has to land at the initial destination. An accurate flight plan is a prerequisite for a succesful redispatch. In order to obtain an accurate flight plan, a few helpful hints are listed:

- The airplane operating empty and landing weights should be accurate.
- The amount of fuel on board should be accurate and so should the fuel gauges.
- Maintenance should take care of airframe and engine deteriorations.
- Proper speed should be maintained.
- The flight plan recommended flight altitudes should be followed.
- Accurate wind predictions are necessary.
- Traffic congestion should be considered (this requires more fuel).
- Taxi fuel planning should consider the distance between the departure terminal and the runway end where the airplane will take off.

- Time of the day (e.g., Friday at 5 p.m.) should be considered because more taxi fuel may be required.
- Air-condition system should be operated per flight plan (e.g., two versus three-pack).
- Airplane trim should be appropriate.

7.1.28 Weight reductions

Airlines established weight-reducing in-house committees whose job is to analyze and reduce aircraft weights. Below is a brief description of how to accomplish this task.

Operating weight empty. Reducing aircraft operating weight empty has a reducing effect on landing weight and trip fuel. Weight reduction in operating weight empty reduces fuel consumption for the rest of the airplane's life. Lower weights permit operations at a higher flight level. The decreases in fuel consumption are listed below related to a 1000-lb weight reduction:

Model	Decrease in trip fuel per 1000 lb of landing weight reduction in percent
B727	0.47
B737	1.00
B747	0.12

Individual weight reductions are relatively small, but adding them up could result in significant savings. There are a few areas where weight reductions are possible.

1. *Passenger service and food items.* As an example, on the B747, based on the number of passengers, meal and beverage service equipment and consumables could run up to 4000 to 8000 lb. Using less heavy equipment and consumables in relation to the number of passengers on board could reduce weights.

2. *Cargo and baggage containers.* Containers used on B707 and B727 aircraft have a total empty weight up to 800 to 920 lb each. On the B747, the empty container weight could be even higher. The use of pallets and bulk loading could save trip fuel consumption by 0.6 to 1.2 percent.

3. *Emergency equipment.* Over-water equipment weighs about 1000 to 2000 lb. For flights that are not over water, it is advisable to remove this equipment.

4. *Potable water.* For the B747, the weight of potable water is about 1800 lb. Filling the tank to the level required for specific flights based on the number of passengers and load could reduce trip fuel by 0.2 to 0.4 percent.

5. *Other items.* Removal of lounges, entertainment equipment, and turbine trust reversers, if approved, could reduce trip fuel by a small amount.

6. *Dirt.* Remove dirt collected on hard-to-reach places and some condensation of water vapors.

7. The airplane should be weighed at certain time intervals. Care should be exercised so that drainage of fuel tanks is done properly.

8. Operating weight empty should be updated when modifications are performed on the airplane. Each modification adding weight should be evaluated carefully because each pound of extra weight increases the fuel burn and cost.

Reducing landing weight. Reducing operating weight empty and removing any weight not affecting the safety of the flight also have decreased the landing weight. To highlight the importance of these saving efforts, we can mention as an example a narrow-body, four-engine airplane requiring 80,000 lb of trip fuel. Reducing the landing weight by 10,000 lb, the trip fuel required would be less with 3200 lb (trip length aroud 3000 nautical miles). As mentioned earlier, depending on the size of the aircraft, the fuel savings could range between 0.3 and 1.0 percent.

7.1.29   Short-trip cruise speed

On short trips, an aircraft flies at light weight, low altitude, and low speed to achieve economical operations. The manufacturer's operating manual recommends the proper airspeed/altitude combinations.

7.1.30   Enroute temperature effect

Temperature has an insignificant effect on fuel mileage. For the B747, the fuel mileage decreases about 0.5 to 0.7 percent for each 10-degree increase in outside air temperature. Selection of a route through a cooler air mass may result in trip fuel savings sufficient to warrant consideration of this route selection.

Recent measurements have indicated small changes in drag due to Reynold's number. Reynold's number is related to air viscosity, velocity, and airplane size. Drag slightly increases when altitude and temperature are higher than reference data. Two more items could be mentioned that have no measurable effect on operations, namely, the Coriolis force that is due to the earth's rotation and the isobar slope where the airplane would maintain constant pressure altitude. In airline daily operation, the airplane cuts across these isobars.

7.1.31 Engine fuel consumption effect

Engine fuel consumption is the function of flight conditions and thrust. It is expressed in pounds per hour per pound or pounds per hour fuel flow for each pound of engine thrust. It is abbreviated as TSFC (thrust-specific fuel consumption). Presently, for modern high-bypass engines, TSFC is in the range of 0.6 to 1.3. After the first few years of operation of a new engine, fuel consumption starts to increase slowly due to engine deterioration to a magnitude of 0.8 to 0.92 percent (or more). After overhaul, it is restored partially, if not fully, to the baseline, and after each overhaul, a slight increase in TSFC is experienced. Airlines conduct engine performance audits and take steps to alleviate engine deterioration before expensive corrections have to be applied.

7.1.32 Mach meter accuracy

A mach meter inaccuracy could adversely affect fuel consumption. Cruising 0.01 mach faster increases fuel consumption by as much as 2.00 percent. For an airline, this 2.00 percent loss could be translated into effective cost dollars. For each 1 million lb of extra fuel (at 50 cents/gal), $74,600 could be wasted.

7.1.33 Derated takeoff

Many times, there is no need to use the maximum takeoff power (EPR or N1); instead, derated takeoff power can be applied. The procedure should be safe, simple, and not degrade airplane performance and/or reduce payload. About 25 years ago, I introduced a derated takeoff procedure by presenting EPR (engine pressure ratio) on the airport analysis charts as a function of elevation and ambient temperature (see Chap. 8 and Figs. 8.4 and 8.5). This presentation, because of its simplicity and accuracy, was widely accepted and used in the industry and was approved by manufacturers at that time. At takeoff, application of the assumed temperature method is always conservative and safe. It is also beneficial in terms of maintenance costs.

By lowering turbine operating temperature, the reduced takeoff thrust operation decreases maintenance cost and increases engine life. This in turn increases engine hot-section life, lowers shop-visit rate, and reduces maintenance material, cost, and labor, as well as any outside vendor cost. It also could affect fuel cost (Table 7.7). By reducing the turbine inlet temperature by 125°F on the JT8D-7 engine, an industry estimate is presented on parts' life improvement, assuming that reduced takeoff power is used half the time.

TABLE 7.7   Improvements by lowering turbine operating temperature

Parts	Failure or distress	Estimated improvement
Burner	Cracking	50%
	Erosion	50%
Transition ducts	Cracking	50%
	Buckling and distortion	50%
First-stage vane	Cracking	50%
	Bowing	30%
First-stage blade	Temperature fatigue	50%
	Creep	50%
	Coating loss	25%

Another approach was the derated engine concept, where, for example, a JT9D-7A could be derated to a -7 or to a lower -3A level. Another solution was the certification of derate I and derate II. This required airlines to furnish guidelines for these derated levels in forms of airport analyses and procedures dealing with derates.

An estimated cost-savings graph is presented for derated takeoff power in Fig. 7.18. Reducing the power setting by 0.03 to 0.05 EPR, the savings could amount to an estimated $25,000 to $45,000 per aircraft per year. The reason is simple: The derated takeoff thrust has its greatest effect on the turbine and hot section. The turbine is the most important engine component affecting fuel consumption. Some airlines have reported that derated takeoff thrust can improve specific fuel consumption by 2 percent. The application of derated thrust reduces the risk of catastrophic failure, improves engine life, and is very cost-effective.

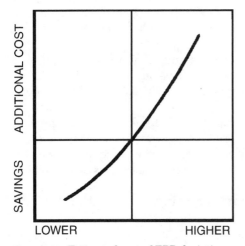

Figure 7.18   Estimated cost of EPR deviation.

7.1.34 Derated climb

Climbing with reduced power slightly increases fuel consumption for overall climb, but there are gains in engine life and lower turbine temperature. The life of the hot section is significantly increased with reduced climb thrust. Furthermore, the fatigue life of turbine airfoils is improved, and oxidation and corrosion of hot-section airfoils are reduced. The combustion chamber liner life is extended. Reduced climb thrust should be used, conditions permitting.

7.1.35 Fuel savings in action

Airlines are closely scrutinizing their operational procedures. Delta Air Lines disclosed its actual figures for the month of June 1996, which showed that the airline saved more than $300,000 in crew costs and almost $90,000 in fuel. Listed below are a few items contributing to these savings.

- Reducing aircraft weights by about 4 percent by eliminating some onboard company materials, cutting the amount of catering supplies, and using lighter service carts.

- Working with the pilot group to develop single-engine taxi procedures.

- Increasing the number of pushbacks and permitting 70- and 90-degrees pushbacks where allowable.

- Tracking and reducing APU usage, for a projected savings of $52 million annually.

- Tankering fuel into destinations such as Florida where fuel is costly.

Only a few items are listed here; however, many more contributed to Delta's success. Another fuel savings approach called NexPlan was developed.

NexPlan does not rely on static routes and calculates preferred routing based on updated wind information. Its variable cost index, tailored for the specific conditions of a given flight, produces accurate fuel, crew, and aircraft costs.

There are many other ways to save fuel. Fuel economy cannot be ignored in today's environment. Practically every air carrier has a policy to obtain the least cost for its operations because fuel is a large part of the direct operating cost. Since 1973, the industry has accomplished significant results in fuel savings procedures, developing hardware and publishing (mostly by manufacturer) information about fuel savings methods.

1. Hardware development
   a. Performance data computers

    b. Flight management system
    c. Full-time automatic throttle
    d. Navigation equipment

2. Airplane/engine condition improvement
    a. Performance analysis
    b. Weight reduction
    c. Engine overhaul improvements
    d. Improved aerodynamic cleanness

New procedures since 1973 that accomplished significant savings in fuel economy include the following:

Start-taxi	Idle and taxi time minimized
Takeoff	Use of reduced thrust
Climb	Best speed for fuel economy
Altitude selection	Best cost/best fuel procedures

	Cruise speed	
Aircraft	Before 1973, Mach	Presently, Mach
B727	0.82–0.84	0.80
B737	0.74–0.78	0.73–0.74
B747	0.86	0.84
DC-8-50-60 series	0.82	0.80
DC-10		0.83
B767		0.80

Descent	Slower speeds, closer to optimum
Holding	Cut down on long holding time
Landing	Increased overweight landing
Reserves	Reduction in reserve fuel
Dispatch	More application of redispatch

The introduction of communication, navigation, surveillance and air traffic management (CNS/ATM) system using satellite technology is a new way to a significant fuel savings. The high accuracy of satellite technology saves about 50 to 100 miles on a 3500- to 4500-mile trip. This can be translated to fuel saved, and a dollar sign can be attached to it.

7.2 Maintenance Department Fuel Conservation

Moving an airplane through air requires power. The energy to move it is in proportion to the power required. It should have a smooth and preferably polished surface, no protruding "bumps," and less drag. This lower drag requires less power and less fuel consumption. The airplane is designed for this reason with a smooth surface and aerodynamically designed contour. The airflow around the airframe is laminar or smooth. Bumps create turbulent airflow and require more power to move the airplane. In addition, turbulence on the wings reduces lift capabilities. Drag penalties can be defined from wind tunnel tests or analytical studies.

In reality, normal wear and tear contributes to elevated drag and higher fuel consumption. Most in-service airplanes experience normal wear and tear that results in degraded aerodynamic "cleanliness." This usually causes no operational problems, just higher fuel consumption. Over a given period, normal wear and tear may include leaking seals, chipped paints, dents, external patches, protruding replacement fasteners, and many other items.

7.2.1 Location of high fuel consumption

Generally, the areas of fuel penalty are classified in four categories:

- Instrumentation
- Seal leakage
- Surface irregularities
- Control surface rigging

The magnitude of the fuel penalty is a function of the location of the problems on the aircraft.

For example, the nose section, the wing leading edge, engine inlet cowls, nose cowl lip, turbine reverser, engine pylon, leading edge of elevator, and rudder are very sensitive locations. Other wear and tear problems may be much less fuel-sensitive. There can be significant performance losses associated with certain types of in-service aerodynamic discrepancies. This may occur randomly within the fleet. The general aerodynamic "cleanliness" degradation is the gradual deterioration of various seals, rigging, and skin smoothness.

It is important to review the degradation in order to make a realistic assessment, and as far as the penalty and restoration are concerned, they could be expressed in dollars and cents. Generally, penalties are expressed in percentages, in drag increase, or in fuel flow increase.

Why is so much attention given to fuel savings? A B737 airplane consumes about 1.9 million gallons of fuel, and at 50 cents/gal, this amounts to $950,000 per year. At the other end of the spectrum, a B747 airplane burns about 13.1 million gallons of fuel (yearly utilization of 3510 hours at 25,000 lb/h fuel flow and a fuel density of 6.7 lb/gal) at a cost of $6,550,000 per year. Assuming only a 1 percent drag-rise in the condition of such an airplane, a significant increase in the operating cost could be avoided.

TABLE 7.8   Fuel and cost penalty for a 1 percent drag increase

Aircraft	Extra fuel gal/yr	Dollar/yr
B737	19,000	9,500
B727	33,180	16,590
B707	49,680	24,840
B747	131,000	65,500

Table 7.8 shows the fuel cost penalties for a 1 percent drag increase. Sometimes drag can increase more than 1 percent, and penalties can be higher. Drag increase is just one of the numerous causes of higher fuel consumption.

Airplane manufacturers conducted extensive studies on airplane maintenance for fuel conservation and made them available for the airline industry. A very significant conclusion of these studies is that the path to in-flight fuel economy lies in accomplishing many seemingly insignificant adjustments or repairs rather than through any single significant fix. The fuel penalties could vary from $100,000 per year to $32 per year, and there are plenty of items where the penalties are between these two cost figures. In each category, only a few items are presented to show the magnitudes of the various fuel penalties.

7.2.2   Instrumentation

When the Machmeter indicates slower speed, in actuality, the airplane flies faster. Flying faster requires more fuel. The relationship between speed and fuel consumption is not linear. Table 7.9 contains a few examples. A 65 cents/gal fuel price is considered.

Note: This range covers both small and large airplanes.

The following list of instrument tolerances is important to know in order to have a realistic evaluation of the accuracy of the data points obtained. A large number of data points averages out inaccuracies. The trends in the items examined should be followed carefully because they point to the problem areas.

7.2.3   Seal leakage

Seal leakage is related to a 10 in² area in order to have a unified basis for comparison (Table 7.10).

7.2.4   Surface irregularities

From more than a few dozen items, only three are presented here:
    Fuel: 50 cents per gallon

TABLE 7.9   Fuel penalties and corrections costs for instrument inaccuracies

Instrument	Fuel penalty, gal/yr	Fuel penalty, $	Cost of correction, $
EPR gauges indicates 0.01 low	10,000–180,000	6,000–108,000	50–150
Fuel flow indicates 1% high	7,000–100,000	4,200–60,000	100–200
Machmeter indicates 0.1 mach low	8,000–250,000	4,800–150,000	150–350

Instrument Tolerances

IAS	2–3 knots	Indicated airspeed
Mach	0.05–0.01	Machmeter
OAT/TAT	2°C	Or Celsius
EPR	0.01	Engine pressure ratio
N1/N2	0.5–1.0%	Engine RPM
EGT	3–4°C	Celsius
Fuel flow	0.5–0.9%	
Altitude		
Sea level	15–30 ft	
2000 ft	70–90 ft	

TABLE 7.10   Fuel penalties, corrections costs for seal leakages

Location	Fuel penalty, gal/yr	Fuel penalty, $	Cost of correction, $
Cargo doors, windows	34,600	17,300	60
Wings/body fairing, flaps	7,800	3,090	60
Ailerons, nose gear doors, aircraft compartment doors, main landing gear doors, seal leaks	3,090	1,545	120

1. Forward nose cargo door, main deck, 0.1 in up or down over entire circumference: Fuel: 60 cents per gallon

Fuel penalty   7500 gal/yr or $4500 per year
Cost of correction $207

2. Rough paint, upper wing and lower horizontal stabilizer critical areas, 10-ft^2 area with equivalent roughness height greater than 0.0015 in:

Fuel penalty   6100 gal/yr or $3660 per year
Cost of correction $53

3. Surface mismatch, upper wing and lower horizontal stabilizer critical areas, 0.1-in step by 10 ft long access panels, doors, etc.:

Fuel penalty   2400 gal/yr or $1440 per year
Cost of correction $125

7.2.5   Control surface rigging

1. Flight with 1-degree sideslip caused by misrigged controls, instrumentation inaccuracy, or asymmetric thrust:

Fuel penalty   67,000 gal/yr or $40,200 per year

Cost of correction $320

2. Rudder misrigged, split rudder, 1-degree difference between upper and lower rudder segments, hydraulic power on:

Fuel penalty   48,000 gal/yr or $28,800 per year

Cost of correction $149

3. Spoiler float, trailing edge of spoiler deflected 1 in up per panel:

Fuel penalty   18,200 gal/yr or $10,920 per year

Cost of correction $107

In these presentations, the penalties expressed in gallon per year values are the same from year to year. The cost of fuel penalties and cost of corrective maintenance change with the then prevailing prices. These figures were based on a large, four-engine airplane.

7.2.6 Auxiliary power unit (APU) fuel penalties

An interesting tabulation is presented for a wide-body, three-engine airplane (Table 7.11). The approximate annual cost of open or missing APU doors in dollars per year per aircraft is tabulated.

Another item is the approximate annual cost of excessive inert weight in dollar per year per aircraft.

Average mission, nautical miles	Penalty/pound, $
500	3.00
1000	5.30
2000	6.80
3000	7.60

These figures are shown in graphic form in Fig. 7.19. For each type of airplane, a similar chart could be developed.

TABLE 7.11 APU Operations

Inlet door	Exhaust door	APU	Penalty, $
Closed	Closed	Off	0
Closed	Removed	Off	2970
Closed	Open	Off	4150
Open or removed	Closed	Off	18,900
Open or removed	Removed	Off	18,900
Open or removed	Open	Off/on	75,600*

*APU fuel cost not included.

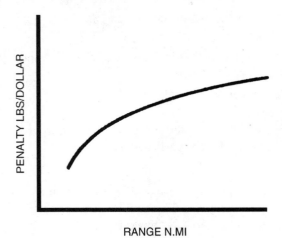

RANGE N.MI

Figure 7.19 Penalty per pound versus range.

Chapter

8

Airport Load Analysis

8.1 Introduction

Airport load analysis is an important presentation for flight operations. It defines the maximum takeoff and landing weights for a given airport, runway, temperature, and wind conditions. Any time a trip analysis is needed, the first step is to obtain the maximum legal takeoff and landing weights from the airport analysis, since these weights are the limiting factors in any further calculations.

8.2 Airport Analysis

At the time of delivery of an airplane, an FAA-approved flight manual or AFM has to be on board. This manual contains the method of calculations to obtain required legal takeoff and landing weights. No weights can exceed the weights extracted from the FAA AFM, since these are the legal maximum weights.

When an airline calculates weights that are lower than those from the FAA AFM, it may lose considerable (passenger or cargo weight) profit potential. If the airline's calculated weights are above those of the FAA AFM, the operation could be illegal and unsafe. For example, if the airline calculates a takeoff weight of 275,000 lb and the FAA AFM method indicates 280,000 lb, the airline is losing a profit potential of revenue generated by 5000 lb of cargo or about 25 passengers. On the other hand, if the airline calculates 281,000 lb, the operation is illegal. One airline was fined $285,000 recently by the FAA for failure to adhere to the maximum allowable takeoff weight requirement.

In order to cope with this complex and tedious calculation, airlines have been computerizing this procedure since the late fifties and early sixties. The FAA AFM is very specific in its description of the requirements of every phase of flight operation, one of which is the takeoff criteria. This is a highly technical subject that requires a good engineering background. Because of its complex nature, it will only be touched on very briefly here.

8.3 Takeoff Criteria

Air worthiness regulations require that the takeoff performance of an aircraft is established for the range of conditions under which it normally operates (Fig. 8.1).

The minimum takeoff field length is the greater of

- One-engine-inoperative takeoff distance to a 35-ft height

- Accelerate-stop distance

- All-engine takeoff distance to a 35-ft height plus a 15 percent margin

It is required that the takeoff distance (with and without engine failure) should be the same as the accelerate-stop distance (Fig. 8.1*b*).

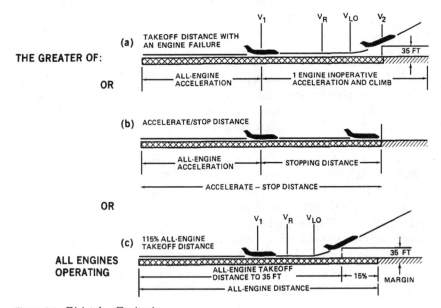

Figure 8.1   FAA takeoff criteria.

This case is referred to as a *balance field length*. Definitions of operational "speeds" include

- V1 is the most critical speed at which an engine failure may occur.
- VR is the rotation speed.
- V2 is the takeoff safety speed (at least 120 percent of stall speed).
- VLO identifies the speed at which the aircraft first becomes airborne (this is not an operational speed).

To ensure safe operations, all obstacles have to be cleared with a prescribed safety margin below the flight path. For each phase of the flight-path profile, certain requirements are described in the FAA-approved AFM that have to be satisfied. A takeoff flight-path profile is presented in Fig. 8.2.

8.4 Landing Criteria

The FAA specifies that the landing distance required for an aircraft be established from a point 50 ft above the end of the runway to the end of landing roll with the use of brakes only (no thrust reversers) and then increased by a margin of 66.7 percent. For runways that are predicted to be wet, the FAA requires an additional 15 percent margin.

All obstacles below the approach path have to be cleared by a certain margin, as defined in the FAA-approved flight manual (Fig. 8.3).

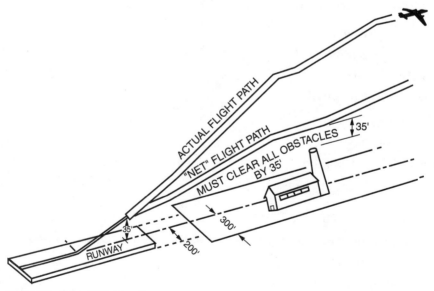

Figure 8.2 Takeoff flight path.

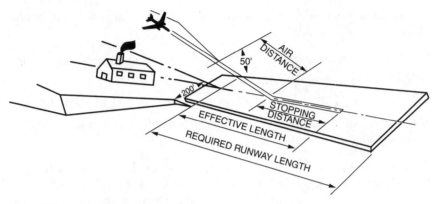

Figure 8.3 Landing flight path.

This description was a simplified presentation of the procedure to calculate takeoff and landing weights. The calculations are much more complex, and it is not the purpose of this chapter to go into detailed analysis.

8.5   Required Airport Data

The input for the takeoff and landing weight calculations constitutes the airport's physical limits and aircraft characteristics:
Airport characteristics:

- Airport elevation
- Airport temperature
- Airport wind
- Runway length
- Runway slope
- Obstructions
- Runway strength

Aircraft characteristics:

- Limiting weights
  Climb gradient
  Runway length
  Bleeds
  Miscellaneous limitations, etc.

- Determine
  - Maximum takeoff weight
  - Maximum landing weight
  - Flap setting for takeoff and landing
  - Miscellaneous other factors

8.6 Airport Load Analysis Presentation

Airlines prepare takeoff and landing weight information from the FAA-approved flight manual. Depending on the airline flight department's philosophy, a sample airport analysis is enclosed here (Fig. 8.4). These presentations may come in many forms and shapes.

The upper part of the sample analysis contains airport information, aircraft name, engine designation, and takeoff flap setting. The following vertical columns are presented:

- The first column is ambient temperature in Fahrenheit.
- The next column contains the performance (sometimes called *maximum*) limit of the aircraft. It is the function of airport elevation and temperature and is independent of runway length. No takeoff weight can exceed this limit. This will ensure a minimum required climb capability for the airplane.
- The next columns, going from left to right, are the weights associated with different runways, as the runway numbers at the top of the columns indicate. The takeoff weights are presented here for the temperature range at this airport at zero wind.

For example, on runway 13L at 10°F, the takeoff weight is 362,400 lb, but the performance limit of 355,000 lb cannot be exceeded. Therefore, the final takeoff weight is 355,000 lb.

- Further on the right-hand side is the engine pressure ratio (EPR) or power setting column.
- The last column shows the temperature in centigrade (celsius). Headwind and tailwind corrections are presented below each runway column.

For the landing chart (Fig. 8.5), similar to the takeoff chart, aircraft and airport data are presented, and 35 degrees of landing flaps are indicated. The airplane operational temperature limit goes up to 120°F (see FAA AFM). At this particular airport elevation, the maximum landing weight is 271,000 lb, and this weight cannot be exceeded. For wet and dry runways, the landing weights are presented in 2-knot tailwind increments.

```
--XABC--                                        --ABC--
ELEV. 3000                                      ANYWHERE AIRPORT1
MAX TEMP 120 F          23 FLAPS                ANYPLACE, ANYLAND
JT3D-7                                          VAR 02 E

                        DC-8-63
                TAKE OFF PERFORMANCE
              NOISE REDUCTION NACELLE
              THE NACELLE CORPORATION
```

RUNWAY NO F	PERF LMT	13L	14L	31R	32R	EPR	C
0	355.0	366.0*	312.9	350.8	332.7	2.05	-17
10	355.0	362.4*	308.9	346.7	328.7	2.05	-11
20	355.0	356.4*	302.5	340.1	322.2	2.02	-6
30	348.4	349.6*	296.4	333.2	315.6	1.99	0
40	339.9	342.0*	290.3	326.0	308.7	1.95	4
45	335.4	338.3*	287.2	322.5	305.3	1.93	7
50	331.0	334.7*	284.3	319.0	302.0	1.91	10
55	326.0	330.5*	280.9	315.0	298.3	1.89	13
60	322.1	327.1*	278.1	311.8	295.5	1.88	16
62	322.1	326.5*	277.6	311.3	294.9	1.88	17
64	322.1	325.9*	277.1	310.7	294.4	1.88	18
66	322.0	325.3*	276.6	310.1	293.9	1.88	19
68	322.0	324.7*	276.1	309.6	293.4	1.88	20
70	322.0	324.1*	275.6	309.0	292.9	1.88	21
72	322.0	323.5*	275.1	308.4	292.4	1.88	22
74	321.9	322.8*	274.6	307.9	291.9	1.88	23
76	321.9	322.2*	274.2	307.3	291.4	1.88	24
78	321.9	321.6	273.7	306.7	290.9	1.88	26
80	321.8	321.0	273.2	306.2	290.4	1.88	27
82	321.8	320.4	272.7	305.6	289.8	1.88	28
84	321.7	319.8	272.2	305.0	289.3	1.88	29
86	319.1	317.6	270.4	303.0	287.5	1.86	30
88	316.2	315.3	268.5	300.8	285.5	1.85	31
90	313.3	313.1	266.8	398.8	283.7	1.84	32
92	310.4	311.0*	265.2	296.9	281.9	1.83	33
94	307.5	308.9*	263.5	294.9	280.0	1.82	34
96	304.6	306.6*	261.8	292.8	278.1	1.80	36
98	301.7	304.4*	260.0	290.8	276.2	1.79	37
100	298.8	302.2*	258.3	288.7	274.3	1.78	38
102	295.9	299.9*	256.6	286.7	272.4	1.77	39
104	292.8	297.7*	254.8	284.6	270.5	1.76	40
106	289.7	295.9*	253.3	282.9	268.9	1.75	41
108	286.5	294.0*	251.8	281.1	267.3	1.74	42
110	283.4	292.2*	250.3	279.4	265.7	1.73	43
112	280.3	290.1*	248.7	277.4	263.9	1.72	44
114	277.2	287.8*	246.8	275.2	261.8	1.70	46
116	274.1	285.4*	244.9	273.0	259.8	1.69	47
118	270.9	283.5*	243.3	271.2*	258.0	1.68	48
120	267.8	281.6*	241.6	269.3*	256.2	1.67	49
RUNWAY LENGTH	11000	80000	10000	9000			
RUNWAY SLOPE	.0	.0	.0	.0			
HEADWIND CORR	696	540	568	543			
TAILWIND CORR	2290	NA	2367	2389			
LIMITED BY	RNWY	RNWY	RNWY	RNWY			

APPLY PERFORMANCE CORRECTION AS REQUIRED

II
                                                DATED 5/5/01

Figure 8.4 Typical airport takeoff analysis.

--XABC--   DC-8-63
ANYWHERE AIRPORT1
                                                              ELEV. 3000FT
ANYPLACE, ANYLAND   35 FLAPS   MAX TEMP 120F
                                                              JT3D-7 ENGINE
                          LANDING PERFORMANCE
                          NOISE REDUCTION NACELLE
                          THE NACALLE CORPORATION

AIRPLANE LIMITS

TEMPERATURE UP TO 120F
WEIGHT LBS   271000

	RUNWAY LIMITS	FOR HEADWIND CORRECTION		ADD: 800 LBS PER KNOT		
RUNWAY NO	EFF LENGTH	ZERO WIND	LANDING WEIGHT		WITH	TAILWIND -LBS
	FT	WEIGHT	2KTS	4KTS	6KTS	8KTS	10KTS
13L WET	10000	262.1	254.9	247.7	240.5	233.3	226.1
14L WET	9000	230.8	223.6	216.5	209.3	202.1	194.9
31R WET	11000	271.0	271.0	271.0	271.0	264.6	257.4
32R WET	8000	199.6	192.4	185.2	178.0	170.8	163.7
13L DRY	10000	271.0	271.0	271.0	271.0	271.0	271.0
14L DRY	9000	271.0	265.8	258.6	251.4	244.3	237.1
31R DRY	11000	271.0	271.0	271.0	271.0	271.0	271.0
32R DRY	8000	237.1	229.9	222.7	215.5	208.3	201.1

	SPOILERS				
RUNWAY NO	EFF LGTH	ANTI-SKID	MAIN GEAR INOP	MAIN GEAR AND	
		INOP TW	NOSE GEAR OP	NOSE GEAR INOP	
			WT UP TO TAILW	WT UP TO TAILW	
13L WET	10000	NA	216.4	0KTS	189.9	0KTS
14L WET	9000	NA	185.2	0KTS	158.7	0KTS
31R WET	11000	NA	247.6	0KTS	221.1	0KTS
32R WET	8000	NA	153.9	0KTS	129.0	0KTS
13L DRY	10000	NA	269.2	0KTS	246.2	0KTS
14L DRY	9000	NA	233.3	0KTS	210.2	0KTS
31R DRY	11000	206.6 NA	271.0	8KTS	271.0	2KTS
32R DRY	8000	NA	197.3	0KTS	174.3	0KTS

                    OBSERVE STRUCTURAL LIMIT

I

                              TAILWIND CORRECTION :
                    SUBTRACT 5000LBS PER EACH KNOT
NA=NOT AUTHORIZED   ABOVE THE LIMITING   TAILWIND

                                                       DATED
                                                       5/5/01

Figure 8.5 Typical airport landing analysis.

On the lower part of the landing plate, antiskid inoperative and certain inoperative spoiler conditions are shown with their associated weights.

A route is planned, e.g., New York to London to Frankfurt to Nairobi. The first step is to prepare an airport analysis for each airport (see Figs. 8.4 and 8.5). The next step is to establish each airport's temperature reliability factor, such as 85 percent of yearly, seasonally, or monthly values. Entering airport analysis charts, the maximum permissible takeoff weights could be extracted at these temperatures and tabulated as follows:

Airport name

Runway number

Runway length

Airport elevation

Airport temperatures and their reliability percents

Maximum permissible takeoff weight based on this information

With these figures, trip analysis can be conducted to evaluate various airplanes for these listed trips.

8.7   Definitions

Various weights used in airline operations are listed below. These weights are used practically everyday by airline personnel. Therefore, it is very important to become familiar with each of them. This information is generally available in the FAA-approved flight manuals, and these are the sources in which to check these figures.

8.7.1   Maximum ramp weight

This is the maximum weight authorized for ground maneuvers by the applicable FAA-approved flight manual (AFM). It is designated in some AFMs as maximum designed taxi or ramp weight.

8.7.2   Maximum structural weight

This is specified in the FAA AFM. This weight cannot be exceeded under any circumstances, even if airplane performance would permit a higher weight.

8.7.3   Maximum takeoff weight

The applicable FAA AFM authorizes maximum takeoff weight at takeoff brake release. The allowable maximum takeoff weight is affected by the most restrictive of several factors that can be grouped into two broad categories: (1) airplane limit and (2) runway limits.

Airplane limit. This limit depends on the airplane's characteristics and not on specific runway conditions. Three separate limitations come into effect:

1. *Structural limit.* This is determined by the strength of the aircraft structure and shall not be exceeded regardless of temperature, wind conditions, or runway lengths (see maximum structural limit).

2. *Climb limit.* This is determined by the maximum weight satisfying a specific FAA takeoff-climb gradient requirement at the existing airport elevation and temperature. It cannot be exceeded regardless of wind conditions or runway length. The climb limit varies with the airport elevation and temperature.

3. *Maximum operating temperature limit.* This is authorized by the FAA AFM due to engine and accessories cooling requirements. No takeoff is permitted above this temperature.

Runway limits

1. Weight based on runway length and slope
2. Weight based on obstacle clearance below flight path
3. Weight based on tire speed limitations
4. Weight based on brake energy limitations
5. Weight based on minimum control speed
6. Weight based on runway clutter (if applicable), etc.

The maximum takeoff gross weight is the lesser weight obtained from airplane limit or runway limit.

Note: Takeoff weight = ramp weight less taxi-out fuel.

8.7.4 Maximum landing weight

Maximum landing weight is authorized by the applicable FAA AFM regulations. The allowable maximum landing weight is affected by the most restrictive of several factors that can be grouped into two broad categories: (1) airplane limit and (2) runway limits.

Airplane limit. This limit depends on the airplane's characteristics and not on specific runway conditions. Three separate limitations come into effect:

1. *Structural limit.* This is determined by the strength of the aircraft structure and shall not be exceeded regardless of temperature, wind conditions, or runway lengths.

2. *Landing climb limit.* This is determined by the maximum weight satisfying a specific FAA landing climb gradient requirement at the existing airport elevation and temperature. It cannot be exceeded regardless of wind conditions or runway length. The landing climb limit varies with the airport elevation and temperature.

3. *Maximum operating temperature limit.* This is authorized by the FAA AFM due to engine and accessories cooling requirements. No landing is permitted above this temperature.

Runway limits

1. Weight based on landing on dry runway
2. Weight based on landing on wet runway
3. Weight based on obstacle clearance below approach flight path
4. Weight based on partially nonfunctioning systems such as antiskid inoperative (if applicable)

The maximum landing is the lesser weight obtained from airplane limit or runway limit.

Note: Landing weight = takeoff weight less trip fuel.

8.7.5   Maximum zero-fuel weight

Maximum zero-fuel weight limit is specified in the FAA AFM. No fuel is included in this weight.

Note: Zero-fuel weight = landing weight less reserve fuel.

8.7.6   Operating weight empty

Operating weight empty consists of the weights of structure, power plant, furnishing, systems, unusable fuel, and other items of equipment that are considered an integral part of a particular airplane configuration. Also included are certain standard items, personnel (cockpit and cabin crews), equipment, and supplies necessary for full operation, excluding fuel and payload. It is described as empty operating weight in some manuals.

Note: Operating weight empty = zero-fuel weight less load (passenger and/or cargo).

8.7.7 Manufacturer's weight empty

This is the weight of the airplane at the manufacturer's location before it is released to the operator.

Note: Manufacturer's weight empty = operating weight empty less operator's items.
To summarize the weights for clarification, Table 8.1 was developed.

8.7.8 Maximum structural payload

This consists of maximum designed payload weight of passengers, passenger baggage, and/or cargo.

8.7.9 Maximum seating capacity

This is the maximum number of passengers specifically certified or anticipated for certification.

8.7.10 Maximum cargo volume

This is the maximum space available for cargo.

8.7.11 Usable fuel capacity

This is the maximum fuel tank capacity.

TABLE 8.1 Misc. Airplane Weights

Manufacturer's weight empty
+ Operators' items
Operating weight empty
Operating weight empty
+ Payload (passenger and/or cargo)
Zero-fuel weight*
Zero-fuel weight
+ Reserve fuel
Landing weight*
Landing weight
+ Trip fuel
Takeoff weight*
Takeoff weight
+ Taxi-out fuel
Ramp weight*

*These weights cannot exceed the maximum weights listed in the FAA-approved flight manual.

8.7.12   Maximum airport elevation

This is the maximum elevation of any airport where takeoff and/or landing is permitted. This limiting airport elevation cannot be exceeded. It is specified in the FAA AFM.

8.7.13   Maximum airport temperature

This is the maximum airport temperature limit. No takeoff and/or landing is permitted above this maximum temperature per the FAA AFM.

8.7.14   Maximum tailwind limit

This is a maximum limiting tailwind of, e.g., 10 knots or as specified in the FAA AFM. This limit cannot be exceeded.

9

Trip Analysis

9.1 Introduction

A *trip* is the basic product of an airline's operation. It is the smallest revenue-generating unit. A *schedule* is merely a set of trips. It consists of a continuous chain of trips by space (geographic) and time. It is important to investigate and evaluate each trip before it is incorporated into the schedule.

9.2 Basic Requirement of Trip Analysis

Applying computer simulation, a schedule could be developed and a real-world environment could be created for airline operation. By adjusting the data and recycling a few times, an effective and profitable operational plan is finalized.

The following departments are involved in the evaluation process of trip analysis because of their economic and operational impact on the company.

1. The Financial Department
2. The Marketing and Sales Department
3. The Flight Department
4. The Maintenance Department
5. The Accounting Department

9.3 Basic Data for Trip Analysis

The trip-analysis simulation report examines the aircraft based on its economics and performance capabilities for a given route structure.

Starting from the top to the bottom, each line is explained in detail. The first part of this trip-analysis simulation report is the basic information on which this analysis is based. It has to list the airplane weight statements. For example:

Maximum ramp weight

Maximum takeoff weight

Maximum landing weight

Maximum zero-fuel weight

These weights generally are listed in the Federal Aviation Administration (FAA)–approved flight manual, and these limitations must be observed.

The operating weight empty has no legal limits like the preceding weights. Each airline builds into its operating weight empty the required items for its particular operation. For example, for overseas flights, there is a need for life rafts, whereas for domestic operation, they are not needed.

As the airplanes are coming off the assembly lines, newer, technically improved items are built in by the manufacturer and/or ordered by the customers, and this can change empty operating weights (see Chap. 4 and Fig. 4.16). This means that for the same type of airplane, the operating weight empty and, for this reason, payload can vary because the zero-fuel weight has a maximum limit as defined in the FAA-approved flight manual that cannot be superseded.

The maximum permissible takeoff weight could be extracted from the airport analysis plates (see Chap. 8 and Figs. 8.4 and 8.5). No higher takeoff weight can be applied at the departure airport elevation and temperature for the selected trip.

Furthermore, the number of seats, the fuel tank capacity, the cargo load in the belly (for passenger operation), its tare weight, and the type of operation (passenger, freight, or combi) have to be specified.

For the revenue, cost, and profit figures, all the necessary elements are listed. For the fuel cost, the prevailing fuel price is considered for that particular station. This is the correct approach instead of using average fuel flows and fuel prices. Yield can be specified for each individual trip if needed.

Table 9.1 presents the trip analysis simulation report part 1 for this particular airplane. The table contains the airplane weights and the elements of direct operating cost and unit costs.

In this example, 195 passengers (60 percent payload) and 11,200 lb of freights in the belly were considered.

TABLE 9.1   Aircraft Specification, Simulation Report Part 1

DC10-30
CF6-50AC
Weight Summary

Ramp wt.	575000 lb
Takeoff wt.	572000 lb
Landing wt.	421000 lb
Zero fuel wt.	391000 lb
Empty opert. wt.	265600 lb
Payload wt.	125400 lb
No. of pass	325
Passenger wt.	165 lb
Baggage wt.	35 lb
Pass. + bag. wt.	200 lb
Ttl pas. + bag. wt.	65000 lb
Addnl. Load wt.	56200 lb
Tare weight	4200 lb
Fuel tank cap.	245600 lb
Configuration	Passenger

Cost Summary
Direct Operating Cost

Flying operations	$/blkhr	852
Insurance	$/blkhr	15
Maintenance & burden	$/blkhr	1419
Depr./rent/interest	$/blkhr	905
DOC less fuel cost*	$/blkhr	3191

*Fuel cost is calculated at each departure station based on fuel price and required fuel; add fuel cost to "DOC less fuel cost" to obtain direct operating cost

Passenger yield used (When not specified)	11.90	c/passmi
Addnl. cargo yield used (When not specified)	32.75	c/tmi
Overhead may be considered: 06/17/1998	3265	$/blockhr

9.4   Trip Analysis

A typical analysis is presented in Table 9.2. It is for a New York to London roundtrip. Each line is explained in detail. The table lists all the various weights, the number of passengers, additional load (if any), weights, altitude, configuration, speed, time, distance, and enroute winds. In this case, a load factor of 60 percent, or 195 passengers, is carried, and a cargo load of 11,200 lb is available.

TABLE 9.2   Trip Analysis. Simulation Report Part 2

Trip Analysis Program		
Route	JFK LHR	LHR JFK
Equipment	DC10-30	DC10-30
Engine	CF6-50AC	CF6-50AC
Season	Yearly	Yearly
Max. takeoff wt. lb	572000	572000
At airport temp. F	70	64
Cruise	M.82	M.82
Cruise altitude ft	33000	33000
Takeoff wt. required lb	450223	471634
Trip fuel lb	105435	125101
Landing weight lb	344788	346532
Alternate station	AMS	BOS
Alternate dist. n.m	161	166
Fuel to alternate lb	9362	9561
Holding (etc.) fuel lb	15426	16971
Reserve fuel lb	24788 I	26532 I
Zero-fuel weight lb	320000	320000
Gross payload lb	54401	54401
Ttl pass. + bag. wt. lb	39000	39000
Number of passengers	195	195
Additional cargo lb	11200	11200
Tare weight lb	4200	4200
Operating wt. emp. lb	265600	265600
Owe is based on	NXXX	NXXX
Configuration	Passenger	Passenger
Trip time h:m	6:14	7:17
Block time h:m	6:44	7:47
Block speed mph	516	446
Block fuel lb	107535	127201
Total fuel lb	131273	152684
Distance n.m	3025 G	3025 G
Wind/reliab. kts/pct	19/85	−54/85
Day of departure GMT	Sun	Mon
Local departure h:m	18:0	7:44
GMT departure h:m	23:0	6:44
Day of arrival GMT	Mon	Mon
Local arrival h:m	6:44	9:32
GMT arrival h:m	5:44	14:32
Ground time min	60	60

Trip analysis: explanation

Route	It is defined by the three-letter airport code.
Equipment	Indicates airplane designation.
Engine	Indicates the type of engine.
Season	Time period expressed in either seasonally or yearly basis.

Max. takeoff weight	The maximum takeoff weight extracted from airport analysis data for this departure airport. It can be specified for any individual trip and for a selected airport temperature.
At airport temperature	Temperature at estimated departure time.
Cruise	Expressed in Mach number, long-range cruise, etc., per airline policy.
Cruise altitude	It can be selected for constant or step-climb cruise altitude.
Takeoff weight required	Weight needed for this trip and enroute wind. It cannot exceed the max. takeoff weight obtained from airport analysis data.
Trip fuel	Fuel required flying from departure runway to destination runway. No taxi-out or taxi-in is included. Enroute wind is considered with specified wind reliability.
Landing weight	Weight when landing at destination airport. It can be specified for any individual trip. For example, when at a certain destination airport, a short runway limits the landing weight.
Alternate station	Alternate airport designated by a three-letter code. If not specified, distance can be furnished.
Alternate distance	Distance between destination and alternate airport. If not specified, 200 nautical miles is considered for international and 150 nautical miles for domestic operations.
Fuel to alternate	Required fuel to fly between destination and alternate airport.
Holding (etc.) fuel	Contingency fuel required for domestic or international operation.
Reserve fuel	Total reserve fuel. Letters D for domestic, I for international operation, and S for island reserve are indicated next to the fuel figure.
Zero-fuel weight	Zero-fuel weight for this trip. It can be specified for any individual trip.
Gross payload	The total payload. For cargo operations, tare weight is included. It can be specified for any individual trip.
Total passenger and baggage weight*	Passenger and baggage weight (e.g., 165 + 35) multiplied by the actual number of passengers. It can be specified for any individual trip.
Number of passengers*	Number of passengers for each individual trip. If not specified, the maximum number of passengers is considered.

Additional cargo*	Additional load above passenger and baggage weight. It can be specified for any individual trip.
Tare weight	Actual tare weight.
Operating weight empty	Actual operating weight empty based on airplane tail number or registration number.
Owe based on:	Airplane identification (tail or serial numbers).
Configuration	This can be passenger, convertible, combi, or cargo.
Trip time	Duration of this trip from departure runway to destination runway.
Block time	Duration of this trip from terminal to terminal.
Block speed	Trip length divided by block time.
Block fuel	Fuel required from terminal to terminal (from engine on to engine off).
Total fuel	Block fuel and reserve fuel.
Distance	Great circle or airway miles designated with a letter G or A.
Wind/reliability	Enroute wind and its reliability factor, expressed in percent, generally 85 percent of not to be exceeded. It can be specified for yearly, seasonally, or monthly.
Day of departure	Day of the week, specified for the first departure only.
Local departure	Departure time expressed in local time.
GMT departure	Departure time expressed in Greenwich Mean Time.
Local arrival	Arrival time expressed in local time.
GMT arrival	Arrival time expressed in Greenwich Mean Time.
Ground time	It can be specified for any individual trip: transit time or turnaround time.

*Applicable for passenger operations.

Note: Operational data are prestored, but they can be changed for any selected trip(s). For example, the gross payload can be specified for the first trip as 59,000 lb, for the second trip as 5511 lb, and so on, but it can never exceed the prestored legal limit—the computer checks this internally.

Prestored data are applied where no actual numbers are specified. For example, if cruise altitude is left blank, a prestored altitude will be entered. If any number is specified at any trip, this will override the prestored data (legal limits must be observed). This presents a very flexible approach to a real-world simulation.

9.5   Financial Report

Table 9.3 contains the financial part of this trip analysis. Listed are factors such as direct operating and indirect operating cost elements, passenger revenue, breakeven load factor, lower-level cargo weight and its potential revenue, and profit potential.
The following list explains Table 9.3 line by line:

Flying operations	Cost based on block time.
Fuel and oil cost	Fuel cost based on fuel requirement for this trip-cost. Cost of oil is added.
Fuel price	Price of fuel at departure station.
Insurance	Insurance cost related to block time.
Maintenance + burden	Cost related to block time.
Depr./rent/interest	Depreciation or rent/lease and cost of interest, based on block time (whichever applies).

TABLE 9.3   Financial Report. Simulation Report Part 3.

Financial data	JFK-LHR	LHR-JFK
Flying operations $	5743	6639
Fuel and oil cost $	9450	15678
Fuel price c/gal	57.00	80.00
Insurance $	101	116
Maintenance + burden $	9566	11058
Dep./rent/interest $	7328	8470
DOC $	32188	41961
IOC $	26583	28773
Total trip cost (D + I) $	58771	70734
Revenue, passengers $	80745	80745
Total pass. + bag wt. lb	39000	39000
No. of passgrs. carried	195	195
Breakeven no. of pass	141	170
Breakeven pass. LDFCR%	72.79	87.60
Potential cargo rev. $	6381	6381
Potential cargo lb	11200	11200
Tare weight lb	4200	4200
Unit cost in c/pas	8.66	10.42
Yield pass. c/passmi	11.90	11.90
Profit potential $	21974	10011

Cargo revenue not included in profit figures (32.75 cents/tonmi)

Reserve fuel: I = international; D = domestic; S = 2 hours fuel
G = great circle; A = airway miles
Total fuel = ramp + trip + reserve fuel
Block fuel includes taxi-out/in fuel
Taxi-in fuel taken from reserve fuel

06/17/1998

DOC	Direct operating cost.
IOC	Indirect operating cost: landing fee, traffic, aircraft service, and system overhead, etc.
Total trip cost	The sum of DOC and IOC.
Revenue, passengers*	Revenue generated by the number of passengers, based on cents/passenger-mile.
Total pass. + bag. wt. in lb*	Total passenger and baggage weight.
Number of passengers carried*	Total number of passengers carried.
Breakeven no. of pass.*	The actual breakeven number of passengers.
Breakeven pass. Loadfactor %*	The breakeven load factor expressed in percentage related to pass. carried.
Potential cargo, revenue*	Additional revenue in passenger operation created by the potential cargo capability of the aircraft. This revenue is not included into the profit potential.
Potential cargo, lb*	Additional cargo potential in pounds in passenger operation.
Tare weight, lb*	Tare weight for the above listed load.
Revenue†	Revenue generated by the payload in freighter operations based on cents/ton-mile.
Breakeven payload, lb†	The actual weight of the breakeven payload in pounds in freighter operations.
Breakeven load factor, %†	The breakeven load factor expressed in percentages in freighter operations.
Unit cost	Unit cost of operation expressed in cents/passenger-mile for passenger and cents/ton-mile for cargo operations.
Yield	Yield of revenue for passenger or cargo operations.
Profit potential	Expressed in dollars: passenger or cargo operations for one trip. Additional cargo revenue in passenger operation is not included. This profit is before taxes.

*Applicable for passengers operations only.
†Applicable for cargo operations only.

The last part of trip-analysis report (Table 9.4) contains the listings of the scheduled stations and the alternate airports. The following items are presented:

Fuel price per station

Landing fees

TABLE 9.4   Costs Per Stations. Simulation Report Part 4.

Station	Fuel price, c/gal*	Indirect Operating Cost		
		Landing fee, $	Traffic svc., $	Handling cost, $
New York JFK	.57	1748	1643	905
Amstrdm. AMS	.82	2227	1388	755
Boston BOS	.56	1243	782	821
London LHR	.80	729	1670	900

*Fuel cost is part of direct operations cost.
06/17/1998

Traffic servicing fees

Handling costs

9.6   Departmental Use of Simulation Report

The use of this simulation program is helpful in achieving a realistic and effective operational schedule. The following departments use this report:

Flight Department. The Flight Department can plan its fuel requirements per trip, station, or geographic area for a certain period of time. It can plan crew requirements based on schedule. It can answer questions from Sales and/or Marketing Departments about individual trips, arrival and departure times, and the number of passengers that the airplane can accommodate for the selected trip or trips, including the amount of load the airplane can carry in its lower deck or, for a cargo airplane, the potential cargo it can carry for a certain distance.

Maintenance Department. The Maintenance Department receives from the Flight Department arrival and departure times for each trip at each station the airplane is visiting. This requires scheduling certain maintenance checks at certain selected stations. After a certain accumulated flight time, the Maintenance Department can plan items that need to be checked, maintained, and/or removed by knowing trip times and ground times. Considering the airplane's position at all times, the Maintenance Department can plan certain maintenance activities and schedule personnel, spares, tools, work orders, materials, and maintenance activities more efficiently.

Marketing and Sales Departments. Both departments receive their information from the Flight Department. They have to know the departure

and arrival times, the number of passengers for passenger operation and the amount of freight for cargo operations, the range of the trip
with full and partial loads, the estimated revenue for a particular operation, and many questions that customers generally are asking about
their trips.

Based on the information received from the Finance Department, the
Sales and/or Marketing Departments can see the profit potential for
each trip and do a better selling job.

Accounting Department. The Accounting Department furnishes all the
interested departments with cost allocation. The availibility of cost
data for each trip makes possible the comparison of cost data with
the simulation-generated cost elements and analyzing the differences
(if any).

Financial Department. The Financial Department can estimate the
profit potential of the whole operation for this schedule and is able to
analyze profit and cost for each trip. It can request schedule changes
to improve profit and conduct sensitivity analyses to improve profit for
the whole system. It can analyze cash flow, return on investment, and
many other types of financial data.

The simulation lets various types of airplanes fly over a given route
structure for a given period, most likely for a week. Projecting this figure for a year period, the following could be established:

1. Revenue for a given year

2. Aircraft miles for a given year

3. Revenue passenger-miles for a given year

4. Available seat-miles for a given year.

The output format contains all the essential information for the
evaluation of a given route structure for various types of aircraft performance and economics, which can be classified into the following
categories:

1. *Route analysis.* Operating revenue, operating expenses, operating income, number of equipment, average load factor, aircraft
   utilization, ground time, trip and block time, and trip and block
   fuel.

2. *Traffic analysis: City pair.* Passengers served, passenger routing,
   number of flights, aircraft movement.

3. *System summary.*   Aircraft type and summary, frequency, load factor, trip time, passengers served, travel times, fuel burned.

4. *Profit and loss.*

5. *Return on investment.*   Cash flow, present value, return on investment summary.

A few trip calculations are listed here. They are presented to enhance your understanding of the logic applied in trip calculation and for clarification of the various weight relationships. Readers may skip the rest of the material in this chapter.

Basic trip data calculations.   Trip data are calculated for a given trip to obtain the expected revenue and cost. This would answer the question: Is this trip economically feasible after all data for the trip are known. Below are relevant data required for certain trip evaluations:

- For revenue determination, the payload has to be known.

- For fuel cost, trip fuel and reserve fuel should be available.

- For scheduling, trip time, block time, and taxi time have to be known.

- For generating cost figures, all the preceding information should be known.

The following examples show quick ways to calculate trip data. The computer can make these calculations faster and more accurately. However, doing it by hand would make it easier to understand the relationship, limitations, and logic of the procedure.

9.7   Sample Calculation of Trip Analysis

Based on the weight statements in Table 9.5, various weight calculations are presented below. The selected airplane is a single-aisle type of airplane.

TABLE 9.5   Airplane Weight Summary

Ramp weight, lb	358,000
Max. takeoff weight, lb	355,000
Max. landing weight, lb	275,000
Max. zero-fuel weight, lb	261,000
Max. payload, lb	100,000
Operating weight empty, lb	161,000
Fuel tank capacity, lb	140,000

Input data. Case No. 1

Taxi-out fuel	3,500 lb
Reserve fuel	13,000 lb
Trip fuel	47,000 lb

Calculate payload and establish takeoff, landing, and zero-fuel weights (Table 9.6).

Input data. Case No. 2

Taxi-out fuel	3,500 lb
Reserve fuel	17,200 lb
Trip fuel	47,000 lb

Calculate payload and establish takeoff, landing, and zero-fuel weights.

This is an example of when reserve fuel is higher than the difference between landing weight and zero-fuel weight. Landing weight is at its limit, and to accommodate this reserve fuel, zero-fuel weight has to be lower. Therefore, payload is reduced by the amount x, as shown (Table 9.7) in this equation:

$$x = \text{reserve fuel} - (\text{landing weight} - \text{zero fuel weight})$$

In this case, $x = 17,200 - (275,000 - 261,000)$, or $x = 3200$ lb.

TABLE 9.6   Trip data for Case No. 1

Operating weight empty, lb	161,000
Payload, lb	100,000
Zero-fuel weight, lb	261,000
Reserve fuel, lb	13,000
Landing weight, lb	274,000
Trip fuel, lb	47,000
Takeoff weight, lb	321,000
Ramp weight, lb	324,500

TABLE 9.7   Trip data for Case No. 2

Operating weight empty, lb	161,000
Payload, lb	96,800
Zero-fuel weight, lb	257,800
Reserve fuel, lb	17,200
Landing weight, lb	275,000
Trip fuel, lb	47,000
Takeoff weight, lb	322,000
Ramp weight, lb	325,500

Input. Case No. 3

Taxi-out fuel	3,500 lb
Reserve fuel	13,400 lb
Trip fuel	126,600 lb

Calculate payload and establish takeoff, landing, and zero-fuel weights (Table 9.8). This is an example of when the high trip fuel reduces payload.

Input data. Case No. 4

Taxi-out fuel	3,500 lb
Reserve fuel	13,400 lb
Trip fuel	126,600 lb

Payload is given as 26,750 lb. Establish takeoff, landing, and zero-fuel weights (Table 9.9).

Input data Case No. 5

Taxi-out fuel	3,500 lb
Reserve fuel	13,400 lb
Trip fuel	126,600 lb

Payload is given as 32,100 lb only. Establish takeoff, landing, and zero fuel weights (Table 9.10).

TABLE 9.8   Trip data for Case No. 3

Operating weight empty, lb	161,000
Payload, lb	53,500
Zero-fuel weight, lb	214,500
Reserve fuel, lb	13,400
Landing weight, lb	227,900
Trip fuel, lb	126,600
Takeoff weight, lb	354,500
Ramp weight, lb	358,000

TABLE 9.9   Trip data for Case No. 4

Operating weight empty, lb	161,000
Payload, lb	26,750
Zero-fuel weight, lb	187,750
Reserve fuel, lb	13,400
Landing weight, lb	201,150
Trip fuel, lb	126,600
Takeoff weight, lb	327,750
Ramp weight, lb	331,250

TABLE 9.10   Trip data for Case No. 5

Operating weight empty, lb	161,000
Payload, lb	32,100
Zero-fuel weight, lb	193,100
Reserve fuel, lb	13,400
Landing weight, lb	206,500
Trip fuel, lb	126,600
Takeoff weight, lb	333,100
Ramp weight, lb	336,600

Note: Sometimes the final numbers are obtained by an iteration process that is not shown here.

Input data. Case No. 6

Reserve fuel	13,400 lb
Maximum takeoff weight	320,000 lb
(due to runway limitation)	
Payload	32,100 lb

Calculate landing weight and trip fuel. In this example, the takeoff limitation has to be observed (Table 9.11).

Similarly, an example can be presented where zero-fuel weight and/or landing weight is limited.

Note: Trip fuel changes slightly with airplane weight. In order to keep these sample calculations simple, trip fuel was kept constant.

9.8   Trip Elements Definitions

A typical trip profile is presented in Fig. 9.1. Each phase of flight is shown and clearly defined. The fuel, time, and distance are calculated and available at any point of the trip. The following paragraphs describe a typical trip:

Taxi-out time. The trip starts from the airline's terminal building. The airplane taxis out to the active runway. The taxi time is a function of the distance to the runway, and sometimes congestion is experienced during taxi-out. This affects the taxi-out fuel and time. For general studies, a taxi-out time of 10 or more minutes could be selected (based on the traffic at that airport). However, at highly congested airports during certain seasons, days of the week, and times of the day (e.g., Friday around 5 P.M.), a higher value should be considered for planning purposes.

TABLE 9.11   Trip Data For Case No. 6.

Operating weight empty, lb	161,000
Payload, lb	19,000
Zero-fuel weight, lb	180,000
Reserve fuel, lb	13,400
Landing weight, lb	193,400
Trip fuel, lb	126,600
Takeoff weight, lb	320,000

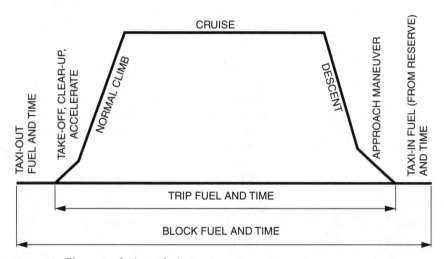

Figure 9.1   Elements of trip analysis.

Climb.   The airplane starts to move on the takeoff runway, and it takes off. After reaching a height of 35 ft, the landing gears are retracted, and flaps and slots move from takeoff position to cruise position. Enroute wind is applied and is accounted for when calculating climb data.

Cruise data.   Cruise data are based on selection of cruise Mach number or long-range cruise. Cruise fuel and time are calculated with the inclusion of enroute wind. Cruise altitude could be at a constant level during the entire cruise regime, or step climb could be applied to improve fuel consumption.

Descent data.   Descent fuel, distance, time, and the wind effects are calculated. In the case of an instrument landing, additional fuel and time should be considered.

Taxi-in time. Taxi-in procedure takes place when the airplane moves from the end of the landing runway at the arrival airport to the terminal building. This is a different phase of operation; taxi-in time generally is shorter than taxi-out time. An average of 5 minutes could be considered for taxi-in time; however, other factors should be considered, as mentioned for taxi-out time.

Block fuel and block time. This is the fuel and time from the terminal building at the departure airport to the terminal building at the destination airport (or between engine-on and engine-off).

Note: Block fuel includes taxi-in fuel taken from reserve fuel.

Trip fuel and trip time. Trip fuel and trip time are sometimes called *mission fuel* and *time* and are calculated from the takeoff runway to the landing runway (without taxi-out/taxi-in fuel and time). Figure 9.1 contains the elements of a typical trip analysis. Based on trip length and traffic conditions, air traffic control could assign airplane to a constant cruise altitude or to a step-climb cruise.

Reserve fuel. The reserve fuel could be based on the following:

1. Domestic rules that consist of the amount of fuel at long range at final cruise altitude for a certain time period (45 minutes per regulation or more). It also includes the amount of fuel based on the conditions at the arrival airport (fog or congestion, accident, or any other reason) and fuel to alternate (generally 150 nautical miles or actual mileage) or missed approach, climb, cruise, and descent.

2. International rules consist of fuel to fly an additional 10 percent of total flight time (per FAA regulation) (or 5 percent of mission fuel allowance, where applicable, e.g., Europe) and fuel to alternate and fuel for 30 minutes at holding speed 1500 ft above the alternate airport. Chapter 7 contains a detailed description of reserve fuel, Table 7.6 and Fig. 7.15 shows the domestic and Fig. 7.16 the international reserve fuel method.

Passenger Research

10.1 Introduction

Passenger research is conducted to identify the characteristics and needs of the passengers choosing the method of air transportation. Interestingly enough, the characteristics of passengers have changed in the past 50 years; in the 1940s and 1950s, business travelers were the mainstay of air transportation. In the year 2000, the situation is just the opposite. Presently, about 80 percent of passengers are tourists, whereas 20 percent are on business trips (Fig. 10.1).

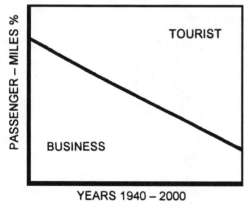

Figure 10.1 Percentage of business and pleasure travel.

10.2   The Significance (if Any) of the Gompertz Curve

In the late 1960s, there was a consensus that the business traveler market was in its mature stage. This assumption neatly followed the forecasters' faith in the "classic" Gompertz curve, which traces the development of any industry as an S curve with identifiable stages, as indicated below. Many traffic analysts assumed that air transportation, at least in the developed countries, had reached the "mature" stage—hence the reliance on GNP correlation—and that traffic growth would slow as a fundamental symptom of approaching market maturity (Fig. 10.2).

There is considerable evidence to suggest that even this basic principle may be open to question as it applies to air transport. In the 1970s, the most obvious support to suggest that air transportation was still in the "high growth" stage came from the remarkable development of air tourism, which added a new dimension to orthodox criteria (Fig. 10.3).

10.3   Objective of Passenger Research

The impact of air tourism has been substantial, and this impact has already changed previous ideas on motivation, traffic patterns, and potential for future growth. The growth of tourism focused attention on the tourist passenger. Many airlines conducted in-depth studies to learn more about their passenger customers. The airlines' customer research departments or independent expert consultants hired investigated any topic of interest with regard to travel behavior, passenger preferences, service evaluation, or any question or problem that might be addressed through research or analytical techniques.

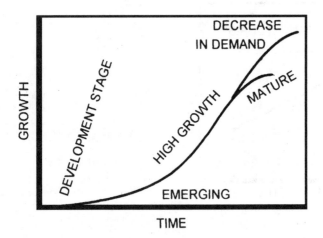

Figure 10.2 Gompertz curve.

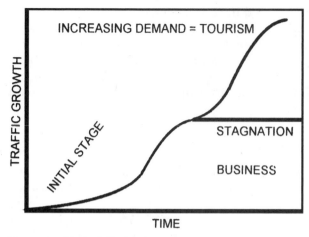

Figure 10.3 Revised Gompertz curve.

The driving force behind this research by the airlines is the idea of knowing their customers better in order to cater to and satisfy their needs, increase their market share, and generate higher profits. Some past and present projects include investigation of the influence of aircraft preferences in flight selection, periodic air traveler attitude surveys, identification of comfort factors and priorities, development of a database on airline passenger demographic characteristics, summaries of passenger survey reports, analysis of wide-body aircraft passenger performance, and so on.

The surveys conducted by the airlines or their hired agencies had to define first the problems and their passenger research objectives as follows:

- Specify and quantify problem
- Apply existing information
- Use alternative approaches, if necessary
- Explore subject areas of interest
- Specify information desired
- Apply results

10.4 Data Collection on Passenger Preferences

Data collection has three basic approaches, namely, mail surveys, telephone interviews, and personal interviews. Each approach has its own advantages and disadvantages. Sampling is important, and random sampling is preferred in certain cases. For survey purposes, the appropriate population should be selected. The design of the questionnaire is very important because if it is poorly designed, it may distort reality. Other research techniques are also applied, such as laboratory research, field observations, analytical methods, and secondary research. Table 10.1 provides a better understanding of these project areas and applications.

10.5   Passenger motivation

Numerous studies have been conducted, and a wide variety of statistical data has been generated regarding passengers. These data are classified into several groups.

In passenger research, many questions arise: Why do people travel? What are their motivations? We may list the following answers:

History

Curiosity and culture

Business and professional

Health

Interpersonal, i.e., visiting friends, relatives (VFR)

Pleasure/romance

Spiritual

Sports

Shopping

The factors affecting tourism are as follows:

TABLE 10.1   Passenger Surveys

Consumer decision making	Factors influencing flight choice behavior
	Travel choice model evaluation
User satisfaction assessment	Passenger preference evaluation
	Aircraft comfort and appeal
Product design	Evaluation of aircraft interior design features
	Exploration of new design features
Product perception	Summary of air traveler surveys
	Evaluation of public and industry opinion
Consumer behavior analysis	Application of consumer research literature to company product lines
	Behavior-attitude relationship

Destination:

Attractiveness

Organization

Accommodation

Social factors

Competition

Origin:

Education

Economy and wealth

Lifestyle

Motivation

Values

Interest

Transportation:

Accessibility

Visa requirement

Customs and immigration

Distance and time

Transport capacity

Increased frequency

Improved customs

Improved airports

Multilingual guides

Cost

The attractiveness is determined by

Good weather

Historical sites

Natural beauty

Different cultures

Recreational facilities

Ground transport

Tour arrangement

Hotel accommodations

Social and political factors could contribute to the growth of tourism. They include

Friendly people

Socioeconomic stability

Acceptable political philosophy

Ethnic ties

Education also has a certain effect on tourism because the increased level of general education (and higher income) stimulates interest in the outside world.

Economy, wealth, and lifestyle all contribute to the growth of tourism:

Increase in discretionary income

Young, multiearner families

People's increased self-concern

Changing role of women

Demand for more leisure time

Easing of currency restrictions

Easier credit

Lower inflation

More vacation time

More retirees with higher income

Sports

Some of these factors listed are interrelated and need to be examined, analyzed, and reviewed together. For example, take the entry "More retirees with higher income." It is true that retirees have more income and time on their hands, but there is also a "dumping" effect—their age. As they grow older, their desire and ability to take longer trips, or any at all, diminish. Their physical condition also deteriorates slowly. They are good candidates for the charter market because charter flights see to it that their baggage is delivered to their hotel rooms because charter operators recognized this need of mature passengers. In other words, we have a market here that has already built in its own limitations. Another factor is that passenger waiting areas are more crowded than 10 years ago, and it is tiresome for an older person to stand there for a longer time period than necessary. In this light, this market segment should be reviewed and evaluated in order to have a good picture of the real world.

10.6   Demographics and Income

The income level of a country reflects the percentage of its population who can afford to travel. To study tourist passengers further, we may set up the following classification:

Age

Sex

Income

Education

Social class

Geographic region

City size

Population density

Family size

Occupation

Lifestyle

We may further classify the tourist group as to

Trip purpose

Reliability

Destination

Availability

Knowledge of alternates

Schedule convenience

Cost

Safety, etc.

Comparing the current tourist volume with the past, we may conclude that increased personal income probably was the most important factor affecting the industry.

Tables 10.2, 10.3, and 10.4 show this increase in income, gender, and age of tourist passengers, respectively.

TABLE 10.2   Median Income in Dollars

1971	1975	1985	1995	2000
$15000	$19600	$22400	$26000	$53300

TABLE 10.3   Passenger Demographics (%)

Gender	1970	1975	1985	2000
Male	74	70	66	63
Female	26	30	34	37

TABLE 10.4   Passenger Age Distribution

Age, years	1970	1975	1985	2000
Male to 49	78	72	71	69
Male 50+	22	28	29	31
Female to 49	68	64	67	62
Female 50+	32	36	33	35

The numerous categories are broken down continuously so as to develop additional statistical data such as the following: From the total number of passengers of a given city-pair (Tables 10.5 and 10.6):

67%	30 to 49 years old
51%	Professional and business field
23%	Retired
55–59%	Persons 30+ years old have family income of $40,000 or more
52%	Pleasure/personal trips
48%	Business trips

Some demographic passenger distributions are presented as a function of age group of passengers in percentage of U.S. population (Fig. 10.4). The new communication technology further erodes the business traveler market.

TABLE 10.5   Frequency of Travel

Frequency of travel	Percent of flyers	Percent of trips
Infrequent	44	13
Moderate	52	56
Frequent	4	31
Total	100	100

TABLE 10.6   Selected Traveler Characteristics

Characteristic	1970	1975	1983	2000
Traveler's age 60 and over	8.8%	11.3	11.5	12.7
Business travelers, women	9	13	15	17
Business travelers making six or more trips	77	75	75	71
Personal/pleasure travelers making six or more trips	25	22	19	21

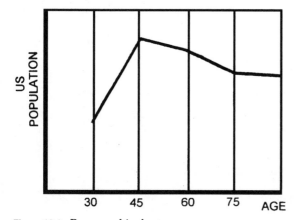

Figure 10.4   Demographics by age.

A chart like Fig. 10.4 can help market researchers concentrate on the age group that makes up the majority of their passengers.

Negative political and social factors could limit the growth of tourism. For example:

Airport congestion

Hotel room limited capacity

Limited transport capacity

Economic downturn

Energy impact

Cost of travel

Changing social values

Destruction of tourist attraction

Overcrowded beaches

Polluted streams

Defaced monuments

Noise pollution

Unfriendly or hostile natives

Possibility of terrorist act

10.7 Passenger Comfort

Some of the subjects surveyed are as follows: airplane preference, convenient arrival and/or departure times, and baggage services.

Passenger ranking of flight comfort is studied quite frequently. Factors that determine flight comfort are as follows:

Seat comfort

Cabin noise

Temperature

Pressure change

Aircraft movement

Lighting

Waiting time in front of the ticket counter

Seating comfort represents another subject that is surveyed, and passenger preferences and suggestions are noted:

Next seat is empty

Wider seats

Wider aisles

More leg room

Safety

Noise control

Note: Passenger preference for certain type of airplanes was always a puzzle for manufacturers and the airlines. Researchers at Boeing Commercial Airplane Group conducted a large-scale study for 3 years, and 90,000 responses were received annually. Airlines were aware that seat pitch could control comfort level. The study found that when a passenger was seated next to an empty seat, his or her comfort level went up. According to this study, an empty seat adds 4½ inches more seat width. This factor alone had the largest effect on passenger comfort. Figure 10.5 illustrates this point vividly.

During a typical survey, a number of passengers were asked to classify certain items as very important in their travels. Some of the answers are listed in Table 10.7.

Figure 10.5 "A pleasant and comfortable seating arrangement."

TABLE 10.7   Survey of Passenger Preferences

Item	Response, %
Convenient arrival time	75
Convenient departure time	63
Seating comfort	59
On-time arrival and departure	62
Meals	35
Pre-boarding	23
Baggage service	61
In-flight entertainment	8
Flight attendants' courtesy	51

Other surveys were conducted about passenger flight comfort. Various categories were listed along with their degrees of importance (little, somewhat, very, and greatest). Some of these categories are as follows:

Convenience

Reliability

Time savings

Cost

On-board service

Several other passenger research surveys have been conducted, but they are not mentioned here. Passenger satisfaction is the primary goal of each airline and the main contributor to revenue and profit. To quantify all the information related to tourism, a host of graphs and tables could be prepared. These data are the basis for forecasting the trends of tourism. Someone may question the importance of having all this information, but forecasting is absolutely necessary because it enables the airlines to ascertain revenue figures and, ultimately, the profits to be expected.

10.8 Passenger Economics

For a given route structure, operating the wrong-sized airplane could spell disaster for an airline. The situation is like building a large hotel without having any idea of the number of guests expected. In order to have a good cross section, one must review all the contributing factors and look at the macro- and microeconomics aspects:

Gross national product

Personal consumption expenditure

Unemployment rate

Fixed investments

Private housing trends

Prime commercial rate

Government purchases

Value of dollar related to other currencies

Fuel prices

Domestic airline revenues

Domestic load factors

Several other items could be incorporated in this list. With today's computer potential, all these data can be digested, and a realistic forecast may be generated. The latest passenger research points in the direction of two class configurations:

Improved business class

Upgraded economy class

Presently, frequent-flyer passengers are eroding the revenues in first class, and in the domestic market. Only two classes of cabins will be available in the future. Furthermore, companies direct their employees toward business class instead of first class. Because business class is close in comfort to first class, this even accelerates the trend toward two classes. This trend generally is applicable to domestic flights and shorter trips. On longer overseas flights, another trend is appearing: First class service is greatly improved due to the fierce competition. The first class seats are exchanged for larger ones by removing a certain number of seats from this class. One international airline installed a free-standing desk and TV in front of each first class seat. This seat could be converted into a comfortable bed. Improved food service and more attention to passenger needs would generate goodwill and more revenue. Power seats have been introduced for first and business class, where computers can be connected to power outlets for the convenience of passengers. At a later date, all seats in economy class will be converted to power seats.

Passenger research explores the needs of passengers that airline management tries to satisfy in order to improve their profit. An interesting comparison is shown in Table 10.8, where past travel speed, time, and cost figures are presented.

TABLE 10.8   Travel Modes Comparison

	1930	1950	1970	2000
Travel modes				
Air	Trimotor	DC-3, 4, 6	B707, DC8, B747, Concorde	B767, B777
Rail	55–65 mi/h	Main	Decreased usage	Same
Ship	8–15 knots	transport		
Bus		Increasing		
Air transport			450–655 mi/h,	500–650 mi/h,
speed	100–130 mi/h	240–310 mi/h	1300 Concorde	1300 Concorde

Note: Airplane speed did not significantly increase between 1970 and 2000. Concordes are still flying, but due to their economic performance, there is no new supersonic transport on the production line. Furthermore, due to airport congestion, travel times during the last one and a half decades were not improved significantly.

Chapter 11

Forecasting

11.1 Introduction

There are two basic definitions that should be discussed here, namely, projection and forecast. A *projection,* as the name implies, is a mathematical extension of historical trends, assuming the same influence of various factors over the life of the projection (Fig. 11.1). A *forecast* may be defined as the expert blending of relevant statistics and economic theory. The procedure uses both quantitative and qualitative information. Forecasting involves the evaluation and application of historical causative and associative factors and patterns and their probable effect and relationship in the future. In other words, forecasting is, and probably will remain, more an art than a science.

11.2 The Purpose of Preparing Forecasts

Why does an airline need to forecast?

- To establish a foundation for scheduling and developing fleet requirements
- To determine financial requirements related to fleet procurements
- To provide a basis for evaluating station staffing and facilities requirements
- To develop marketing strategies and promotional programs
- To establish the public's future needs, desires, and ability to travel and transport goods by air
- To plan and be ready for the preceding requirements

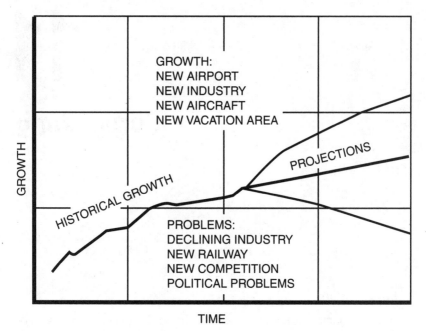

Figure 11.1 Projection and forecasting.

Forecasting methodology are based on

- Relevance and availability of historical data
- Degree of accuracy desired
- Time period of forecast
- Costs versus benefits of forecast for end user
- Time and resources available to develop forecast

11.3 Types of Forecasts

Three basic types of forecasts are available:

1. Qualitative techniques
   a. Delphi
   b. Market research
   c. Managerial judgment
2. Time-series analyses
   a. Trends
   b. Smoothing
3. Causal models
   a. Single equation
   b. Simultaneous equation

11.4   Time Range of Forecasts

Forecasting is an integral part of airline operations; it tries to quantify such future operational factors as revenue and demand. Planning will enable management to attain the goals predicted by forecasters. Generally, there are three types of forecasts: short, medium, and long term.

Short-term forecasts.   These are within a time frame of 1 month to $1\frac{1}{2}$ year and mostly concern daily operations.

Medium-term forecasts.   These normally cover up to 5 or 6 years. They are used for market and route planning.

Long-term forecasts.   These extend to 15 or more years and involve fleet planning, type of equipment needed, and routes. Once a forecast is in place, management can set goals, policies, and planning strategies and determine how to obtain the goals based on the forecasted figures.

Forecasts deliver answers to such questions as

How fast can the market be penetrated?

What aircraft types should be used?

What is the potential market?

How can an airline plan for future expansions?

What types of expansions should be made—new stations, advertisements, new city-pairs, personnel, etc.?

There are various methods to generate forecasts, depending on the objectives. These objectives can be the purpose of the forecast, the time frame involved, and available data and its accuracy. These methods raise the following questions:

How sophisticated should the approach be?

How much time is available for preparation?

Are there ample resources to obtain the necessary data?

How many variables should be used and/or are available?

11.5   Forecast Models

Some models apply a sophisticated approach to traffic forecasting. The following advantages are associated with these types of models:

- Forecast factors affecting traffic demand rather than traffic itself

- Systematic explanations of the relationship between traffic and causative factors

- Mathematical models that facilitate control and reproducibility of results once models are estimated and quickly evaluate alternative scenarios for the independent variables (e.g., sensitivity analysis, impact on revenues, cost, profit, etc.)

To generate a workable forecast, several steps should be taken. Figure 11.2 indicates the elements that are part of this complex procedure.

Once a traffic forecast is established, a schedule is developed. This would affect airport development and station personnel, facility planning, promotions, etc. All these items would establish future invest-

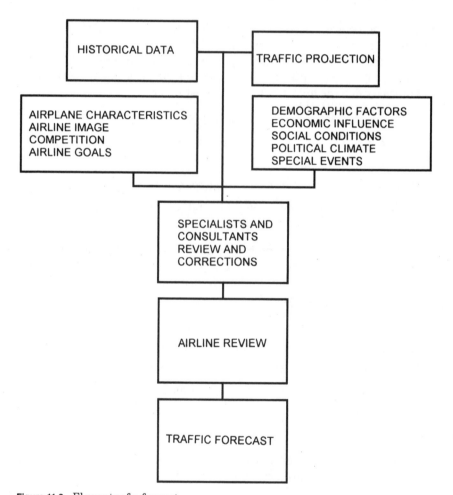

Figure 11.2   Elements of a forecast.

ment programs. Traffic forecasting has a far-reaching effect. This is why airlines are making serious efforts to have a well-researched base for traffic forecasts.

Forecasting starts with historical data and traffic projections. The airline's Traffic Department has to consider the following:

- Political trends
- Socioeconomic influences
- Demographic factors
- Special events
- Strategic plans
- Airline goals
- Aircraft capabilities

Once a traffic projection is available, it has to be cycled over to fine-tune the forecast. After review, a final forecast is adopted.

Based on this traffic forecast, the airline has to estimate its market share. Market share relates individual parts to the whole. Growth of a part cannot be analyzed realistically without detailed knowledge of the whole (Fig. 11.3).

City-pair forecasts have to be established as part of the microanalysis, based on the available and necessary information. A flowchart type of representation is presented in Fig. 11.4.

Based on the economic forecast, the level of service, and the historical traffic data, a preliminary forecast is developed. When reviewing the market forecast, it is sometimes necessary to recycle the data to

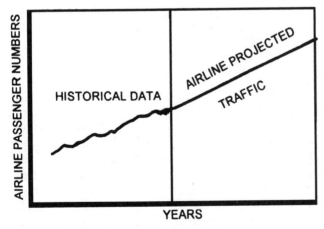

Figure 11.3 Specific airline growth versus time.

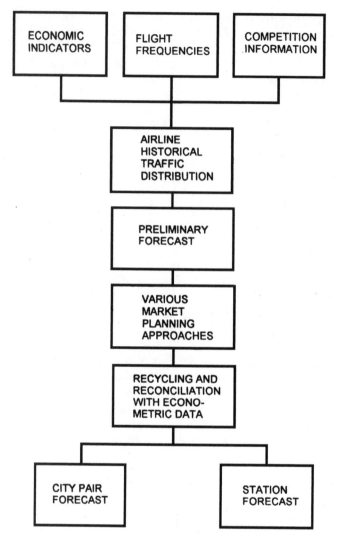

Figure 11.4   City-pair forecast.

improve the city-pair forecast and/or the station forecast. Various companies develop econometric models. The Chase and Wharton econometric models are very well known in the industry. These models can provide forecasts of employment, prices, economic indicators, and many other factors.

One of these models for traffic and revenue forecasting is presented in Fig. 11.5. Based on various input factors, an econometric model is developed that enables the prediction of passengers and revenue.

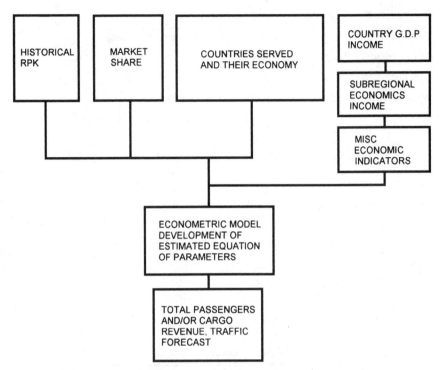

Figure 11.5   Econometric model.

11.6   Historical, Industry, Passenger, and Cargo Data

The historical data include airline origin and destination traffic and market share. The projection reflects total industry growth. For the airline, market share is established.

It is desirable to obtain a 5-year history of the data. Information can be collected from organizations such as the Department of Transportation (DOT), INS, ICAO, and the United Nations.

Various methods are available to obtain the required information for a traffic projection. Projection in general terms means an extrapolation of historical data based on a calculated trend equation. It assumes that the forces operating in the past will continue to function in the future in the same fashion. Tables 11.1 and 11.2 show elements necessary to generate a forecast. RPK and ASK signify *revenue passenger-kilometers* and *available seat-kilometers*. Load factor is the relationship between available seats and occupied seats. Load factor can be applied for a single trip, round trips, sections, regions, or the whole operation. In this case it applies for the whole operation.

TABLE 11.1   Passenger History and Forecast

The following elements are applied to establish Passenger Forecast:

-Minimum of 5 years data should be available

-Scheduled Operation

-Non-scheduled operation

Necessary yearly data:

Revenue passenger mile

Annual growth rate of passenger mile in RPM

Available sent mile ASM

Available growth rate of seat mile ASM.

Load factor

NOTE: RPM = Revenue Passenger Mile
      ASM = Available Seat Mile

TABLE 11.2   Cargo Traffic History and Forecast

The following elements are applied to establish cargo traffic forecast

-Minimum of 5 years data should be available

-Scheduled service

-Non-scheduled service

-All services

Necessary yearly data:

Freighter RTM

Annual growth rate of freighter in RTM

Total freight RTM

Annual growth rate of total freight capacity in ATM

Load factor

NOTE: RTM = Revenue Ton Mile
      ATM = Available Ton Mile

For cargo forecasts, RTK signifies *revenue ton-kilometers* for cargo operations.

The forecast can be broken down for a specific region, for certain well-defined sections, or for the whole operation. Figure 11.6 shows that historical data were available from 1965 until 1971 for the now-defunct Seaboard World Airlines.

This chart indicates that Seaboard World Airlines' market share was around 16 percent of the total market. The data were presented until the year 1975. Once the market share is established—in this case 16 percent—it can be presented in tons per year. The market share can be broken down for each city-pair, e.g., JFK to LHR. It can be further broken down for weekly and daily available load. Obtaining the results, a good estimate can be forecasted for fleet planning procedures. This would answer the questions of how many airplanes should be purchased and what kind of schedule should be planned. Forecasting can further establish a retirement schedule for older (and maybe noisier) airplanes.

11.7 Market Share

There is a quick way to calculate the market share for a given city-pair. It requires a list of airplanes and their seating capacity for all the airlines serving the city-pair. A preference factor can be established for nonstop and multistops flights. More stops decrease the desirability of such flights for some passengers. Table 11.3 indicates

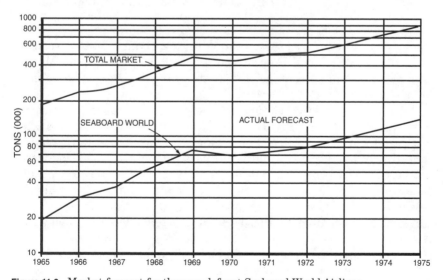

Figure 11.6 Market forecast for the now-defunct Seaboard World Airlines.

TABLE 11.3   Market Share Calculation

Alpha Airlines	AVRO RJ85	Seats = 95
Beta Airlines	B727-200	Seats = 140
Gamma Airlines	757-200	Seats = 186

Stop factor:

Nonstop = 1.0

Onestop = 0.5

Twostops = 0.25

Alpha Airlines   8 flights per week

Beta Airlines   6 flights per week all onestop

Gamma Airlines   4 flights per week

Alpha Airline market share:

$$\frac{(8 \times 1 \times 95)}{(8 \times 1 \times 95 + 6 \times 0.5 \times 140 + 4 \times 1 \times 186)} = 39.5\%$$

Alpha Airline market share is 39.5 percent

---

the method of this procedure. Other factors defining market share include fares, flight times, departure and arrival times, equipment (turboprop or jet), and many others not listed here.

$$\text{Market share} = \frac{(\text{airline seats} \times \text{no. of flights} \times \text{stop factor})}{(\text{all airline seats} \times \text{no. of flights} \times \text{stop factor})}$$

For passenger operations, market share is shown in Fig. 11.7 for three trunk lines (indicated by the letters A, B, and C) for a selected market (e.g., JFK to ORD or LHR to Dulles) for this city-pair. Market share is shown in percentage, and at any time, the three airlines' market share adds up to 100 percent.

Figure 11.8 shows the relationship between seats available, number of passengers, and (implicitly) the load factors of the flights for a selected city-pair for a large airline. For example, in November 1997, reading from the graph, seats = 57,000, and passengers = 30,000. This would represent a 52.6 percent load factor. In order to achieve an economically sound operation, management needs a breakeven load factor that is below this value (assuming that all other factors are satisfactory).

Figure 11.9 is similar to Fig. 11.8 but is extended in such a way that it shows seat capacity for 100, 80, 60, and 40 percent load factors. The chart also shows the breakeven and actual number of occupied seats.

For example, in February, the maximum number of seats available was 40,000. The breakeven number of seats is 20,000, and seats

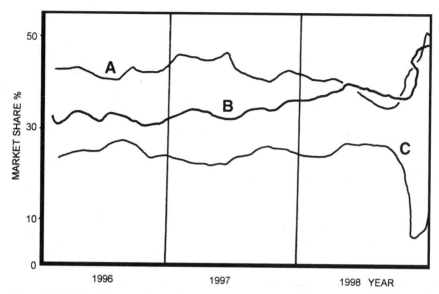

Figure 11.7 Market share presentation.

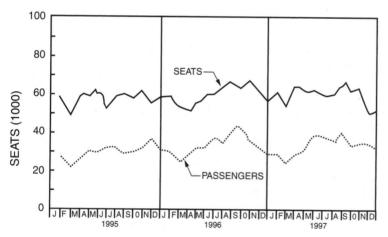

Figure 11.8 Seats available and number of passengers on board.

occupied were 15,000, which is a 38 percent of load factor. In order to minimize losses, the total number of seats should be reduced. The black area on the chart for this month indicates that the actual revenue seats are below the breakeven number of seats. Taking the month of December, the total number of available seats was 85,000 and seats occupied were 78,000, for a (very) high load factor of 91.7 percent. This would require a higher number of available seats. This

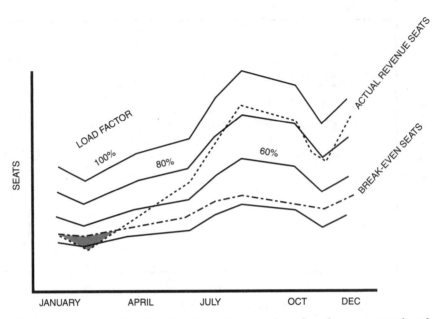

Figure 11.9 Seats available for selected load factors and number of passengers on board.

figure shows at a glance the load factor, the breakeven load, and the losses (see the black region around February). Furthermore, it indicates where capacity could increase or decrease and in what month this could occur. It shows where overcapacity or undercapacity exists. It also shows the time period where capacity cutbacks should be planned and where an option such as leasing an aircraft would be advisable. In addition, it shows where the operation is not profitable or has questionable profitability. The black area indicates a situation where the number of passengers carried is less than the breakeven number of passengers.

Figure 11.10 presents the monthly passenger distribution for a popular trip (this could be Seattle to Honolulu or New York to London). By viewing this graph, the following can be concluded:

- The passenger volume increases from year to year.

- The lowest and highest passenger volumes are experienced generally at the same time of the year.

- The distribution is cyclic.

- Passengers take their vacations in the middle of the year, and management should expand its marketing share during that time period.

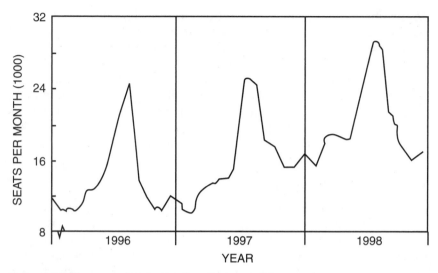

Figure 11.10   Passenger distribution for a 3-year period.

11.8   Methods of Forecasting

11.8.1   Consensus method or the Delphi* Technique

This method is based on subjective evaluation, intuition, educated guesses, and past experience. Mostly, it is used when sufficient data, resources, and time are not available. With this method, forecasting can be done in-house, by outside consultants, or by both. Surveys can be sent out to experts, and their forecasts can be consolidated and then sent back for reevaluation. Cycling the information over a few times, a fairly good forecast can be established. This forecast then should be evaluated in-house by a committee and routed through the Sales and Marketing Departments. Polls also can be taken. After several reviews, a final forecast can be prepared.

Since the forecast is based on experience and intuition, and maybe on poor judgment, it could be susceptible to a number of errors. Nevertheless, if applied for a shorter period of time, it may be useful to obtain a workable forecast. Sometimes it is combined with other forecast methods to improve its accuracy.

11.8.2   Market research

Market research is very useful when historical data are not available in certain areas or are inadequate. This need generally arises when

---

*Delphi is a location in Greece where in ancient times oracles were given.

an airline intends to open new routes or move into developing countries where no previous data are available. Market research also will show where to expect traffic flow. For example, a certain city opens up a Monte Carlo or Las Vegas–type of resort with casinos or other tourist attractions. Market research studies hotel facilities, entertainment, and other tourist attractions affecting the volume of business. Sometimes market research is the only way to forecast trends for a certain region and/or city-pair. For cargo operations, market research examines the industrial background.

11.8.3 Managerial judgment

Managers or directors who are not involved directly in forecasting but are familiar with the region and/or city-pairs are drawn into the project of forecasting based on their experience and detailed knowledge of the area. These employees generally are involved in operations, sales, and marketing. This type of forecast can be developed in a short time. It is not based on any kind of theory or mathematical background. This forecast also can complement the more sophisticated ones and may serve as a useful tool in short-term forecasting.

11.8.4 City-pair method

Based on historical average annual growth rates, this method calculates projected future growth. Assuming that enough information is available, growth rates can be established.

11.8.5 City-pair systems average

When the preceding method is not viable because of unsatisfactory information related to the city-pair, another approach is to examine the growth rates of both cities compared with other cities and base the projected growth on the average.

11.8.6 Systems growth rate

This method is used when historical data are insufficient.

11.8.7 Growth-rate estimate

If a city-pair is anticipated to undergo significant changes in the future for economic or any other reasons, an estimated growth rate has to be used that considers the historical data adjusted by the expected changes.

11.8.8   Deflated or inflated growth rate

Sometimes growth rates obtained from historical data can be irrationally low or high. In order to get realistic figures, a lower or upper limit would have a controlling effect on the growth rates.

11.8.9   Database correction

During a given historical period, certain events, such as strikes, unusual weather conditions, and other irregularities, may grossly affect data. In order to have a more accurate projection for the future, these distortions should be adjusted for accordingly.

11.8.10   Traffic seasonality options

The fluctuating behavior of a time series can be classified into four basic categories:

Secular trend

Cyclic variation

Seasonal variation

Irregular variation

The *secular* (or *long-term*) *trend* of a time series generally is described as a smooth curve over a longer period of time; it is a gradual trend. The cycle has a normal growth trend.

The *seasonal variation* is shorter than a year and has a relatively regular pattern, as shown in Fig. 11.10. Sometimes, a random fluctuation takes place in a time series, attributable to strikes, weather, and other irregular events. By smoothing out the irregularities for a period of a few years, the only remaining items are the seasonalities of the trend. By establishing index values, a projected seasonalized traffic can be calculated.

11.8.11   Exponential smoothing

Exponential smoothing is applied for short- and medium-term traffic projections. The data values are decreasing with age. Such a smoothing requires city-pair historical traffic data, exponential weight ranges, and a minimum of 3 years of historical data.

11.8.12   Least-squares method

Least squares is a method that fits a computed arithmetic trend line to observed data. The sum of the squares of the deviations is at a minimum. The method of least squares is applicable for intermediate- and long-term projections based on historical time series. The least-squares routine assumes that the average relationship between traffic demand and time and can be described by a linear equation.

11.8.13   Geometric progression

The geometric progression method calculates a geometric trend line based on historical time series. The sum of the squares of the deviation is at a minimum. This type of progression is used for medium- and long-term predictions.

11.8.14   Multilinear regression

The multilinear regression analyzes the average causative and associative relationship of one or more independent variables and the dependent variable. A linear relationship exists between the variables. Multilinear regression is applicable for medium- and long-range projections and forecasts. The calculation is used for variable data of traffic demand, such as political, social, economic, and demographic factors.

11.8.15   Polynomial regression

A polynomial regression calculates the linear or nonlinear causative or associated relationship between a dependent and an independent variable using least-squares equations. A nonlinear relationship could be a curve with an exponent of 2 (parabola) or any exponent of higher degree. This method predicts traffic data for medium- and long-range projections.

11.9   Econometric Forecasting

The econometric routine consists of three elements:

- Macro analysis is a time-series analysis based on elements of the macroeconomy and is a top-down approach.
- Secular trend is based on a geometric progression.
- Micro analysis is a time-series analysis based on elements of the microeconomy and is a bottom-up approach.

11.10   Accuracy of Forecasting Methods

There is a certain relationship between the different forecasting methods and their accuracy (Fig. 11.11). Each airline aims to prepare an

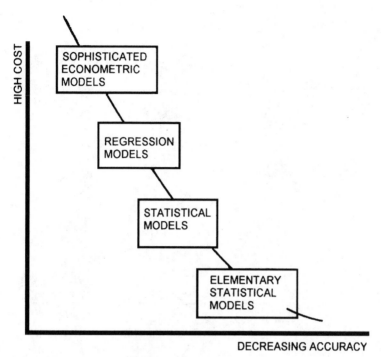

Figure 11.11   Cost versus accuracy of forecast.

optimal forecast with the least cost and within the shortest time. This is not an easy task, and a lot of effort is required for a realistic forecast (Fig. 11.12).

In conclusion, many forecasting models are available with various levels of sophistication. Forecasts can be generated within just a few days or within a longer period, and their accuracy varies with their complexity.

Figure 11.12   Scientific market strategy session.

Chapter

12

Scheduling

12.1 Introduction

Scheduling is one of the most important tasks in a well-functioning airline. The purpose of scheduling is to serve passenger demand for an organized, convenient way to travel. A good schedule attracts passengers, thus helping to increase revenues and, ultimately, profits. It also has to satisfy the requirements of the company's various departments to enable a smooth operation at a minimum cost with maximum profits.

12.2 The Scheduling Department

The Scheduling Department has the key to implement the airline's goal to present a viable schedule, observing all the constraints and limitations to obtain the highest revenue. It collects all the data necessary to present a profitable schedule. A few items are listed in this section merely to indicate the type of information required from other departments. In addition to this information, common sense, economics, costs, benefits, and experience should be the guiding factors in the preparation of a good schedule. There are no two airlines in this world having the same aircraft type, routing, and/or conditions that would generate the same schedule. Thus a schedule cannot be copied or used as a uniform solution. The scheduler needs to know a multitude of information before considering how to build a new schedule or modify an existing one. Such information includes

- Segment average daily, weekly, or seasonal traffic demand
- Segment yield per traffic mile

- Segment maximum allowable load factor
- Segment minimum allowable average daily, weekly, or seasonal one-way frequency level
- Segment distance
- Segment block time for each aircraft type
- Average DOC per flight per aircraft type
- Seats and/or cargo space available per aircraft type
- Maximum load for each segment per cargo aircraft type
- Maximum operating profit for each segment and for each type of aircraft
- Directional imbalances
- Whether to integrate new equipment, if any, into an existing schedule
- Aircraft modification program schedule
- Maintenance requirements:
  - Main base
  - Line station
- New routes
- Equipment changes
- Schedule changes due to competition, etc.

When operational changes occur, the first step is to evaluate their financial impact on seat-mile, load factor, and passenger and/or cargo volume and to analyze the consequences in revenues and, ultimately, profit.

A hypothetical schedule is shown for Alpha Airlines in Table 12.1. This schedule is prepared for the winter of 2002. The following conditions are furnished:

Load factor:

Planned 65%

Maximum 70%

Utilization: maximum of 13 hours daily

Aircraft type: Astra-797 airplane with 270 passengers (fictitious airplane type)

DOC: $1220 per block-hour

Maximum range: 4900 nautical miles

Operational information:

Curfew at LHR: 23:59 to 05:59 hours

Ground time for:

Minimum transit: 0:45 minutes

Minimum turnaround: 1:00 hour

At EWR minimum turnaround: 1:45 hours

Maintenance: each aircraft has to be rotated to Washington (IAD) for 12 hours maintenance at least once every 150 hours (EWR-WASH ferry may be added to the schedule).

The marketing forecast is shown in Table 12.2. Only a few stations are listed only with the expected market share of Alpha Airlines. Block times are listed for a few stations in Table 12.3.

12.3   Aircraft Rotation Schedule

An aircraft rotation summary is shown for a certain schedule (not shown here) in Fig. 12.1. Due to limited space, only some of the flight numbers and stations are shown (time of day lines are for 6:00/12:00/18:00/24:00). What the summary does is it follows the schedule segment by segment, indicating the station names, flight numbers, the day of the week, and the time of the day.

12.4   Equipment Routing

Equipment routing is shown on Fig. 12.2. Here, an airplane is followed in time and space. The chart shows the airplane's scheduled route from one station to the other, its flight number, the day of the week, and the time of the day.

This schedule was developed for a small hypothetical airline. This is the first step for a graphic presentation of a schedule. (Based on the resources of the particular airline, a computerized approach may be used.) The next step is to analyze the operational feasibility of this schedule. An economical evaluation also is in order to check its profitability. Since each airline has different route structures, aircraft types, cost structures, operational policies, and market shares, each airline has to do its own evaluation based on its own unique characteristics. This is only the beginning. Subsequently, many other problems will arise while a schedule is being prepared, and they will have to be solved in order to have a profitable operation.

TABLE 12.1 Alpha Airline Schedule for Winter 2002

Eastbound Read Down

STA		X71	X73	X75	X77	X79	X81	Z*
		Dly	Dly	Dly	2	6	4,5	
EWR	Dep	1801	2002	2159				+4
IAD								+4
LGW	Arr	0541	0740		2200	2202	2201	+1
	Dep	0641	0840	0940	1000	1000	0959	
CDG	Arr	0845	1045		1100	1100	1059	+2
	Dep	0930	1131		1304	1304	1304	
FRA	Arr	0940			1349	1349		
FCO	Arr		1320		1400	1542		+1

Westbound Read Up

STA		X72	X74	X76	X78	X80	X82
		Dly	Dly	Dly	3	7	5,6
EWR	Arr	1814	2126	1436	2306	2327	1911
IAD	Arr				2010	2031	1614
LGW	Dep	1540	1851	1202	1911	1932	1514
	Arr	1441	1752		1854	1916	
CDG	Dep	1424	1736		1808	1832	1603
	Arr	1338	1651				
FRA	Dep	1130			1600		
FCO	Dep		1503			1543	

Eastbound Read Down Z^* = Time Zone   Westbound Read Up

Airport Code   Airport Name
CDG   DeGaulle-Paris
EWR   Newark
FCO   Fiumicino-Rome
FRA   Frankfurt
IAD   Dulles
LGW   Gatwick
ORY   Orly-Paris

TABLE 12.2   Marketing information: Number of passengers

Origin-destination	Industry total	Alpha Airline's market share
EWR-LGW	759,400	125,200
EWR-FRA	378,100	85,300
EWR-ROME	412,400	76,205
etc.		

TABLE 12.3   Block Times

	Block times	
City-pair	East	West
LGW-CDG	1:05	1:15
FRA-LGW	1:50	1:53
FRA-ORY	1:10	1:14
EWR-LGW	6:40	7:42
etc.		

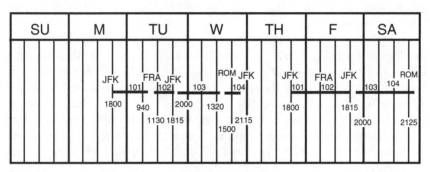

Figure 12.1   A typical aircraft rotation schedule (partial).

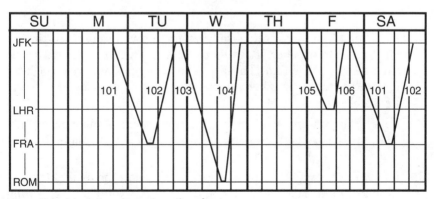

Figure 12.2   A typical equipment routing chart.

The two major categories of scheduling are scheduling by revenue requirements and scheduling by the operational needs and constraints of the airline. These two, sometimes opposite requirements have to be reconciled and optimized.

12.5 Schedule by Revenue Requirements

Scheduling should satisfy the needs of the airline's (revenue-generating) customers. Carrying passengers and/or freight generates revenue. To obtain maximum revenue, an attractive, convenient, and competitive schedule should be developed.

It is not enough to select departure and arrival times; the transported volume also has to be known for a given city-pair, as well as the connecting traffic. Similarly, the proper equipment has to be planned, based on traffic volume. When the load factor is too low and below the breakeven load factor, the operation is not economical. If the load factor is high, a higher profit margin is ensured, but not all the available revenue-generating passengers may be seated on a flight, and some would have to be left behind and possibly lost as customers forever.

Aircraft utilization is measured in trips per day, month, or year or block-hours per day, month, or year. If utilization is high with a low load factor, the generated profit is also decreasing. The frequency is different for a domestic airplane flying an average trip length of 400 nautical miles or for an international flight with a range of 3500 nautical miles.

The market predicts a certain number of passengers for a set of trips. The airline scheduler's job is to match this market to airplane size, estimated utilization, estimated load factor, estimated frequency, and estimated revenue or profit.

To point out some aspects of scheduling, consider passengers related to aircraft capacity versus time of day. Three conditions could exist for a certain city-pair:

1. The available number of passengers for a given departure time is more than the number of seats on the airplane, and the extra passengers are left behind. These passengers may depart on a later flight or switch to another airline. Steps should be taken to avoid this problem.
2. The available number of passengers and the airplane capacity are equal, no passengers are left behind, and the airplane is fully loaded. This is the ideal situation.
3. The available number of passengers is less than the available seats in the airplane. The load factor and the breakeven load factor relationship have to be examined. If the latter is above the airplane

operational load factor, the trip does not generate enough revenue, and steps should be taken to correct this situation.

The scheduler's task is to select the optimal result. Generally, airlines have a schedule already working for them, and the task is to constantly incorporate necessary changes such as new departure times or adjust current departure times to meet changing demands.

12.6   Time Zones

Time zones are another factor that causes concern. Eastbound flights from the United States often depart in the evenings, whereas airplanes flown into the United States from Europe usually arrive in the morning hours. From a marketing point of view, departure and arrival times are essential when the flights cross time zones.

Time zones are limiting factors when schedules are prepared. A trip from the West Coast to the East Coast requires 8 hours due to time zones (3 hours), and the best time for departure is in the early-morning hours, until 1 or 2 P.M. The window of opportunity is about 6 or 7 hours when flights are to be scheduled from the West Coast to the East Coast.

12.7   Departure Time Distribution for Passenger Operations

It should be noted here that taxi-in and taxi-out times should be scheduled for optimal conditions. Taxi time varies with time of day, weeks, seasons, and peak-hour congestion. If it is possible, by coordinating departure time and the associated taxi-out time and arrival time and the associated taxi-in time, considerable savings in block-time and trip cost can be realized. However, given the many constraints, it is very hard to obtain the optimum.

Taxi-out and taxi-in times consist of three factors. Generally, the first two are present, and the third is present only sometimes. The first factor is constant and is a function of distance to or from the terminal building to or from the active runway. The second factor is a variable that is the function of time of the day, week, or season and depends on traffic and congestion. For some airports, the peak congestion is in the morning or afternoon, but this could be different for another set of airports. The third factor comes into play when the airplane is assigned for any reason to another (frequently further situated) runway. This does not occur on a regular basis. These factors define taxi time and taxi fuel.

Data are collected by some airlines for certain airports' taxi-out and taxi-in times as the function of season, month, weeks, days, or hour of the day. To show the significance of these data, let us assume the following:

1. Alpha Airlines operates out of JFK and Newark airports.
2. It has flights from both airports to London. Trip time is 6 hours and 45 minutes (no taxi times included).
3. Departure time from JFK is 5:00 P.M. and from Newark is 11:00 P.M. on Friday.
4. Taxi-in time at arrival at LHR is 10 minutes.
5. Establish block times for both flights.

Based on the airline database, taxi-out time at JFK at 5:00 P.M. is 22 minutes and at Newark at 11:00 P.M. is 11 minutes. Thus JFK to London:

$$6:45 + 22 + 10 = 7:17 \text{ block time}$$

EWR to London:

$$6:45 + 11 + 10 = 7:06 \text{ block time}$$

Assuming an operating cost of $5000 per block-hour for the airplane and 300 trips per year, the Newark operation is shorter by 11 minutes per trip. This would generate a cost savings of $275,000 per year per airplane. (A similar situation could be experienced in London, where Heathrow Airport is more congested and has a higher taxi-out time than Gatwick Airport. It may be assumed that operating from Gatwick Airport would show a similar magnitude of cost differences versus Heathrow.)

The purpose of this exercise is to highlight the effect of every factor, however small, when deciding any kind of change in schedule.

Business travelers prefer early-morning departures between 7 and 8 A.M. in order to arrive at their destinations at the beginning of office hours. They also like to return to their base, if possible, the same night. The general public favors later departure times in the morning and arrivals around 5 or 6 P.M. Normal passenger traffic takes place between 7 and 8 A.M. until 4 to 10 P.M.

A passenger daily departure time by the hour is shown for a large U.S. airport in Fig. 12.3. As was described previously, it reflects a typical distribution for passenger departure times.

12.8 Departure Time Distribution for Freight Operations

The same type of departure distribution for cargo flights is shown in Fig. 12.4. Both Figs. 12.3 and 12.4 are typical for their respective operations, and both demonstrate the operational characteristics.

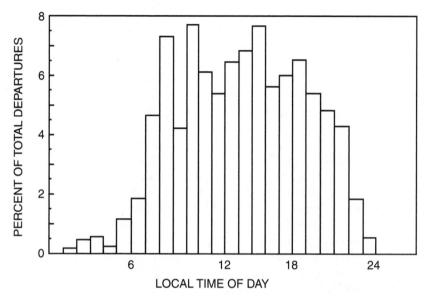

Figure 12.3 Percentage of departures for passenger operations versus local times.

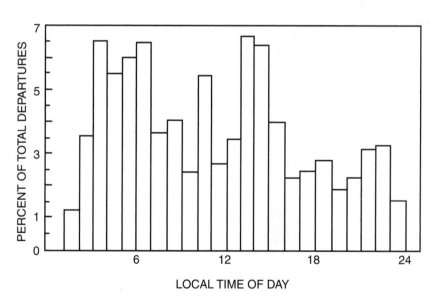

Figure 12.4 Typical cargo airplane departure pattern at a large airport.

Freight shippers have entirely different demands from those of passengers. For cargo departures, the preferred time is late afternoon, after close of the business day. The Postal Service also requires 5 to 6 P.M. departure times up to midnight to ensure an early-morning arrival time (between 4 and 7 A.M.). However, night and early-morning flights bring up noise problems. Certain communities in densely populated areas have curfews to avoid noise above a certain level at certain times of the day.

Another important consideration when preparing a schedule is weighing multiple stops against nonstops. Stops could generate higher revenues, but operation costs may increase due to additional expenses such as extra landing fees and additional block times and block fuel, which may exceed the additional profits.

Flights can be classified as scheduled flights based on a published schedule, charter flights, and extra sections. An *extra section* is a particular flight in which an unusually large number of passengers are to be carried. Flights can be further categorized as nonstop, multiple-stop, and hub operations. Hub operations are not too convenient for passengers because of such problems as changing equipment, rushing from one gate to another, and standing in line again. In addition, passengers may not reach their connecting flights on time in the case of a late arrival into the hub. So far, only a few requirements for preparing a good schedule based on passenger needs have been discussed.

12.9   Schedule Requirements Based on Airline Operational Constraints

Once a schedule is established based on revenue demands, the airline has to consider how to attain that schedule based on the limited resources it has to deal with. To mention a few:

- Flying operations
- Station scheduling
- Maintenance scheduling
- Fuel purchase policy

In order to achieve an optimal schedule to coordinate the needs of these various departments as well as to obtain maximum revenue, the airline must prepare a very detailed and carefully conducted study. Nevertheless, since revenue enhancement is the primary goal, the operational requirements should be subordinated to maximize profit based on the airline's capabilities and limitations.

12.10   Crew Scheduling

Crew scheduling is a complex task; federal aviation regulations (FARs) and an airline's contracts with its crew have to be observed. To mention just a few limitations, there is a daily limitation of 10 flight hours under certain conditions described in the FAA regulation; a flight crew member has to have a minimum of 16 hours' rest after completing his or her last flight; and maximum weekly flying hours are limited to 40 hours (applicable to a consecutive 7-day period).

To demonstrate the importance of observing time limits on a given trip, if when the schedule is under development it is realized that the crew's duty time is 10 minutes shorter than trip time, this schedule would not be acceptable, and a new crew schedule would have to be generated. The contract generally calls for a maximum of 80 hours of flight time per month. Depending on many other factors, this 80 hours of flying time is sometimes reduced to 55 to 60 hours per month. To complicate the matter, crew seniority has to be taken into consideration. Not all crew members are based in the same city, so the expenses involve the cost of traveling, lodging, and food. All basic elements should be scrutinized and integrated into crew scheduling to achieve the most efficient solution.

12.11   Environmental and Operational Limitations

Several conditions influence on-time performance. An unusual enroute wind can have a great effect on arrival and/or departure times, and on arrival, no ramp, gate, or personnel may be available beyond scheduled hours. Air traffic control (ATC) restrictions also should be mentioned in this category. For example, ATC selects a longer routing, causing a longer flying time or diversion.

Operating in a hot climate adds another restriction factor. In the FAA-approved flight manual (AFM), when temperature climbs above the maximum operating temperature, operation is not permitted (Fig. 12.5).

In Fig. 12.5, a daily temperature distribution chart is presented to a city located close to the Equator. This chart reveals that takeoff and landing are not permitted between 11:50 A.M. and 5 P.M. due to the maximum operating temperature limitation described in the FAA-approved flight manual. At this location, early-morning, late-evening, or nighttime operation should be applied.

Another limitation to be observed, as shown in the FAA-approved flight manual, is airport elevation, depending on the aircraft-engine combination. No takeoff and landing are permitted from any airport with an elevation above the established legal limit.

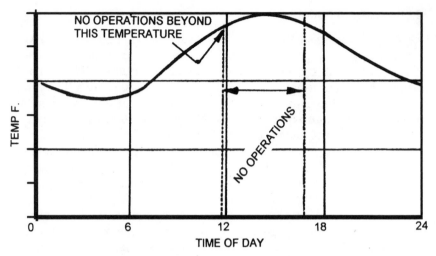

Figure 12.5   Operational limits versus time of day.

The pavement strength limitation hardly plays any role in curtailing operations, but it is still an item to be checked. Another limiting factor could be the seasonal meteorologic conditions. Sometimes ground fog cannot be ignored completely (Fig. 12.6).

The figure shows the probability of limited operation (e.g., fog) as a function of time of day. Since delays have a ripple effect on the system, the scheduler has to schedule the flight around this time of day.

Sometimes runway lengths are shortened by unscheduled runway closings or for a short time to conduct maintenance work. This may introduce an unscheduled fuel stop and/or additional delays.

Sometimes the airport listed in the proposed schedule has a short runway. Short runways require a lower takeoff weight, which may prevent carrying enough fuel for the trip. The scheduler has a choice:

1. Take off with less load (losing partial revenue).

2. Introduce a one-stop operation. Operating cost would be higher due to a second landing fee and additional takeoff climb and descent cost figures.

3. Select a close-by airport that has a satisfactory runway length that supports the required takeoff weight for that trip.

Winter flying into northern cities adds another problem for operations: The airplane has to stay outside during night, when the temperature may fall below zero. Snowfall and/or ice could cause delays for morning departures. For this reason, it is advisable to provide hangar space for the night.

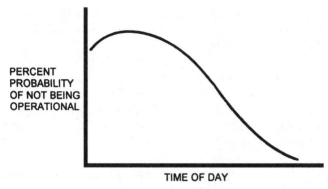

PERCENT
PROBABILITY
OF NOT BEING
OPERATIONAL

TIME OF DAY

Figure 12.6   The probability of not being operational versus time of day.

Winter operation generates further problems. The FAA has established regulations for defined icing conditions for winter operation. These place additional burden on operations because in certain weather conditions the deicing procedure is unavoidable. In the case of snow, sleet, and/or ice, a reduction in takeoff weight may be necessary relative to dry runway performance. The data for operating on runways covered by 0.25 to 0.5 in of slush and ice present the limiting weights due to contamination and are published in the aircraft operations manual. This limitation affects mostly cargo operations because cargo airplanes have higher takeoff weights.

12.12   Station Schedule

There is more than one way to generate an efficient schedule between stations; no fixed formula exists. Each airline has to evaluate the traffic, airplane types, and available resources to plan an economical schedule. No airline can imitate another airline's schedule because factors such as cost structure vary from airline to airline.

Airlines developed a variety of statistics about the volume of their passengers because this defines their schedule, size of airplanes, ground services, and other required equipment and expenses. Let's say that for a given city-pair, the average number of seats available per month is 68,000, but only 37,400 seats are taken, so the average load factor is 55 percent. If the breakeven load factor is 48 percent, there is a potential for revenue. Should the breakeven load factor be 58 percent, management should take actions to avoid losses. As a remedy, it could bypass this city-pair, change to airplanes with a different seat capacity, or introduce one-stop operation versus the current nonstop operation. Management has to search for and study the best available solutions to this problem.

Nonstop service. If the airline market can support a nonstop operation, then this is the preferred way between two cities. It is fast, cheaper, and is preferred by passengers to multiple stops.

One-stop operations. Sometimes a one-stop operation is introduced into the schedule. This is done when the number of available passengers is such that one of the segments is not profitable on its own. However, combining the two segments could generate enough revenue to make the trip economically feasible.

Local service. A smaller aircraft connects low-traffic stations and delivers passengers to a larger city. This service may prove to be economical to the airline; however, changing airplanes inconvenience passengers.

12.13   Linear Route Systems

A *linear* route structure may be in the form of a single line or a combination of straight lines in a T or Y shape. It is characterized by a limited number of turnaround points and no crossover flights that bypass the hubs (Fig. 12.7).

12.14   Hub and Spoke Operations and the Role of Regional Airlines

A good example of hub and spoke scheduling was a U.S. trunk line routing for New York to Atlanta with the designated hub Raleigh-Durham Airport. The advantage to the airline was that a large volume of passengers could travel on a larger airplane to the hub. From this hub, a smaller airplane could carry a smaller number of passengers to any other destination.

Regional airlines fit very well into this type of operation. They fly smaller airplanes into a hub, and they operate them more economically than a large airline using larger aircraft. This fact was recognized by the regional, as well as the major airlines, and this is how the so-called code-sharing method was implemented.

The use of smaller airplanes is more economically feasible than having a nonstop flight; nevertheless, it is inconvenient for passengers to keep changing airplanes in a very large and busy airport. More cities can be served with this approach. A nonstop service between cities would require more airplanes and less efficient transportation. A hub and spoke type of route structure has one or two hub cities from which all flights originate to outlying points (Fig. 12.8).

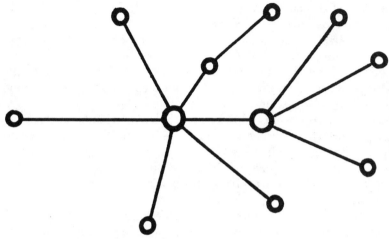

Figure 12.7 Linear route structure.

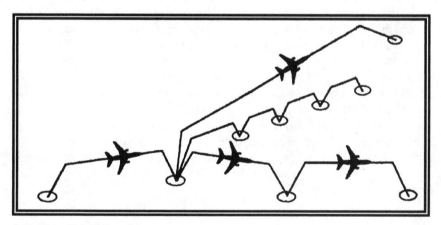

Figure 12.8 Hub and spoke.

Considering revenue and cost, the hub and spoke operation would generate higher revenues with a moderate escalation in cost as opposed to nonstop service for the same set of cities. It is hard to establish for another airline a similar hub and spoke system for the same cities, not to mention the high amount of investment with questionable return. From an operational standpoint, hub operations, after reaching a certain level of traffic, could result in congestion, delays, and increasing passenger travel time and, for the airline, more personnel and higher operational costs.

For the airport, the hub operation would generate higher airplane traffic. When one large jet lands and discharges 200 to 300 passengers, a great many of the passengers board smaller airplanes. These smaller airplanes have to taxi to the runways and take off. As a consequence of the hub operation, instead of one landing and one takeoff, several other, smaller airplanes take off with the passengers discharged from the larger airplane. The number of passengers passing through the airport does not change, but the additional airplane movements may cause significant congestion at the same time.

Curfews may represent additional problems at hub operations. They may limit the number of operations or limit or ban aircraft movements during certain time periods. They may require lowering takeoff weights when prescribed noise limitations have to be observed or banning certain airplane types for a given period or for any operations.

With hub operations, as business is growing, additional problems arise. The increase in traffic requires additional gates because a limited number of gates would stunt growth. The scheduler's problem is to use these gates more efficiently. In other words, he or she must plan more departures per gate per day. The average daily departures per gate is generally between 5 and 8, but some airlines at certain airports are able to squeeze out up to 12 departures per day.

Another constraint is the availability of the runways. During takeoff and/or landing, due to the vortex generated by larger airplanes, smaller airplanes following the larger ones have to observe a safe distance. This again slows down the operation, limits airport capacity, and contributes to additional congestion.

Heavier traffic increases taxi time, block time, and block fuel; could further erode revenues; and could prevent baggage from being loaded, sorted, and transferred in due time. Ground time has to be extended due to connecting flights.

Note: For hub operations, three phases could be observed. The initial phase occurs when the airlines select a hub, and (in-house and/or hired consultants) studies indicate that a hub operation could result in increased profit after a certain amount investment. The mature phase begins when an airline starts to show profit for this type of operation. The last phase is when many of the problems mentioned before start to erode profit.

The life span of the hub operations could last from a short time period to many years, and it depends on the local characteristics of the airport, airplane, market, industry in the area, and many other factors. Hub operation requires airline management to constantly evaluate the operations, economics, and the hub environment and be very alert to every change in any of these factors.

12.15   Selected-Stops Operation

When providing service between multiple stations (e.g., eight stations), the scheduler could provide flights for each even-numbered station and for another flight connect all the odd-numbered stations. In this way, it is possible to cover all stations, requiring less time, and still provide a certain level of service. The only disadvantage to this approach is that there is no service between the odd- and the even-numbered cities.

12.16   Routing

In the 1970s, Omega Freight Airlines had the following routing pattern: JFK-BOS-DTW-ORD-Europe. Three daily cargo flights covered these stations, but this routing schedule was not profitable. By reworking the schedule, after a careful analysis of this case, the following changes were initiated:

JFK-BOS-Europe

JFK-DTW-Europe

JFK-ORD-Europe

The same load was delivered to Europe. The airplanes covered only once a day BOS, DTW, and ORD, saving landing fees, fuel cost, and station cost and increasing profit.

12.17   Traffic Flow Distribution

An uneven distribution of traffic flow represents another significant problem in scheduling. Considering the traffic distribution from one city to another, such as from SFO to DFW, at 7 A.M., 180 passengers are predicted to fly; at 8 A.M., 200 passengers; and at 9 A.M., 88 passengers only. The first two flights can use the same aircraft, whereas the latest flight would have a very low load factor. The airline should choose a smaller type of aircraft (if possible).

Now, there is another situation that has to be discussed, namely, the spill and recapture problem. Let us assume that airplane type A has a seating capacity of 287. At a high load factor, there will be times when a certain number of passengers cannot be accommodated. Selecting a higher-seating-capacity airplane (with higher operating costs) results in fewer passengers being turned away. The airline has to make a decision:

1. Using a smaller airplane with lower operating costs and turning away a certain number of passengers

2. Using a larger airplane with higher operating costs and turning away fewer (or no) passengers

The answer is whichever solution is more economical. It should be noted that a turned-away passenger could be lost forever.

Related to this problem is the recapture of those passengers who could not be seated on their original flight. In the case of a nonstop flight, such passengers could depart on a later flight, and maybe their goodwill can be retained. The issue gets complicated when a passenger has a one-stop flight. In this case, the second-segment seat is lost too. When placing this passenger on a later flight, the second leg has to be available for that passenger.

Sometimes location has a great influence on the load factor. For example, passengers in New York—specifically in Manhattan—prefer to depart or arrive at LGA instead of JFK or EWR. LGA is closer to Manhattan and Queens, not as crowded as JFK, and closer than EWR. In addition, ground transportation cost has an impact on selecting the most convenient airport. Thus, the closer the airport, the lower is the cost of ground transportation. Many passengers consider total transportation cost as the sum of the airline ticket plus ground transportation costs.

Another uneven traffic flow distribution problem has to be addressed in the schedule, namely, a drastically fluctuating load factor experienced between each subsequent station. Some trips between two adjacent stations have very low load factors—sometimes even below breakeven load factor—and others show high load factor and profit. If the total revenues are not acceptable, the schedule has to be rearranged.

12.18 Frequency versus Load Factor

It is advisable to obtain certain statistics related to load factors for a specific route and a fixed time period. Knowing the number of passengers (P_t) for the same time period and the number of passengers per airplane (P_s) and the scheduled frequency (F_t), the available mean load factor (LF_a) is calculated as follows:

$$LF_a = \frac{P_t}{P_s \times F_t} \tag{12.1}$$

And the individual load factor (LF_n) is

$$LF_n = \frac{P_n}{P_{sn}} \tag{12.2}$$

where P_n = number of passengers on a given flight
P_{sn} = passenger seats on a given flight

The relationship between mean load factor and frequency is decreasing, corresponding to diminishing returns (Fig. 12.9). At low frequency, the load factor is high. When frequency increases, the load factor decreases. If the stations with the low load factor are bypassed, the profitable trips may be affected because the passengers from the unprofitable stations would not be available.

12.19 The Effect of Changing Departure Time

Marketing plays a significant role in defining departure time because the Marketing Department can design the schedule based on passenger preference. Sometimes the desired and preferred departure time is not available. The schedule has to be reworked and all negative ripple effects eliminated. Let us examine a hypothetical case where a flight has to be scheduled later at a certain station than originally would have been preferred by the airline.

To visualize this problem, a graphic presentation is helpful (Fig. 12.10). A station with three flights are shown, and a schedule is prepared. In this figure, on the upper time line three airplanes are presented with their arrival and departure times. They are well positioned, and there is enough time between the departure and arrival times. Let us assume that for some reason the second airplane's arrival time has to be shifted. The positions of airplanes 1 and 3 do not change. The graph clearly indicates that in the case of rescheduling, airplanes 2 and 3 are on the ground, and their times on the ground are partially overlapping.

The impact of the newly proposed schedule would cause numerous changes in the operations. This is best described with a practical example: it takes one hour for a ticket counter to process a planeload

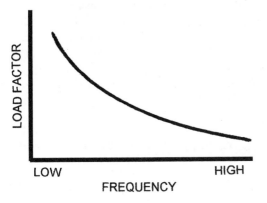

Figure 12.9 Load factor versus frequency.

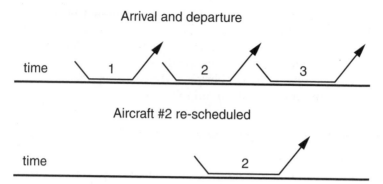

Figure 12.10   Station schedule.

of passengers. By changing the departure time by only half an hour, the facilities would not be able to handle ticketing on time, and passengers would have to stay on line much longer because another flight is already scheduled for this period.

Passenger comfort should be considered in the waiting area, and at least 15 ft² per passenger is required. Furthermore, when two airplanes are departing at about the same time, the passengers should not be confined to a more densely populated waiting area, not to mention that it is hard for senior citizens to stand in crowded areas.

In addition, passenger transfer time may be too short or impossible in this station or at the next station. Food service would have to be rescheduled, baggage handling overtaxed, and delays would be experienced. Very likely, another shift would have to be hired to handle the workload. In the case that outside contractors are used for aircraft cleaning, food service, or other aircraft services, the contractors also would have to reschedule their activities. Sometimes, contractors may decline to render their services, or they may slow down because of lack of personnel and/or equipment. Changes in the schedule may require additional ground equipment such as baggage vehicles, aircraft starters, and forklifts. These should be purchased or rented, or another airline should be engaged to do the job.

In conclusion, a ¹/₂-hour shift in departure time for a certain flight for a certain station would have the following consequences:

■ Additional ticket counters

■ Additional facilities (lavatories, waiting rooms, gates, etc.)

■ Additional personnel

■ Additional ground equipment

- Review of outside contractors contracts for food service, aircraft cleaning, aircraft fueling, baggage handling, other miscellaneous services

Not necessarily all these problem areas would be affected; maybe only a few. Still, the existing problems have to be solved, but at what price? Considering the ripple effect of this change in departure time, all other stations involved in this routing have to be reevaluated. This may cause certain flights to experience difficulties due to noise curfews or reduced takeoff weights and/or introduction of fuel stops, if any. Maybe some flights have crew scheduling problems or maintenance difficulties due to a lack of personnel or spare parts. Furthermore, the airplane cannot leave the station because of awaiting the connecting flights.

If schedule changes are too numerous, a new schedule should be printed, published, and distributed. Advertisements and other related expenses also should be considered.

This $\frac{1}{2}$-hour change in schedule could affect all the preceding items. Attaching a price tag to all or some of these items could amount to a considerable sum of money. When all the stations are affected by this change, it could become a substantial amount and should be available as needed. Having listed the potential expenses, the question arises: Would this $\frac{1}{2}$-hour change in departure time generate enough revenue to justify the additional expenses?

Departure times have to be examined carefully for each station in order to enable the stations to satisfy the planned traffic conditions and the designated times. Taxi times at both stations should be considered because they vary as a function of (daily) times.

Slot allocation. Another important factor to be observed in station scheduling is the slot allocation. Larger airports have a slot allocation plan. For example, Alpha Airlines has been allocated three slots, which are usable at a predetermined time. In the case of a schedule change, these slots may not be available.

12.20   Airplane Ground Time Activities

Ground time has to be established based on the activities related to airplanes, passengers, and/or cargo. Ground time varies according to

Airplane type

Maintenance checks

Equipment availability

Crew changes

Personnel availability

Connecting flights

Line station

Turnaround station

The minimum ground time for a through flight is about 25 minutes for domestic flights and about 1 hour for international flights. For connecting flights, turnaround time naturally is longer. Trans-Atlantic turnaround time is about 1 hour and 20 minutes or maybe more depending on aircraft type, location, and service available. Table 12.4 shows various ground times.

12.21   Schedule Development

After discussing scheduling, the following figures show a hypothetical scheduling problem. It starts with a station where scheduling for departure and/or arrival is not limited by any regulations or problems, as shown in Fig 12.11.

12.22   Curfew

In the next figure (Fig 12.12), a night curfew is indicated from 11 P.M. until 6 A.M. This can be a total curfew with no airplane movements at all or a

TABLE 12.4   Ground Times

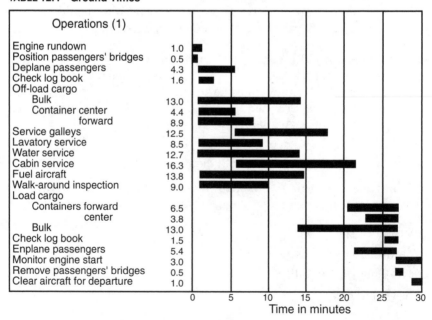

Operations (1)		
Engine rundown	1.0	
Position passengers' bridges	0.5	
Deplane passengers	4.3	
Check log book	1.6	
Off-load cargo		
Bulk	13.0	
Container center	4.4	
forward	8.9	
Service galleys	12.5	
Lavatory service	8.5	
Water service	12.7	
Cabin service	16.3	
Fuel aircraft	13.8	
Walk-around inspection	9.0	
Load cargo		
Containers forward	6.5	
center	3.8	
Bulk	13.0	
Check log book	1.5	
Enplane passengers	5.4	
Monitor engine start	3.0	
Remove passengers' bridges	0.5	
Clear aircraft for departure	1.0	

0   5   10   15   20   25   30
Time in minutes

MIDNIGHT

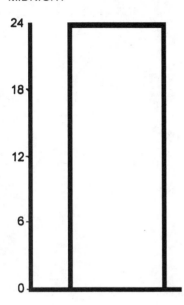

Figure 12.11 Schedule development: No constraints.

MIDNIGHT

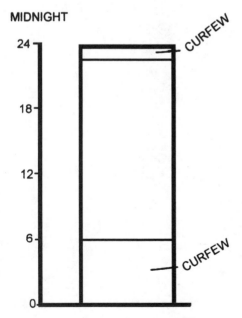

Figure 12.12 Schedule development: Curfew.

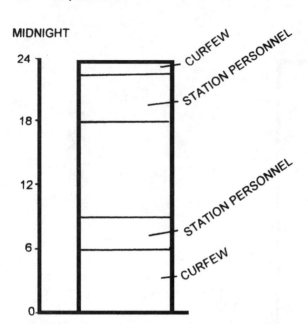

Figure 12.13 Schedule development: Curfew and availability of station personnel.

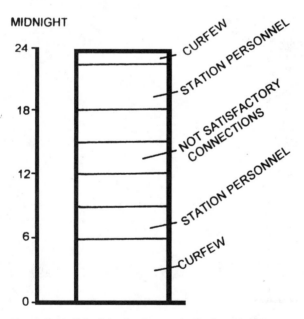

Figure 12.14 Schedule development: Curfew, station personnel, and satisfactory connections.

curfew limited to a specific number of movements, limited by a certain noise level, or limited by takeoff weight because of noise level. If the type of airplane cannot satisfy this noise-level limit, the operator can dispatch it with less takeoff weight (less payload or one fuel stop) or not take off at all. For example, JFK, LHR, FRA, and ZRH airports have noise limits.

12.23   Limitation of Station Personnel

In Fig. 12.13, an additional problem exists, namely, station personnel. Maybe one shift is enough and available to handle departure and/or arrival of the airplane.

12.24   Connections

Once these problems are solved, the connections (Fig. 12.14) to and from this station should be realistic. There will be a certain period, however, when good connections are not available.

12.25   Gate Availability

Adding another problem to the previous one, gate availability is now examined. In Fig 12.15, when an airplane arrives before scheduled arrival time, sometimes a few minutes delay may be experienced due to nonavailability of the assigned gate. This also clearly indicates the sensitivity of scheduling.

Note: In actual operation, not all these constraints act together; however, they are here to stay in some combination.

12.26   Short Summary of Schedule Activities

Schedule development is a very complex activity. From the numerous conditions affecting scheduling, only a few items are listed here:

- Departure times:
  Marketing
  Maintenance
  Traffic service
  Ground service
  Assigned slots
  Flight and cabin crews

- Airport:
  Day of week
  Time of day

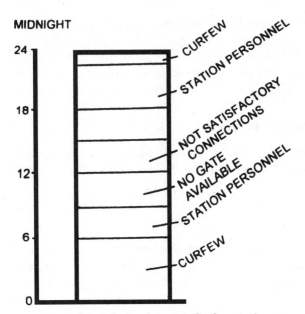

Figure 12.15   Schedule development: Curfew, station personnel, satisfactory connections, gate availability.

Station personnel
Ground crew

- Arrival times:

Connections
Gates
Curfews
Customs

- Unique to airline:

Traffic and traffic flow
Schedule changes
Operational considerations
Scheduling maintenance activities
Seasonal changes
Local service carrier
Trunk or long-haul carrier
Flight crew regulations

- Airport considerations:

Curfews
Ground facilities
Gate facilities

Equipment turnaround and transit
Late arrivals
Connections
Weather
Availability of maintenance facility

- Equipment changes:

Operating times
Payload restrictions
Service patterns
Type of aircraft
Ground support
Operating temperature limit

- Financial impact:

Seat miles
Load factors
Passengers
Profit and loss
Etc.

12.27 Maintenance Scheduling

The maintenance organization within an airline has one single goal: to keep the airplanes flying and keep the airplane safe by accomplishing all prescribed maintenance services within the shortest time on the ground and with optimal use of personnel. These are opposite requirements, and the scheduler has to compromise to obtain the best and most economical solution.

There are, along the airplane routes, designated stations able to handle certain maintenance jobs. For some components, the maintenance program describes removal, overhaul, and inspection based on predetermined flying time. These predetermined flying times cannot be exceeded. Part of the maintenance can be done at line stations, whereas some can only be done at the maintenance base. The proper personnel have to be present at the right place and (scheduled) time with the necessary tools and spare parts on hand. The work flow should be scheduled without delays.

The component flying time limits have to be observed, and inspections, overhauls, etc. must be carried out according to the maintenance schedules. Different airplanes reach their designated flying times (e.g., inspections are scheduled) at different times and locations; therefore, maintenance service should take the least time and employ the most efficient utilization of personnel. Personnel consists of highly skilled specialists. Since there is a very large investment in facilities, spare parts, tools, and the remuneration of these highly skilled personnel, an even work flow is required.

The maintenance scheduler also has to accommodate the handling of mechanical breakdowns. To provide for these unforeseen breakdowns, the following problems should be solved:

Is the station able to repair this breakdown while the airplane is on the ground?

Is there another airline with maintenance capabilities available? Should the airplane be ferried to another station or to the main base?

Is there another airplane available to continue the original flight?

Can the substitute airplane be maintained according to its own schedule?

The repaired airplane should enter service again with an updated maintenance schedule. A constantly changing scene should be updated to achieve optimal utilization. To keep up these activities, the airline has to spend large sums of money; therefore, every decision should be based on a cost-benefit relationship with the strictest consideration for safety.

In this chapter, some of the problems associated with schedule building have been highlighted. The conclusion is that a good schedule must satisfy all external and internal requirements in order to maximize profit and minimize cost.

13

Fleet Planning

13.1 Introduction

Fleet planning is one of the most important activities of an airline. It will determine the future of the airline and its fleet requirement.

13.2 Elements of Fleet Planning

Numerous airlines went out of business because of their poor fleet planning procedures. Fleet planning defines an airline's structural buildup and characteristics by

- Aircraft type
- Number of aircraft
- Timing of purchases and retirements of aircraft
- Deployment of aircraft
- Attainment of financial goals

A typical fleet planning model requires the following input:

- Current fleet
- Cost
- Network characteristics
- Passenger demands
- Alternate options

The components of fleet planning are the following:

Traffic forecast

- Economic and financial data
- Performance:
  - Aircraft
  - Aircraft/engine combination
  - Airport
  - Route analysis
- Environmental factors:
  - Noise
  - Smoke
  - Pollution
- Airline objective
- Schedule

In case an additional aircraft is incorporated into the fleet, expenses would be higher as far as spares, training, and new employees are concerned. Additional expenses could be expected if this aircraft is not the same type of airplane in the existing fleet. The dollars per flight cycle cost would influence maintenance cost. Short trips generate increased cycle costs.

All these items contribute to generate an operational fleet plan with a simultaneous processing of financial data. By reviewing the figures obtained, the airline can change and recycle the input data to finalize the planning for a profitable operation.

Generally, two fleet planning approaches are considered, namely, the macro approach and microanalysis (Fig. 13.1). This figure illustrates the relationship between microanalysis and the macro forecast.

13.3   Macro Fleet Planning (Systems Approach)

The macro technique is based on top-down or aggregate methods. Typically, it is performed at the system or subsystem level.

Note: A subsystem exists when an airline can divide its operations into regions, e.g., domestic and overseas operations.

The macro technique uses revenue-passenger-mile (RPM) data, not origin-destination-passenger data. The elements of a macro-level analysis are as follows:

- Demand based on system RPM

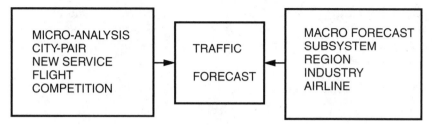

Figure 13.1 Traffic forecast.

- Trip distance
- Departures
- Estimated load factors
- Available seat-miles (ASM)
- Number and type of aircraft
- Cost elements

All these items are combined into the fleet selection process that results in a fleet planning strategy, financial results, and operating data. Application of this approach has the advantage of a quick evaluation. In the case of new schedules, updated traffic forecasts, different aircraft types, new cost data, and a revised airplane retirement plan, this procedure can result in a rapid update.

13.4   Micro Fleet Planning Approach

Micro-level analysis is a bottom-up approach. For the identical routes and different types of aircraft, an economic comparison is conducted. It examines the schedule and airplanes identified by their tail numbers, and it lists route assignments and requires traffic flow determination. For example, it examines the passenger or cargo origin-destination traffic. In relation to traffic, the following information should be supplied:

- Fares
- Realistic load factor limits
- Ground times
- Frequency and airplane capacity
- Reasonable connection of flights
- Preferred times and days of departures and arrivals

A typical information flow starts with schedule, traffic forecast, aircraft parameters, costs, and fares. By processing the data, figures relating to traffic allocation, itineraries, operating costs, and profit are generated. Reviewed by the airline, the results may be accepted or changed if necessary. This analysis has a wide application and is used in the following ways:

- Obtains financial figures
- Follows tail routing (refers to airplane tail number)
- Furnishes revenue and cost figures
- Prepares statistics for fleet planning
- Presents alternate fleet plans
- Reports segment traffic loads
- Presents airport activities for review
- Furnishes a schedule for evaluating its practicality

13.5   Aircraft Economic Comparison

Economic comparison could be used when a route-by-route detailed analysis is required. The elements of this economic evaluation are presented in the flowchart in Fig. 13.2.

Marketing requirement. This should contain passenger and cargo demands, apply a predetermined load factor and limitations of frequency, and on each route determine frequency for each aircraft.

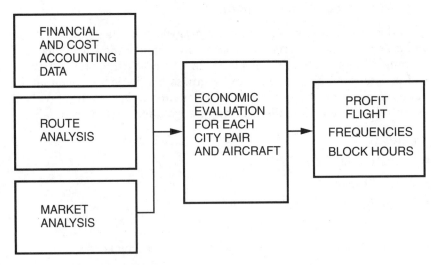

Figure 13.2 Economic evaluation.

Trip analysis. This should generate trip distances, payload, block fuel, and block and trip times.

Financial data. This should contain fares and rates for revenues. To obtain operating costs, interest, insurance, depreciation, station cost, flight frequency and block time, and fuel and other factors should be available.

Information obtained from this approach for each route and for each aircraft type would provide:

- Profit figures
- Optimal frequencies
- Block-hours
- Costs

The aircraft economic approach evaluates aircraft performance and financial capabilities, and it contributes to an efficient planning.

13.6   Fleet Planning: Case History

13.6.1   Introduction

This is an example of a study for Omega Airlines; it is prepared for aircraft capacity planning for a given set of cities and routes. This report is strictly for teaching purposes; therefore, the numbers from this sample report should be considered accordingly.

In this study, Omega Airlines' expansion of its fleet will be based on its forecasted market share. It will be shown how to approach this forecast and consider its fleet productivity.

This type of report always should contain an introduction with a historical background and an indication of the report's purpose. The discussion should describe the method of presentation and handle the exceptions. The conclusion should summarize the subject.

13.6.2   Fleet requirement and marketing analysis

The board of directors' decision to expand Omega Airlines' route network between the United States and Europe required additional aircraft to meet future cargo market potential. This decision would influence Omega Airlines' domestic cargo market in route structure as well as in actual tons of cargo generated. The North Atlantic cargo market meanwhile would be increasing at a relatively steady average annual growth rate.

It is the purpose of this study to forecast both Omega Airlines' domestic and North Atlantic cargo markets and to compare the forecasted potential with Omega Airlines' fleet productivity at various stages in the forecasting period (1990–1995). The forecasts for the North Atlantic and domestic markets were done separately, but they could be combined at the end of this study to obtain the total system outlook (only the domestic market will be shown here in more details).

It was assumed that Omega Airlines would maintain the same amount of passenger miles flown, roughly equivalent to four DC-8-73 aircraft. A constant allocation of four aircraft accommodates passenger transport.

13.6.3   Aircraft productivity analysis

It is necessary to know aircraft productivity. The capacity (tons), utilization (hours per day), block speed (mi/h), and time period (aircraft months) involved for each type of aircraft are well known.

The capacity of Omega Airlines' fleet was determined by the analysis described on the following pages. The composition of Omega Airlines' fleet at any given time period throughout the forecast was obtained from airline annual reports. The remaining factors were supplied by other sources. Productivity is then calculated by multiplying utilization, capacity, block speed, aircraft months, and a composite factor of 26 days per month. This is presented in Table 13.1.

As a first step, productivity (for 1993) has to be defined as follows:

$$\text{Productivity} = \text{utilization} \times \text{block speed} \times 100\% \text{ capacity} \times \text{days/months} \times \text{number of months}$$

$$= 13.0 \text{ hours/day} \times 460 \text{ mi/h} \times 53.1 \text{ tons} \times 26 \text{ days/month} \times 144 \text{ aircraft month}$$

$$= 1188.9 \text{ ton-miles, in millions}$$

Note: Considering 60 percent load would result in 713.3 ton-miles in millions.

The total productivity of the present fleet consisting of 12 airplanes at 60 percent load factor is 713,317,363 ton-miles. Four of the airplanes were dedicated to passenger service, with an equivalent of 238,070,966 ton-miles.

In Table 13.1, the effect of decreased service time is reflected in the decrease from 13 to 12.5 hours per day utilization of the DC-8-73 aircraft in the year 1995.

TABLE 13.1   Omega Fleet Productivity Analysis (Ton-Miles in Millions)

	1988	1989	1990	1991	1992	1993	1994	1995
Type of aircraft	ATM	ATM	ATM	ATM	ATM	ATM	ATM	ATM
DC-8-55[1]	425.1	215.8	0	0	0	0	0	0
DC-8-73[2]	107.3	883.4	1,188.9	1,188.9	1,188.9	1,188.9	1,188.9	1,143.1
DC-8-61[3]	125.4	31.3	0	0	0	0	0	0
Total								
100% LF	657.8	1,130.5	1,188.9	1,188.9	1,188.9	1,188.9	1,188.9	1,143.1
60%								
LF	394.7	678.3	713.3	713.3	713.3	713.3	713.3	685.9

Productivity (ATM) = utilization × block speed × available tons × composite factor
(=26 days/month) × aircraft months

	1988	1989	1990	1991	1992	1993	1994	1995
$^1U =$	13.0	13.0	—	—	—	—	—	—
Blk spd. =	450	450	—	—	—	—	—	—
Avail. Tons =	43	43	—	—	—	—	—	—
$^2U =$	13.0	13.0	13.0	13.0	13.0	13.0	13.0	12.5
Blk spd. =	460	460	460	460	460	460	460	460
Avail. Tons =	53.1	53.1	53.1	53.1	53.1	53.1	53.1	53.1
$^3U =$	11.0	11.0	—	—	—	—	—	—
Blk spd. =	450	450	—	—	—	—	—	—
Avail. Tons =	48.7	48.7	—	—	—	—	—	—

To see the relationship presented in the tables and figures, the year 1993 has been selected to explain the charts and tables and correlate the numbers. Based on Table 13.2, the following information is available for the selected year of 1993:

Domestic forecast in ton-miles:	112,469,405
International forecast in ton-miles:	850,000,000
Total forecast in ton-miles	962,469,405
Omega Airlines' total fleet productivity:	713,317,363
Omega Airlines' equivalent passenger ton-miles:	238,070,966
Omega Airlines' available cargo service ton-miles:	475,246,397
Total forecast in ton-miles:	962,469,405
Omega Airlines' cargo capacity:	475,246,397
Omega Airlines' cargo capacity required:	487,223,008

The DC-8-55 and the DC-8-61 aircraft will be phased out of service in 1990, either by lease arrangement or by selling them. A shortage of 487,223,008 ton-miles or an equivalent of nine additional airplanes would be necessary to satisfy demand for the year 1993.

Figures 13.3 and 13.4 present these data as a function of time and productivity.

Omega Airlines' domestic and international market forecast is presented in Table 13.2.

Table 13.3 presents the aquisition schedule of additional aircraft.

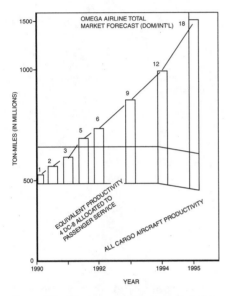

Figure 13.3   Fleet requirements and productivity analysis.

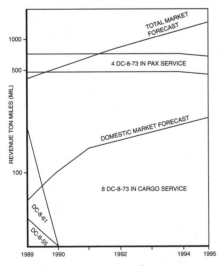

Figure 13.4   Revenue ton-miles versus years.

TABLE 13.2 Omega Airlines' Productivity Forecast (Ton-Miles)

	1989	1990	1991	1992	1993	1994	1995
Forecast							
Domestic	28,378,844	50,504,593	79,859,988	94,763,718	112,469,405	133,615,560	158,909,793
International	321,000,000	466,000,000	578,000,000	699,000,000	850,000,000	1,057,000,000	1,336,000,000
Total	349,378,644	516,504,593	657,859,988	793,763,718	962,469,405	1,190,615,560	1,494,909,793
Fleet productivity at 60%							
LF	678,335,440	713,317,363	713,317,363	713,317,363	713,317,363	713,317,363	685,882,080
Equivalent ton-miles per year for 4 aircraft allocated to passenger service	238,070,966	238,070,966	238,070,966	238,070,966	238,070,966	238,070,966	238,070,966
Ton-mile capacity available for cargo service	440,264,474	475,246,397	475,246,397	475,246,397	475,246,397	475,246,397	447,811,114
Additional ton-miles required	0	41,258,196	182,613,591	318,517,321	487,223,008	715,369,163	1,047,098,679
Number of additional (actual) DC-8-73 (whole aircrafts)	—	0.7	3.07	5.35	8.19	12.02	17.59
(At 59,517,744 ton-miles/yr)	—	1	3	6	9	12	18

$4 \times 59,517,744$ ton-miles/yr $= 238,070,966$ ton-miles

Utilization 13.0 hours/day $\times$ Block speed 460 miles/hour $\times$ 60% capacity 31.9 tons $\times$ 26 days/month $\times$ 12 aircraft months

TABLE 13.3   Acquisition Schedule of Aircraft

Year	Service passenger present	Service air freight present	Service air freight additional	Total fleet
1990	4	8	1	13
1991	4	8	3	15
1992	4	8	6	18
1993	4	8	9	21
1994	4	8	12	24
1995	4	8	18	30

13.6.4   Domestic market forecast

This forecast consists of five elements:

1. Historical trends
2. Economical indicators
3. Origin and destination forecast
4. Tons forecast
5. Ton-mile forecast

In preparing any forecast, the effect of uncertainty on the outcome increases as the forecast period is further projected into the future despite the fact that short-term variables may be averaged out in the process. The air freight industry in particular is extremely susceptible to fluctuating conditions in the economy, thereby making reasonable forecasts difficult to formulate. It is because of the close relationship between cyclic variations in the economy and variations in the development of air freight that the approach used in this study places emphasis on these economic indicators that show high correlation with air freight. The study involves certain selected city-pairs.

13.6.5   Economic indicators

A downturn might be explained by examining general economic indicators in the U.S. economy. There were significant decreases in certain economic indicators correlating with total freight movement that showed a market downturn in the late 1980s:

- Industrial production
- Manufacturers' inventories
- Retailers' inventories
- Corporate profits (before and after taxes)

- Business expansion into plants/equipment
- Average monthly earnings

Notable among these indicators was the decrease in retail inventories and the corresponding increase in manufacturing inventories. This would indicate that goods were held up at their point of manufacture, manifesting itself as a general decrease in air cargo. In addition, GNP experienced an actual downturn around that time.

Another useful economic indicator, the principal diffusion index, was examined to obtain an insight into the cyclic nature of the economy. The index is composed of about 20 business indicators classified and summarized into industrial and nonindustrial subgroups. A few are mentioned here:

Industrial indicators:
- Computer and related industries
- Communications
- Building industries
- Textile, paper, and iron industries
- Production
- Unemployment rate
- New orders of durable goods

Nonindustrial indicators:
- Stock market indicator
- NASDAQ
- Personal income
- Number of bankruptcies
- Manufacturers' inventories

The fluctuations in these indices or indicators correspond to the expansions or contractions of the economy, and these changes have an influence on the cargo market.

13.6.6 Growth pattern for selected cities

Six cities (Chicago, San Francisco, Los Angeles, Detroit, Philadelphia, and Baltimore) were examined for the domestic market. Certain growth patterns in each city's outbound tons emerged, as seen in Fig. 13.5. Los Angeles, San Francisco, Detroit, and Philadelphia witnessed a high initial growth but slower growth later. Economic factors for each city were analyzed.

The annual outbound air freight growth rate for these years was selected as the representative base growth period for forecasting purposes. Baltimore is the exception due to the fact that a major domestic airlines applied hub concept to Baltimore. Thus the result-

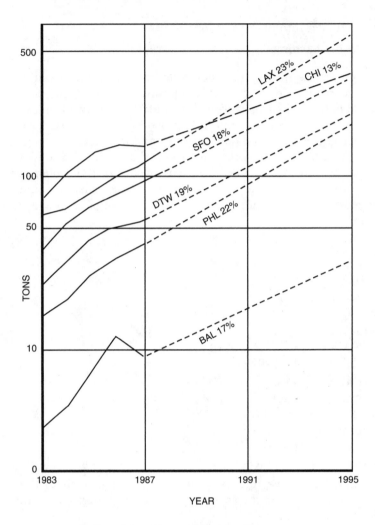

Figure 13.5 Outbound tons forecast for a few cities.

ing service through a nearby larger terminal effectively reduced Baltimore's outbound tons considerably. From 1983 to 1985, all cities had a good air freight growth with the exception of Baltimore. The year 1983 was selected as the beginning of the historical series because prior years involved a certain fluctuation.

Air freight growth during 1983–1986 had a high rate of increase. In order to be realistic, this high rate is adjusted and lowered to obtain a fairly good forecast.

Table 13.4 lists the adjusted and unadjusted values. For Baltimore, the unadjusted and adjusted rates are shown in Table 13.5.

TABLE 13.4   Growth Rate for a Few Selected Cities

Cities	Unadjusted 1985–1986 growth rate, %	Adjusted for forecasting rate, %
Los Angeles	23	23
Philadelphia	23	22
Detroit	19	19
San Francisco	18	18
Baltimore	56	17
Boston	14	14
Chicago	14	13

TABLE 13.5   Baltimore: Growth Rate

	1985–1986 Growth rate, %	1986–1987 Growth rate, %	Average growth rate, %
Baltimore	56	−23	(56 − 23)/2 = 17

Table 13.6 contains the historical outbound tons and their growth rates for the period from 1983 to 1987.

For each city, a growth rate was established. Taking the 1987 tons and their growth rates for 1988, a forecasted tonnage was established. A similar forecast was made for each year up to 1995. For example, Los Angeles' growth rate of 23 percent was applied to its 1987 tons of 114,239, and for 1988, 140,514 tons were calculated. For 1993, the forecast showed 395,590 tons (Table 13.7).

13.6.7   City-pair market forecast

City-pair market distribution of tons originated were calculated from the airline's source material and represent the best judgment. Omega Airlines' fair share of a given city-pair market cannot be fully realized until a few years after entry. This phenomenon has been identified as the *initial growth factor,* and a formula for its calculation was developed from historical data. Normally, during the first year of operation, an airline can realize 50 percent of its fair market share, 75 percent the second year, and a full fair market share the third year. Under these circumstances, total output will increase, but beyond some point, the resulting output increases will become smaller and smaller (diminishing returns). Omega Airlines' full market shares (1991–1995) of selected city-pairs were calculated and are listed in Table 13.8.

In 1993 for LAX, 395,589 tons originated outbound (Table 13.9). From this, only 12.6 percent (49,844 tons) terminated in Chicago.

TABLE 13.6 Outboard Scheduled Air Freight History

Community	1983	Percent change	1984	Percent change	1985	Percent change	1986	Percent change	1987
Chicago	69,316	47.0	101,895	28.0	130,398	13.6	148,144	(4.1)	142,096
San Francisco	36,331	39.8	50,773	30.6	66,310	18.3	78,436	13.3	88,890
Los Angeles	58,848	10.8	65,194	21.5	79,191	23.2	97,608	17.0	114,239
Detroit	22,493	39.1	31,288	37.8	43,100	18.8	51,210	5.2	53,861
Philadelphia	15,075	27.1	19,161	40.0	26,911	23.3	33,176	16.9	38,782
Boston	20,057	24.7	25,011	30.8	32,727	13.5	37,149	(2.5)	36,207
Baltimore	3,472	33.2	4,624	65.2	7,639	55.5	11,877	(23.1)	9,131
Total	225,592	32.1	297,946	29.6	386,276	18.5	457,600	5.6	483,206
Average	32,227	32.1	42,564	29.6	55,182	18.5	65,371	5.6	69,029

TABLE 13.7 Scheduled Air Freight Outbound Forecast (Tons)

Community	Annual forecasted growth rate	1988	1989	1990	1991	1992	1993	1994	1995
Chicago	13%	160,568	181,442	205,030	231,684	261,803	295,837	334,296	377,755
San Francisco	18%	104,890	123,770	146,049	172,338	203,359	239,963	283,157	334,125
Los Angeles	23%	140,514	172,832	212,584	261,478	321,618	395,590	486,576	598,488
Detroit	19%	64,094	76,272	90,764	108,010	128,532	152,953	182,014	216,597
Philadelphia	22%	47,314	57,723	70,422	85,915	104,816	127,876	156,008	190,330
Boston	14%	41,276	47,055	53,642	61,152	69,713	79,473	90,599	103,283
Baltimore	17%	10,683	12,499	14,624	17,110	20,019	23,422	27,404	32,063
Total		569,339	671,593	793,115	937,687	1,109,860	1,315,114	1,560,054	1,852,641
Percent change		18.0	18.1	18.2	18.4	18.5	18.6	18.8	
Average		81,334	95,942	113,302	133,955	158,551	187,873	222,865	264,663

Table 13.8 indicates that Omega Airlines' market share for LAX-ORD (Chicago) would be 14.3 percent, or 7128 tons per year. This multiplied by the trip length (LAX-CHI) of 1742 statute miles would result in 12,416,976 ton-miles (Table 13.10). Using the same calculations for other city-pairs for the year of 1993, a total of 112,469,405 ton-miles is predicted. This is the same amount predicted in Table 13.2. An additional nine aircraft were required in 1993.
. The purpose of this exercise was to show the methodology of market share development and productivity requirements. Exceptions are also considered in this case.

13.6.8 Conclusion

There are many approaches to forecasting cargo and/or passenger potentials. This was one way to approach this task.

The sophistication of these types of reports is based mostly on available time, personnel, and database and financial resources.

TABLE 13.8   Omega Airlines' Penetration into Selected City-Pairs (1991–1995)

City-pairs	Share, %	City-pairs	Share, %
LAX-PHL	14.3	SFO-BAL	25.0
PHL-LAX	14.3	BAL-SFO	20.0
SFO-PHL	16.7	LAX-CHI	14.3
PHL-SFO	14.3	CHI-LAX	14.3
LAX-BOS	14.3	SFO-CHI	16.7
BOS-LAX	16.7	CHI-SFO	16.7
SFO-BOS	16.7	LAX-DTW	16.7
BOS-SFO	16.7	DTW-LAX	16.7
LAX-BAL	20.0	SFO-DTW	20.0
BAL-LAX	25.0	DTW-SFO	16.7

TABLE 13.9 Omega Airlines' Destination Forecast

LAX-CHI	1987	1988	1989	1990	1991	1992	1993	1994	1995
Industry tons origination (LAX)	114,239	140,514	172,832	212,583	261,477	321,617	395,589	486,574	598,486
Percent terminating in Chicago	12.6	12.6	12.6	12.6	12.6	12.6	12.6	12.6	12.6
LAX-CHI tons	17,705	—	21,777	26,785	32,946	40,524	49,844	61,308	75,409
Market share (%)		—	7.2	10.7	14.3	14.3	14.3	14.3	14.3
Annual tons		1,568	2,866	4,711	5,795	7,128	8,767	10,783	
Average weekly tons		30.2	55.1	90.6	111.4	137.1	168.5	207.4	

TABLE 13.10 Omega Airlines' Share of Domestic Ton-Mile Market Forecast

	1989	1990	1991	1992	1993	1994	1995
SFO-CHI	2,707,367	4,812,073	7,585,788	8,949,608	10,562,233	12,462,366	14,705,297
CHI-SFO	2,276,105	3,873,986	5,847,839	6,607,155	7,465,993	8,437,254	9,533,839
LAX-CHI	2,731,456	4,992,572	8,206,562	10,094,890	12,416,976	15,272,114	18,783,986
CHI-LAX	3,140,826	5,273,034	7,964,424	8,999,172	10,169,796	11,491,974	12,984,868
SFO-DTW	1,207,650	2,137,250	3,361,500	3,967,400	4,681,200	5,523,650	6,517,575
DTW-SFO	2,587,525	4,637,625	7,320,600	8,775,175	10,441,400	12,425,100	14,786,450
LAX-DTW	1,076,375	1,994,750	3,276,525	4,030,975	4,957,250	6,098,800	7,501,050
DTW-LAX	2,012,525	3,608,325	5,735,400	6,825,600	8,121,200	9,665,650	11,502,400
LAX-PHL	1,162,545	2,126,139	3,494,826	4,300,218	5,287,782	6,505,458	8,001,186
PHL-LAX	997,152	1,807,338	2,945,913	3,593,103	4,384,113	5,347,707	6,524,634
SFO-PHL	956,080	1,700,816	2,679,540	3,160,096	3,731,228	4,403,000	5,195,540
PHL-SFO	1,159,876	2,103,376	3,431,824	4,186,624	5,107,480	6,229,616	7,633,544
LAX-BOS	1,329,060	2,431,398	3,994,998	4,914,916	6,043,314	7,434,918	9,144,454
BOS-LAX	1,362,938	2,340,188	3,565,008	4,065,360	4,633,468	5,282,362	6,022,466
SFO-BOS	971,640	1,724,661	2,717,893	3,209,111	3,786,697	4,466,845	5,271,147
EOS-SFO	701,740	1,376,490	2,094,424	2,388,615	2,723,291	3,103,850	3,538,389
LAX-BAL	481,068	890,092	1,459,472	1,794,128	2,205,476	2,714,432	3,337,264
BAL-LAX	639,100	1,124,816	1,749,972	2,047,444	2,393,720	2,802,744	3,279,164
SFO-BAL	532,084	941,568	1,478,556	1,745,824	2,059,680	2,429,932	2,868,840
BAL-SFO	345,732	608,096	948,924	1,108,304	1,297,108	1,517,788	1,777,700
Total	28,378,844	50,504,593	79,859,968	94,763,718	112,469,405	133,615,560	158,909,793

Scheduled Maintenance Program

14.1 Introduction

In 1968, a maintenance program was developed for the B747 aircraft by the airlines and the manufacturer, and this became the *Handbook of MSG-1* (Maintenance Steering Group). In 1979 it was updated to MSG-2 by an ATA task force, and in 1993 ATA and the airlines developed the MSG-3 revision.

The last revision incorporated damage tolerance and considered fatigue, corrosion, aging airplane problems, multiple failures, etc. A task-oriented logical flowchart was developed. The tasks in the scheduled maintenance program consisted of lubrication, servicing, operational and visual checks, inspection, functional checks, restoration, and discard.

The logic developed considered the item's function, its functional failure, the failure effect, and the failure cause. The logic asks questions, and when such questions are answered with yes or no, it goes on to the next task until all subsequent questions are asked and answered. For each task, the criteria are applicability and safety and operational and economical effectiveness.

Maintenance program documents are approved by the maintenance review board and accepted by the Federal Aviation Administration's (FAA's) Flight Standard Service. A maintenance program includes aircraft structural maintenance program development. Without going into detailed description, certain repeat inspection intervals are established, and the tasks are described in a logical fashion.

Scheduled maintenance programs provide operators with flexibility because after the minimal in-service operating time, inspection frequency changes may be requested and submitted to regulatory agencies. In this fashion, operators may schedule their maintenance intervals more efficiently.

There are numerous ways to schedule maintenance checks. They are divided into categories with the lightest listed first and the heaviest, last. For each phase of maintenance, identifying letters are designated, such as A check, B check, and so on.

14.2 General Description of Various Maintenance Checks

1. *Service check.*   General visual inspection of aircraft interior and exterior for condition. This includes engines (cowling closed), tires and brakes, fuel tanks, and some specific items depending on fleet type.

2. *A check.*   This includes all items on service checks and more checkout of systems. Some fleets require engine inspection.

3. *B check.*   This includes service and A check items and additional system checks, some lubricating of components, detailed inspection of certain areas, and some minor opening.

4. *C check.*   This is a major structural inspection of the airframe. It includes lots of opening of access panels and lubrication and servicing of systems. It also includes all structural requirements. Depending on fleet, additional replacements of cables, etc. are called out.

5. *D check.*   This is a major structural inspection of the airframe that requires the almost complete opening of the airframe, interior and exterior.

Operators sometimes eliminate C checks by splitting them into B and D checks. Some operators have introduced trip checks, K checks, etc.

The FAA Advisory Circular (AC) 121-1A classifies them as "A, B, C, and D. But some operators use their own terminology. Table 14.1 contains these checks and the average frequency of inspection for the well-known B727 airplane. Each airline has its own schedule of inspection, and with time and experience, the service frequencies are extended.

The checks can be split, and personnel hours are varied for each airline. Another consideration is the nonroutine-to-routine work relationship. This ratio could vary from 1.5 to 2.5 depending on factors such as the age of airplane, the geographic area where the airplane is operating, the climate, and the quality of maintenance work. The aging of aircraft resulted in introduction of programs that helped to extend the useful life of such "geriatric" airplanes.

TABLE 14.1   Inspections and
Frequencies

Inspection	Average frequency
A	155 hours
B	500–700 hours
C	3000 hours
D	20,000 hours

The repair assessment program evaluates fuselage pressure shell repairs and considers damage tolerance. The service action requirement is covered by service bulletins. Some of these requirements were included on aircraft on the production lines, and they concentrate on structural modifications.

Corrosion prevention and control programs are covered in airworthiness directives. Such programs require inspection at certain intervals. The inspections are calendar-driven and describe repairs and procedures.

Service action requirements were initiated by airworthiness directive. They require the inspection of structures that can experience stress that could cause them to fail. The approach applies statistical methods and sampling programs. *Fail-safe life* is defined by the number of cycles the structure can accumulate before failing. The first period of fail-safe life is called the *safe life*. Beyond this, service life is called damage tolerance.

Certain components after a certain amount of time in service are inspected and could be discarded, repaired, or reconditioned and reintroduced into a spare's pool. A few definitions are listed below.

14.3   Hard Time Limit

This is a maximum interval for accomplishing maintenance tasks. The intervals usually apply to overhauls as well as the total life of parts or units.

14.4   Time Between Overhauls (TBO)

This is the maximum time permitted for an item between overhauls.

14.5   Condition Monitoring (CM)

This is the maintenance process for items that have had neither a hard time nor are on condition maintenance as their primary maintenance process. Condition monitoring is accomplished by appropriate means available to an operator for finding and solving problem areas. These may range from notices of unusual problems to special analysis of unit performance. No specific monitoring system is implied for any given unit.

14.6   On Condition (OC) Maintenance

Items and appliances listed as "on condition" must be restricted to components on which determination of continued air worthiness may be made by visual inspection, measurements, tests, or other means without teardown inspection or overhaul. These OC checks are to be performed within the time limitations prescribed for the inspection or check. Performance tolerances and wear or deterioration limits shall be contained in the air carrier's maintenance manual. If any item or appliance cannot be maintained in a condition of air worthiness by use of OC procedures, it must be placed on specific time-limitation control or be controlled by an acceptable standard for determining time limitations, e.g., hard time or condition monitoring.

14.7   Structurally Significant (SS) Items

These are local areas or primary structures that are judged by the manufacturer to be relatively the most important from a fatigue, corrosion vulnerability, or failure-effect standpoint. Boeing introduced the BITE (built-in test equipment) to improve the accuracy of fault isolation and reduce component removals caused by a less accurate maintenance. The introduction of the OC and CM programs enabled operators to reduce cost and optimize utilization of equipment, thus ensuring higher profits. Here are the advantages of these procedures:

- Elimination of unjustified removal of working components to affirm condition
- Reduction of component mortality rates by avoiding intrusion except when an item has malfunctioned
- Reduction of spares required
- Elimination of contamination that is carried into good systems by opening such systems for removal of good equipment
- Reduction of shop maintenance costs by virtually eliminating time-controlled component overhaul

14.8   Engine Maintenance

For engine part replacements, frequencies are established. Table 14.2 shows engine maintenance unit replacement frequency for a large size commercial jet engine. Any item exposed to high temperature is assigned a low flight time limit. The parts time limits are shown for 4000, 8000, and 12,000 hours.

Next (Table 14.3) is an example of an airline maintenance schedule for airframes and engines. In order to distribute more evenly the lengthy D checks, they are divided and partially combined with C checks.

14.9   Airplane Checks (Example)

A check after every 25 flight hours

B check after every 300 flight hours

C check after every 3600 flight hours

Landing gear after every 25,000 flight hours

D check built into C checks

Note that the quoted flight hours vary by airline and by aircraft.

14.10   Engine Checks (Example)

During A check

During B check

Borescope after every 900 flight hours

X-rays after every 4500 flight hours

Turbine parts after every 10,000 flight hours

Compressor parts after every 12,000 flight hours

Burner after each 20,000 flight hours

TABLE 14.2   Replacement Frequency

Replacement	4000 h	8000 h	12,000 h
Fan rotor			X
Fan stator			X
Compressor			X
Combustor	X	X	X
H.P. turbine stator	X	X	X
H.P. turbine rotor	X	X	X
L.P. turbine stator		X	X
L.P. turbine rotor			X
Gearbox			X
Control and accessories			X
Service bulletins	X	X	X
Rev. spoiler system			X

TABLE 14.3 A Typical Aircraft and Engine Maintenance Schedule

					Engine			Engine	Engine	Engine	Engine	Landing
A	A	A	B	B	Borescope	C	X-ray	Turbine	Compress	Burner	Gears
25	50	75	300	600	900	3600	4500	10,000	12,000	20,000	25,000

Note: The A, B, and C checks are shown only at the beginning of this time line for easier legibility.

Note that flight hours may vary from airline to airline.

The intervals between various maintenance checks were improved due to increased reliability of the components and systems. This is demonstrated in Fig. 14.1, prepared for the first 14 years of operation for the B707 airplane. New operators for this aircraft derived an immediate benefit because they did not have to start at the bottom of the scale. During the subsequent years, maintenance parts and system reliability improved significantly, reducing the cost of maintenance.

A good maintenance program is based on reliability, simplicity, and accessibility. These three elements are the basis for an economically sound maintenance program.

Another measure—to be touched on only very briefly—is the component unscheduled removal rate per 1000 component hours. Figure 14.2 shows a typical component history. In the beginning, the removal rate is relatively high, but as the system matures, the removal rate improves in an asymptotic way.

Similarly, there is an improvement in the service bulletin outputs, where, on reaching about 3 years of operations, the rate diminishes significantly (Fig. 14.3).

14.11   Maintenance Location

Maintenance is performed on the maintenance base, which has the capability to overhaul and conduct all the major functions described in

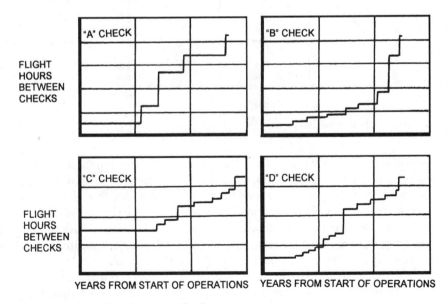

Figure 14.1   Aircraft maintenance checks.

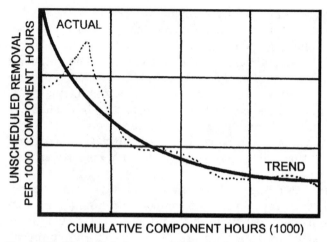

Figure 14.2 Unscheduled component removal versus component hours.

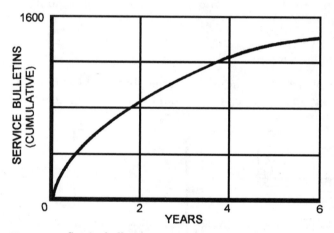

Figure 14.3 Service bulletin output.

the maintenance program. It has the personnel, spare parts, tools, and reserves to satisfy maintenance needs. Some airlines assign certain maintenance functions to outside contractors in order to reduce their maintenance base. Others, mostly smaller air carriers, completely farm out their entire maintenance program. In practice, this means that they sign up with larger airlines and adopt their maintenance program. Each airline originally establishes its maintenance program, and it is approved by the FAA. This program then becomes mandatory to its originator. Major stations have their own maintenance base, personnel, spare parts, tools, and facilities, so they are able to conduct line maintenance. Service stations are on a smaller scale. In some other stations, no maintenance is performed at all.

For the major stations and service stations, the same practices are applicable as mentioned before such that partial or complete maintenance jobs can be performed by outside contractors. The decision to use outside contractors or do the maintenance in-house is based strictly on economic factors.

14.12   Inventory Management

In short, an efficient inventory management system can save the airline industry millions of dollars. The trend in inventory management follows such concepts as

- Just-in-time delivery (specifies when and where an item should arrive)
- Pooling to reduce inventory and conserve capital
- Improved cycle time
- Leased inventory
- Elimination of nonproductive inventory

In the recent past, the airline industry had an inventory composed of high-value rotables of about $23 billion and repairables and expandables of about $27 billion, to make a total of $50 billion. About 4000 spare engines were included in this figure. Airline operators owned about 80 percent of the total spare parts inventories. The reason for keeping such a large amount of inventory on hand was the fear of having an expensive aircraft-on-the-ground situation, and for this reason, airlines were slow to introduce new inventory management techniques common to other industries. Today, the picture is completely different. Airlines constantly review their spare parts inventory and take steps to reduce their volume and expenses.

One step is to eliminate the obsolete and excess parts and scrutinize new purchases.

In the past, the cycle of ordering and delivering parts took about 30 days. Most of this cycle time was spent on processing the orders. In this field, there have been significant improvements, such as Boeing's 24-hour turnaround time for more than 400,000 standard parts. With quicker response times, inventory volume can be reduced and less money spent.

Pooling is another way of saving money. Many airlines are members of the Airline Technical Pool. Larger airlines are opening their warehouses to smaller operators that lack the funds, time, and resources to have their own inventories. Some airlines choose certain suppliers who come on-site and manage inventory for them. Furthermore, suppliers establish partnerships with each other to serve better their airline customers and share costs.

Effective inventory management is important for efficient and timely operation and cost savings.

Chapter
15

Airline Operational and Economic Comparison

15.1 Introduction

Airlines are using several indices to measure productivity and efficiency in their operations. It is a complex task because each airline has different types of airplanes in its fleet, two or three flight crews per airplane, short- and/or long-range trips, different philosophies in their operations, and so on. Presented below are a few of these indices.

15.2 Flight Operations

One measure is block-hours versus number of flight crew members. This could be subdivided by cockpit crew only, cockpit crew and cabin crew, and cabin crew only. Instead of block-hours, flight-hours could be used. In this case, no distortion is caused by the long taxi times resulting from congestion at certain airports and at certain times. For the number of flight crew per airplane, a comparison can be developed, and another can be done for cabin crew per airplane. Another comparison could show the number of passengers versus flight crew and/or cabin crew (distortion could occur if an airplane has two or more flight crews). Available passenger-miles or ton-miles could be compared with the number of flight crew and/or cabin crew. Cabin crew should be related to revenue-passenger-mile (to be more realistic). Another index is the flight crew compensation in cents per available passenger-mile or ton-mile. In case of cabin crew, cents per revenue-passenger-mile should be used for comparison. Another comparison could be trip fuel per number of passengers for a given trip or for the whole system. Revenue can be compared and/or divided by the number of flights or flight-hours and/or cabin crew or the number per employee in the Flight Department. The same could be applied to the relationship between profit (before and/or after taxes) and number of flight crew, number of airplane types, and so on. Flight operations cost could be presented in percentage of the direct or total operating cost.

15.3 Maintenance and Burden

The Maintenance Department can issue similar indices and compare them with the figures of other airlines. Distortions should be corrected when outsourcing is being applied in maintenance operations. One measure is the number of maintenance personnel per airplane. Similarly, other measures include materials used, expressed in dollars per airplane, total maintenance budget per airplane, and maintenance personnel remuneration per airplane. Another basis for comparison is the total maintenance budget as a percentage of direct or total operating cost. Other comparisons are spare parts values per airplanes, per personnel, amount of maintenance cost, and per maintenance burden. Another frequently used relationship is the direct maintenance cost related to the maintenance burden.

15.4 Depreciation and Amortization

Depreciation and/or amortization can be shown per aircraft or as a percentage of direct and total operating cost.

15.5 Relationship Between Various Direct Operating Cost (DOC) Factors

Another way to compare various indices is the relative direct operating cost (DOC) factor comparison. Airlines compare their DOC elements with the DOC expressed in percentage. Table 15.1 presents the average industry percentage figures.

The figures in this table are applicable for the present economic situation. Their relationship changes as a result of tax laws, fuel price changes, interest-rate changes, and so on (Fig. 15.1).

For example, a few years ago, airplane depreciation was changed from 15 to 20 years. This changed the ownership percentage from 44 to 46 percent to the present 35 to 39 percent. In essence, this is a good way to compare the airlines' cost-element relationship. International airlines' cost distribution is slightly different from domestic distribution due to different laws and some different expenses (navigation fees).

TABLE 15.1   Direct Operating Cost
Elements in Percent

Flight and cabin crew	24–27%
Maintenance	18–23%
Fuel cost	14–17%
Ownership	35–39%
Miscellaneous	2–4%

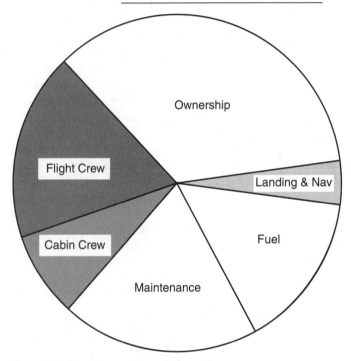

Figure 15.1   Direct operating cost elements.

15.6   Indirect Operating Cost

Assessment of indirect operating cost is a difficult task because each airline's operation and cost assignment are not necessarily identical. Operation cost for passenger service, including sales and ticketing, aircraft servicing, and general administration costs, can be compared. The number of scheduled passengers per ticket and sales personnel could be presented, and/or the renumeration of ticket and sales personnel versus the number of passengers could be developed. Station costs could be compared with the number of transits per month, employees at the station, passengers processed, fixed costs, variable costs, revenue, sales costs, and so on. Furthermore, airplane handling and servicing costs can be evaluated and various indices developed for comparison purposes.

15.7 Miscellaneous Comparisons

Passenger and/or cargo load factors can be compared with those of various other airlines. Generally, airlines flying longer trips have higher load factors than those flying shorter trips. Another measure is the daily utilization of the aircraft. Here again, the longer the average trip is, the better the daily utilization becomes. Another comparison is the relationship between revenue-passenger-mile divided by revenue or by airplane or number of employees. Another comparison is the relationship between revenue-passenger-miles divided by revenue-ton-miles and the percentage of total revenue versus freight revenue. Revenue is compared and indices are established per airplane, passengers, and pounds of cargo and per total number of employees.

Yield in cents per revenue-passenger-miles and/or ton-miles and operating expenses in cents per available passengers and/or ton-miles could be compared with the performance of other airlines.

To illustrate the productivity of capacity per employee, Fig. 15.2 shows the number of employees and available seat-mile relationship for a few major airlines. The highest productivity was reached by Delta Air Lines in 1996. These data were taken from DOT Form 41.

Furthermore, total employee productivity expressed in revenue-passenger-miles or ton-miles can be compared with the figures of other airlines on a yearly basis or, in the same fashion, employee compensation per available passengers or ton-miles.

Developing and comparing these figures with those of other airlines will furnish a good picture of the airline's operation. Some indices with higher figures than the industry average may indicate an expensive or less efficient operation. By scrutinizing these higher figures, there is a chance to discover some inefficiency in the operation related to that particular index or realize that the conditions of comparison were not identical.

Total labor cost per employee, passenger-miles, or ton-miles and airplane could be compared with the figures of other airlines. In order to achieve a fair comparison, the possibility of outsourcing also should be considered. An additional decisive factor is the number of employees per airplane and passengers or tons carried. For certain items, the yearly trend is important, for example, the yearly trend of employee compensation related to that of other airlines.

An airline management's job is to find out the causes of large discrepancies between identical indices. The causes may be different kinds of operations, less motivated employees, less training, poorly organized

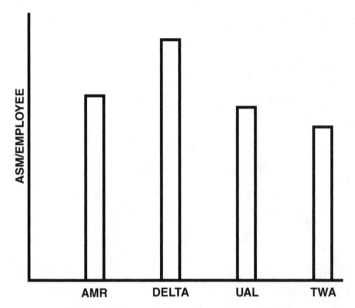

Figure 15.2 Productivity per employee for various airlines.

work process, jobs being farmed out, etc. Once differences are explained, the necessary corrective steps should be taken to improve productivity.

Airlines lease a certain percentage of their fleets. Table 15.2 lists the percentages of leased aircraft related to airline fleets that are carried off the balance sheet. This information is valid for a given time period and therefore should be dated.

Other comparisons could be developed. A few are listed below:

- Operating revenues
- Operating costs
- Labor
- Materials
- Fuel
- Depreciation

Per employee, seats, or any other variable, these comparisons would be good indices for evaluation of the economic measurement of operations.

An interesting comparison is given in Table 15.3, which shows the average daily trading volume (ADTV). The year in this case is 1998.

Stock analysts are interested in this presentation. Various indices could be developed based on specific indices, such as cents per avail-

TABLE 15.2   Percent of Leased Airplanes
in the Fleet of a Few Selected Airlines

Airline	Percent leased
American	27
Delta	30
UAL	42
British Airways	29
KLM	22

TABLE 15.3   Average Daily Trading Volume
For a Few Selected Airlines

Airline	ADTV
American	1600
Delta	1300
United	750
Southwest	2300

Note: ADTV is presented in multiples of 1000.

able passenger-mile or ton-mile or revenue cents per passenger-mile or ton-mile, etc. However, this subject is beyond the scope of this book. Figure 15.3 indicates for a given time period the airline industry unit revenue and unit cost figures. There is a steady growth from 1988 for the next 11 years. The chart shows the years when there were no profitable operations. All in all, the industry is profitable, especially in the last few years.

Figure 15.4 shows the relationship between industrial production and domestic passenger revenue.

Theoretically, many more meaningful indices could be developed for comparison purposes. Each airline could and should select only those indices which provide it with a good comparison that could improve its operations.

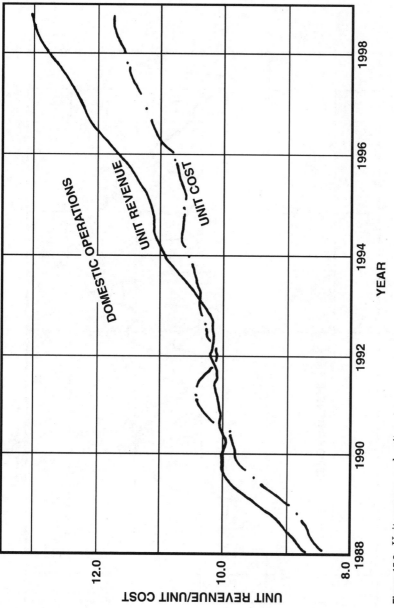

Figure 15.3 Unit revenue and unit cost versus years.

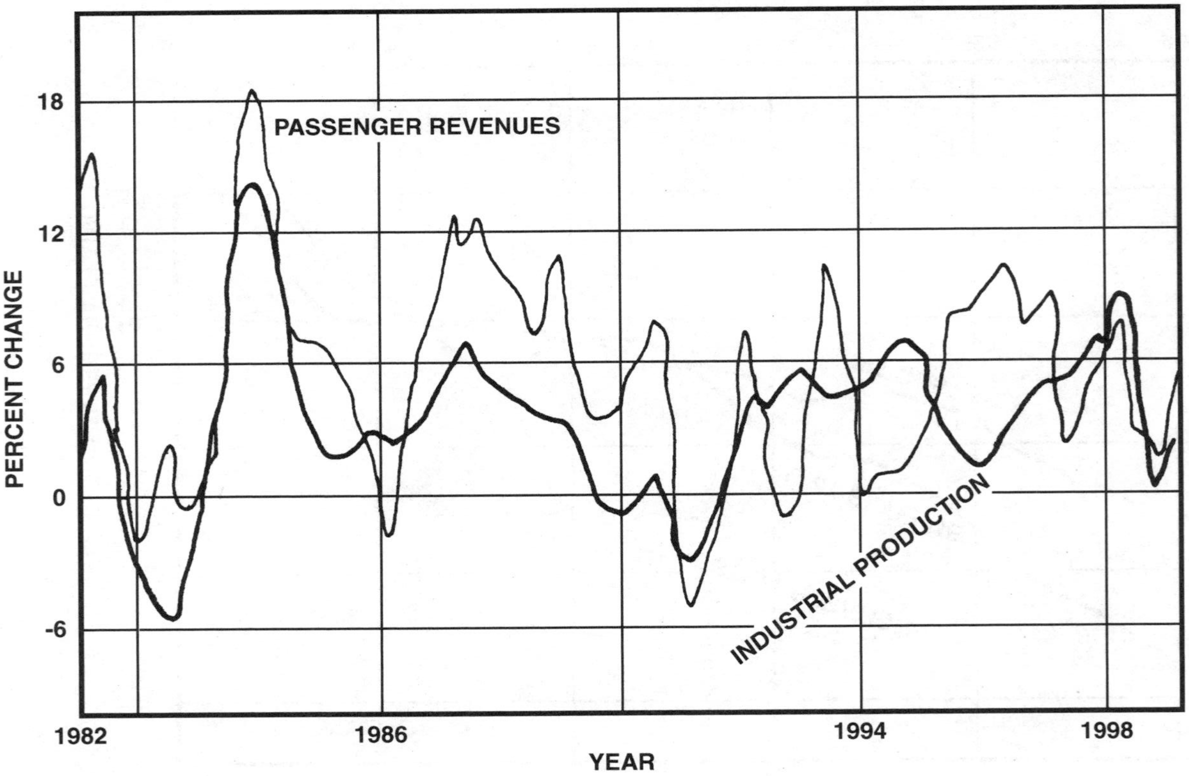

Figure 15.4 Industrial production versus airline passenger revenue.

16

The Environment and Airline Operations

16.1 Introduction

The development of air transportation has had an impact on environment, just as the environment has had an effect on air transportation. With the introduction of environmental regulations, the economics of the airline business changed. The airline industry acted in a responsible way in response to the environmental requirements and has taken steps to conform to regulations. This chapter discusses the problem areas and expenses involved in meeting environmental requirements (e.g., noise, smoke) and operational problems (e.g., bird strikes, runway contamination, wind shear, and volcanic ash).

16.2 Noise Impact and the Environment

The introduction of larger jet engines had a significant impact on environmental, economic, social, and technical issues. The environmental effect encompasses such factors as noise, air quality, ecology, and wild life preservation (e.g., at JFK).

Aviation also has economic effects on surrounding communities. Improved communications between the aviation industry and communities is imperative.

Technical impacts include the satisfactory design of airports and their environments, runway design, terminal design, and so on.

The outstanding issue between the aviation industry and communities is the noise generated by the operation of jet aircraft. Aviation has had a far-reaching effect on improving our lives and strengthening the local and national economy. Since noise is a constant factor that reminds people, practically daily, of an airport's existence, corrective actions have been taken to alleviate the noise problem.

Several lawsuits have been filed, and many communities have tried to limit airport operations or expansion or have attempted to prevent building of new airports. Naturally, all these events affect many people and sometimes cause severe economic hardships. About half a million people are exposed to extreme noise levels, and because of the close proximity of their homes to the runways, the value of their properties have decreased by about $6 billion. In addition, about 14 times more people are subjected to a normally unacceptable noise exposure. Because of the high level of aircraft noise, many properties lose most of their value, and these losses may be as high as $50 billion to $60 billion.

16.3   Definition of Noise

Before going further into noise-related problems, let us define what noise really is. *Noise* is a pressure wave with frequency and amplitude. The audible range for the human ear is from 20 to 20,000 hertz (Hz, or cycles per second). The range of amplitude is between 1 and 10^8. Pressure is defined as Newtons per square meter. Related to reference pressure, the dimensionless ratio of pressure is known as a *decibel* (dB). To have a general idea of noise level, Fig. 16.1 shows various noise sources and their decibel values.

16.4   Typical Noise Levels

The scale in Fig. 16.1 goes from the threshold of hearing to the threshold of pain. In the middle of the scale is the conversation level, which is around 60 dB.

The approximate velocity of sound is about 345.6 m/s, and it is a function of temperature. It is important to know that the acoustic pressure is proportional to the inverse of the distance from the source.

16.5   Airplane Noise Regulations

The Federal Aviation Administration (FAA) issued a noise regulation, and it is well known as FAR 36. The FAR 36 noise-compliance stages are described briefly in Table 16.1. From these three stages, the first was accomplished by elimination of the fleet of stage 1 aircraft by U.S. airlines. For stage 2 airplanes, the limit is the end of the last century. By "hush-kitting" stage 2 airplanes, the owners will still have old aircraft with old engines. Placing new stage 3 engines on old aircraft brings up the question of aging airframes and the additional costs of maintenance.

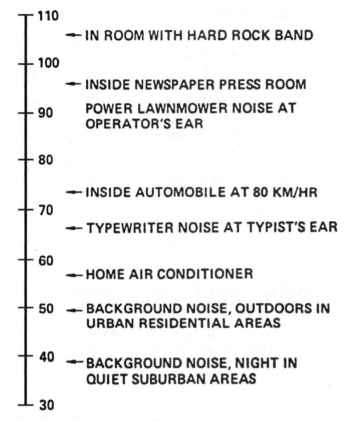

Figure 16.1   Noise levels (in dB)

TABLE 16.1   FAR 36 Noise-Compliance Stages

Stage 1	Aircraft cannot meet original 1969 FAR 36 (1971 Annex 16) noise requirements for new-type aircraft. These are the older designs.
Stage 2	Aircraft meets original 1969 FAR 36 requirements for new-type aircraft. Applies to aircraft with type application between Jan. 1, 1969 (Jan. 1, 1967 if bypass ratio > 2) to Nov. 4, 1975 (Jan. 1, 1969, March 1, 1972* BPR > 2 to Oct. 6, 1977).
Stage 3	Aircraft meets more stringent 1978 FAR 36 requirements for new aircraft. Applies to aircraft with application for type design on or after Nov. 5, 1975 (Oct. 6, 1977).

*Individual airplane certificate of airworthiness.

16.6   Airplane Noise Measurement

During the certification of an airplane, noise measurements have to be conducted at three arbitrary points. Noise limits are set, and the airplane cannot exceed these limits. The three points are shown in Fig. 16.2.

The takeoff reference point is 3.5 nautical miles from the start of roll. The sideline reference line is 0.25 nautical miles. The approach reference point is 370 ft below the approach flight path and 1.0 nautical mile from the approach end of the runway. For stage 2 and 3 airplanes, Figs. 16.3 through 16.5 present the noise limits. The horizontal scale (logarithmic) is the takeoff weight scale, and the vertical scale is the noise limit in decibels. These charts show the noise limits for three reference points. Measurements must be made at three arbitrary points, and the airplane cannot exceed the established noise limits.

The prescribed noise levels may be exceeded at one or two of the measuring points if

1. The sum of the exceeding decibels is not greater than 3 EPNdB.

2. No exceeding value is greater than 2 EPNdB.

3. The exceeding values are completely offset by reduction at other required measuring points (trade provision of FAR Part 36, Appendix C).

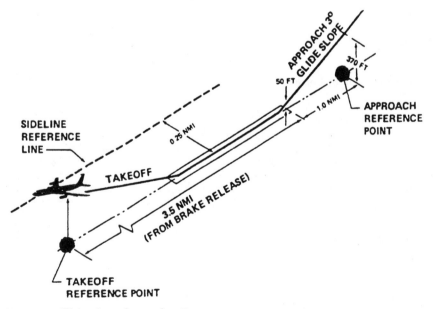

Figure 16.2   FAA noise reference locations.

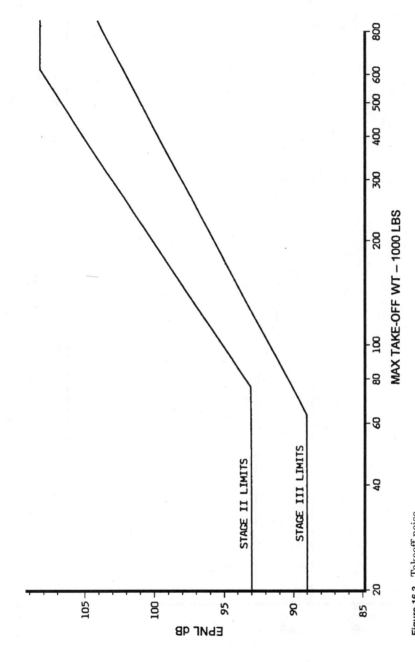

Figure 16.3 Takeoff noise.
Note: EPNL = Effective Perceived Noise Limit in dB

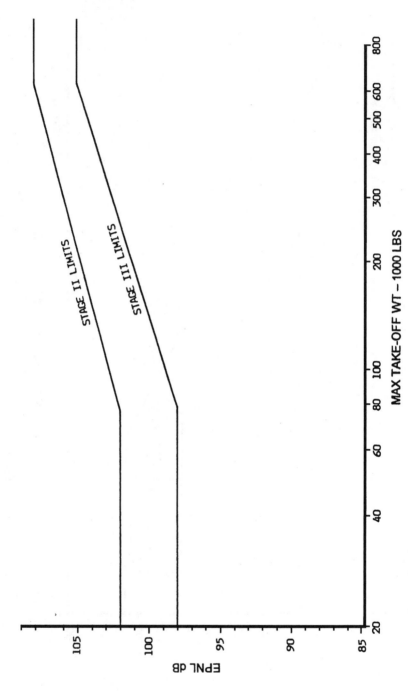

Figure 16.4 Approach noise.

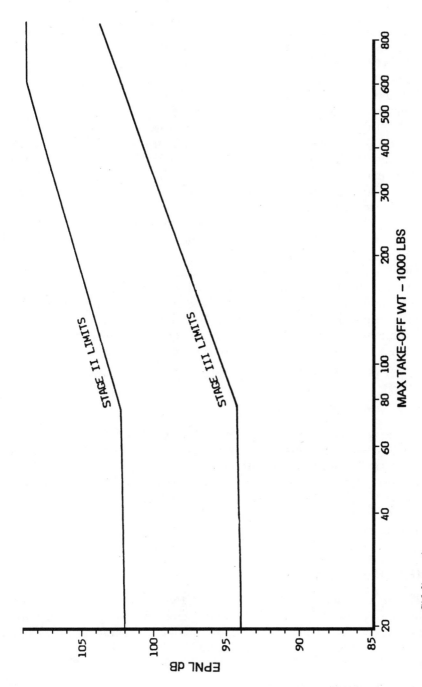

Figure 16.5 Sideline noise.

Note: The International Civil Aviation Organization (ICAO) has a similar noise regulation with 4 EPNdB and 3 EPNdB trade. (EPNdB = Effective Perceived Noise in decibel. Noise is measured in various dB units.)

The purpose of this test is to ensure that each airplane is measured against a preset "yardstick."

The European Economic Community (EEC) also has certain noise criteria governing their member airlines' operations. London (LHR), for example, has nighttime operation quotas. By keeping the number of nighttime operations constant but reducing the night quota for noisier airplanes and increasing it for quieter airplanes, a lower noise level is achieved. Certain requirements, e.g., for takeoff 4 mi^2 and for landing 2.5 mi^2 contour at 95 EPNdB, had to be satisfied.

The noise regulations had a profound effect on the operational and economic situations of the airlines. To comply with the regulations, some airlines exchanged their older and noisier airplanes for newer and less noisy equipment. Other airlines kept their older airplanes but applied noise suppressors to their existing engines or reengined with quieter equipment. By their compliance with regulations, they established a healthier, friendlier, and quieter environment.

Some airports took various steps for noise reduction. A few are listed below.

16.7   Airport Noise Restrictions

The decision of the New York Port Authority at JFK Airport, for example, imposed a daytime noise limit of 112 EPNdB and a nighttime limit of 102 EPNdB. This was intended to reduce noise around the airport. To observe these noise limits, takeoff weight sometimes had to be reduced, and for this reason, passengers and/or cargo were left behind. If it was possible, some airlines shifted their operations to another, less restrictive airport.

16.8   Night Curfews

The introduction of night curfews at certain airports reduced the number of takeoffs and/or landings or completely restricted any movement during certain time periods. This affected mostly cargo airlines having older and noisier aircraft and frequent night departures.

16.9   Preferential Runways

Noise limits affected airline operations due to delays (if any) because only certain preselected (by the port authorities) runways were permitted for use. Tucson airport, based on a recommendation, shifted most aircraft movements $\frac{1}{2}$ mile to the less populated southeast, resulting in a higher approach altitude over populated areas and thus lowering the noise level over the community.

16.10   Preferential Treatment of Quieter Airplanes

Smaller airlines with older and noisier airplanes, in the process of complying with the noise regulations, may experience some delays in their daily operations or decrease their departure frequencies. London Heathrow Airport has a noise quota system; each aircraft is classified into QC1, QC2, QC4, and QC8 levels. This QC is related to a point system; a noisier airplane uses more points, and in this way, the available slots are limited for it.

16.11   Airport Noise Monitoring Stations

Monitoring stations at certain strategic points on and around an airport keep track of each airplane's takeoff noise. Explanations must be submitted to airport authorities as to why certain airlines using the same type of aircraft are noisier than others. Specific noise levels are set for daytime and/or for nighttime operations.

16.12   Noise-Abatement Procedures

In order to avoid various airport noise regulations imposed on airlines that may be unreasonable from a technical point of view and/or questionable in terms of safety, the FAA, ICAO (International Civil Aviation Organization), ATA, and ALPA (Airline Pilot Association) established precise noise-abatement procedures. The reason to have standard procedures rather than a different one for each airport is to avoid confusion and errors and to improve the safety of operations. Climb thrust cutback is applied for lower noise levels where applicable. Figure 16.6 shows a standard noise-abatement departure profile.

Recommendations arising in December 1995 from the Committee on Aviation and Environmental Protection (CAEP) came as a result of a 3-year effort by a group of technical experts representing aircraft manufacturers, airline pilots, airport authorities, and CAEP member states. The objective was to identify operational strategies to improve aircraft-generated noise around the airports. For this project, all aspects of safety were considered, examined, and reported. The recommended procedure for noise abatement—when the communities are close to the end of a runway—consists of the following segments:

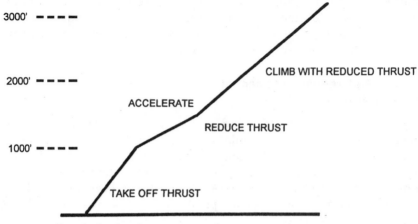

Figure 16.6 Standard noise-abatement procedure.

1. Liftoff and climb to at least 240 m or 800 ft above airport elevation.

2. Reduce engine thrust.

3. Retract flats or slats on schedule.

4. At 900 m or 3000 ft above airport elevation, accelerate to enroute climb speed.

The procedure changes when communities are some distance from the end of the runway, namely, step 3 becomes step 2 and step 2 becomes step 3. This procedure is aimed at achieving noise reduction for communities either close to or more distant from the end of a runway. It permits the operator the flexibility to choose a procedure for maximum noise reduction at different airports.

16.13   Engine Retrofits and Engine Noise Suppressors

In order for older airplanes to satisfy FAA noise regulations, retrofit was a way to stay in operation. For smaller airlines, purchasing new aircraft was impossible because the cost would drive them out of business. By installing noise suppressor materials on the engines, they were able to extend the life of older airplanes and at the same time satisfy the FAA noise regulations. The treatment of engine inlets and tailpipes with noise-attenuation materials, impervious sheets, and honeycomb and porous sheets also reduced noise to an acceptable level. For some airplanes this caused a 1.0 to 2.0 percent increase in trip fuel and/or a reduction in the allowable takeoff weight by sometimes as much as 10,000 lb. Another solution was an alternate flap setting for landing. On some engines, blow-in doors on inlet cowls were replaced with fixed-inlet cowls with acoustic material treatment (Fig. 16.7).

For some airplanes, a complete reengining was the answer to noise limitations. The DC-8-60 series airplanes were reengined with CFM-56 engines and became DC-8-70 series airplanes. This procedure had two advantages, namely, satisfactory noise levels and lower fuel flow. The reengining extended the useful life of the airplane by 8 to 12 years or more. The only downside to this solution is the slowly increasing maintenance cost due to the age of the airframe.

In this case, the reengining affected the payload versus range curve. It extended the range of the DC-8-73 (CFM-5b engines) by about 800 nautical miles and increased the payload by 17,500 lb. Comparing a JT3D-7 engine on a DC-8-63 airplane with a DC-8-73 airplane powered with a CFM engine, a significant noise reduction and improved fuel burn of 15 to 17 percent were experienced. Reengining shortened the takeoff field length requirement by 700 to 900 ft and reduced the noise level as shown in Table 16.2.

16.14   Airplane Noise Contour

The 100-dB noise contour was reduced from 6 mi^2 for an early DC-8 airplane to less than 1 mi^2 for a DC-10 airplane. At certain airports, the noise footprint area was prescribed, and the airplane had to satisfy this requirement. (Fig. 16.8)

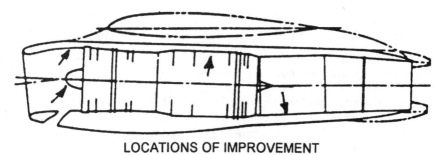

LOCATIONS OF IMPROVEMENT

Figure 16.7   Locations of noise-suppressing materials.

TABLE 16.2   Noise comparison (in dB)

DC-8-63/73	Requirement	JT3D-7 Engine	CFM-56 Engine
Takeoff	104.2	112.0	100.0
Approach	106.5	114.1	100.1
Sideline	106.5	100.0	89.0

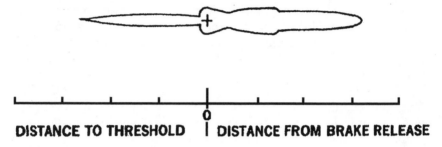

DISTANCE TO THRESHOLD | **DISTANCE FROM BRAKE RELEASE**

Figure 16.8   Typical noise contour.

16.15   Airport Noise and Effects

As more and quieter new airplane models are being built, the noise problem is beginning to come under control. It will take some time, however, until older airplanes disappear from the scene.

To sum up the status of noise, the following can be stated: The noise generated by individual airplanes is decreasing, but airplane movements are increasing significantly, and for this reason community noise is increasing. People living around airports today are more sensitive to noise, and their attitude is more aggressive as far as noise is concerned. Any attempts to build new or extend present airports are running into resistance, environmental studies, and/or lawsuits.

Considering all the preceding when building new airports, more emphasis should be placed on the following items in the planning stage:

- Control of the land in and around the airport to minimize the noise impact

- Zoning to minimize the residential dwellings

In conclusion, first secure enough land at the time of a master plan and the approval of a new airport. Furthermore, only a minimum number of dwellings (if any) should be in the noise-sensitive area.

16.16   Jet Engine Exhaust

Figure 16.9 shows a typical jet engine exhaust velocity distribution in top and side views. These contours correspond to the jet exhaust temperature distribution, and the contours should be considered as guidelines only. Winds can have a considerable effect on these lines. Operators have to make sure that their employees or anyone else is not exposed to jet engine exhaust because temperature, noise, and exhaust speed can have deadly effects on people. This type of chart could help operators develop an environment that is safe from this problem. For example, many airports have blast fences for safety reasons.

16.17   Engine Smoke Emissions

Global warming induced law makers and environmental groups to focus more attention on airplane engine smoke emissions. The main targets are nitrogen oxide (NO_x), carbon dioxide (CO_2), and unburned hydrocarbon (HC).

For engine manufacturers, it is a difficult task to reduce gas emissions, improve fuel efficiency, and comply with worldwide noise standards. The Committee on Aviation Environmental Protection (CAEP) is composed of 14 different industrialized nations. In 1980 the ICAO Committee on Aircraft Engine Emissions developed standards that became applicable in 1981. In 1991, the original limit was reduced by 20 percent. An additional 16 percent reduction is contemplated for the year 2008. Table 16.3 shows the nitrogen oxide generated by highway vehicles, aircraft, and railroads. The airline contribution is really quite low when compared with other means of transportation.

Besides the NO_x, there are other aviation pollutants such as carbon dioxide, water vapor, and sulfur oxide. The reduction in NO_x on advanced, medium-sized engines was accomplished. On a more recently certified large engine with a very high compression ratio, the improvement was less than expected. The reason probably is that not enough air is passing through the engine to adequately cool the combustion gases to reduce the generation of NO_x and cool the walls of combustion chamber. This increases the pressure ratio and NO_x and decreases fuel efficiency. Our present state of knowledge cannot answer the question of which one of these pollutants should be limited first.

General Electric is in the research stage of a new catalytic method to reduce its engines' NO_x.

Aircraft operations are a growing source of emissions into the atmosphere. Some studies indicate that the global civil aircraft fleet consumed 120 million tons of fuel in 1992 and generated approximately 1.3 million tons of NO_x, 380 million tons of carbon dioxide, and 150 millions of water vapor. Forecasts were prepared for the years of 2015 and 2040.

Traffic rate is estimated to grow by 3.6 percent, and fuel consumption will be 2.5 times more, producing more NO_x and carbon dioxide. Table 16.4 shows this forecast.

It was discovered recently that contrails generated by airplanes (white clouds of straight lines that airplanes leave behind at high altitudes) could form clouds sometimes a few hundred miles long. These contrail clouds are under investigation to find out their effect on the weather.

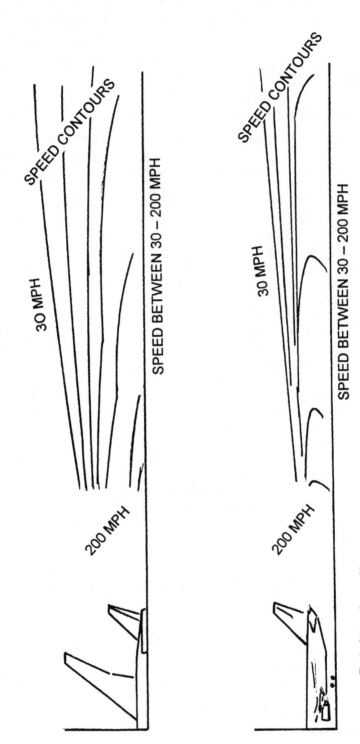

Figure 16.9 Typical jet exhaust diagram.

16.18   Ozone Distribution

Studies indicate that ozone likely would be destroyed in the atmosphere where supersonic airplanes are flying. On the other hand, the same emittance from subsonic aircraft would produce ozone. Ozone is measured in milligrams per cubic meter. It is a complex relationship, and it takes time to discover the interaction between all the factors. There are some operational considerations that may reduce air pollutants.

There are measures that can reduce fuel consumption and emissions. A few items are listed on this subject:

- Air traffic control (ATC) improvement

- Improvements in load factor

- Reduced airplane taxiing and more towing at airports

- Lower cruise speeds

- Optimization of climb and descent speed schedule

- Avoiding tankering

- Reducing flights

16.19   Smoke Emission Improvement At
and Around the Airport

There are some other methods that can be applied on the ground to achieve improvements. Various methods to reduce smoke emission include the following:

1. *At the airport.*   The taxiway could be optimized to reduce fuel burn. Other measures include increased towing of aircraft, replace-

TABLE 16.3   U.S. Emissions of Nitrogen Oxides (in million short tons)

Year	Highway vehicles	Aircraft	Railroads	Total transportation	Nontransportation
1970	7.39	0.07	0.50	8.09	12.54
1980	8.62	0.11	0.73	9.70	13.58
1990	7.49	0.14	0.93	8.90	14.14
1991	7.37	0.14	0.93	8.78	13.89
1992	7.44	0.14	0.95	8.88	13.97
1993	7.51	0.15	0.95	8.96	14.32
1994	7.67	0.15	0.95	9.14	14.52
1995	7.61	0.15	0.99	9.12	12.68

Source: U.S. Environmental Protection Agency.

TABLE 16.4   Aircraft Fuel Consumption and Emissions

	1992	2015	2040
Fuel	120	302	420
NO_x	1.3	3.0	24.44
CO_2	380	953	1326
H_2O	150	372	517

Note: Items are shown in millions of pounds.

ment of auxiliary power units by fixed air systems, the use of electric vehicles, etc.

2. *Surrounding the area of the airport.* Emissions can result (besides airplane emissions) from road traffic, industry, agricultural activities, cooling and heating units, etc.

3. *On the roadside.* Measures include enhancement of the public transportation system and development of alternate fuel-powered vehicles and promotion of ride-sharing and car-pooling systems for airport employees.

Engine manufacturers are continually researching to achieve an environmentally friendly, nonpolluting solution to conform to environmental requirements while maintaining the fuel efficiency of their engines.

16.20   Bird Strikes

This is a growing concern for airports and airline management. The presence of numerous birds around airports and runways is a serious risk to flights. In the summer of 1995, some 180 bird strikes were experienced at JFK Airport. Bird ingestion may lead to engine failures. In the past 18 years, bird strikes have damaged 13 engines. The FAA issued monthly statistics for the years of 1993, 1994, and 1995 about the number of bird strikes. Transport Canada issued statistics about aircraft parts struck and damaged in 1994. The economic impact must be high because, besides parts, labor costs should be added. In addition, when an airplane has to be out of service, this could mean an additional loss of revenue.

The most serious incident at JFK took place a few years ago when a DC-10 collided on takeoff with a flock of gulls. The airplane was destroyed by fire. Luckily, there were no fatalities.

The ICAO recently published a bird strike report. It included information from 45 states. For the year 1994–1995, some 11,000 bird strikes were reported. Bird strikes were classified into different categories. For the year 1995, various classifications are presented:

Bird strikes: Flight phase

Takeoff run	1248
Climb	1012
Cruise	174
Descent	107
Approach	1918
Landing roll	1005

Effect on flight

Number of bird strikes with no effect	4930
Aborted takeoff	132
Precautionary landing	234
Engine shutdown	16
Forced landing	16
Other effects	136

Height of bird strikes

0–100 ft	3698
100–500 ft	730
500–2500 ft	667
Above 2500 ft	369

Aircraft parts struck

Radome	478
Windshield	625
Nose	561
Engine	639
Fuselage	512

16.21   Airplane Damages Due to Bird Strikes

Considering the damage done to equipment and lost time, bird strikes could run into losses of a few million dollars, not to mention the lives at stake too. Bird strikes endanger flights and have to be taken seriously.

Besides gulls, the Canada goose is also a growing concern. According to Transport Canada, where these geese were involved in accidents, the damage in 1995 was over $200 million Canadian, while 24 fatalities were recorded. These birds migrate south to the United States each year. In the same year, Northwest Airline's Airbus A320 collided at La

Guardia Airport with a flock of Canada geese, and the damage amount-
ed to $2.5 million. A similar fate was experienced a few months later by
a B747 operated by Polar Airlines. Airport authorities took steps to solve
this environmental problem. Tapes of bird distress calls were played,
and the noise was directed toward these birds. All these measures, however, had a very low success rate. At London's Heathrow Airport, a person was hired to scare birds away from the airport by wrapping himself
in a black cloth. For better visibility, he walked up a small hill and
flapped his arms, imitating a big, black bird's movements. This procedure's effects lasted only for a short time. Presently, airport management at JFK employs full-time personnel for the fight against bird
strikes. The following steps have been taken to improve the situation:

- Proper planning for vegetation management

- Improvement in solid waste management

- Elimination of freshwater sources and wetlands and better drainage

- Effective insect control

- Establishment of a falconry program

Three falconers were hired at JFK with extensive experience working with trained falcons and hawks. They had their own trained birds.
The upkeep of these trained birds was quite costly because they
required a special diet. The falcons and hawks would fly around and
above the airport in a certain pattern such that they would be visible
and scare away other birds. In the future, more working birds will be
considered to scare away gulls and other birds that endanger the operations in and around airport areas. Naturally, this is and will be an
additional expense that will be included in the user fees of customers.

16.22   Airport Pavement

Pavements were developed to support airplanes rolling on the ground.
For commercial operations, two types of pavement are available:

- Flexible pavement surfaced with asphalt (10 cm typical)

- Rigid pavement surfaced with concrete (25 cm typical)

Note: Some B737 airplanes operate from gravel runways.
The ACN/PCN system was developed to measure the use of pavement to avoid pavement failure that could be costly for airline operations for the following reasons:

- Deteriorating breaking coefficient

- Discomfort to passengers
- Increased takeoff distance
- Engine foreign object damage (FOD) or engine ingestion

Engines can easily pick up broken pieces of runway. Naturally, it is helpful when an engine is mounted high above the ground. Engine repairs due to ingestion can be costly, and this may increase maintenance costs considerably since it may require removal of the engine and replacement and/or repair of the fan and compressor parts. During engine change, the aircraft is on the ground, not generating cash flow and thus losing profit.

Similarly, overuse or excess airplane weight can cause pavement failure, and this is quite costly for airport operators.

For each airplane type, an airplane classification number (ACN) is established. For each runway, a pavement classification number (PCN) is assigned (reference ICAO Aerodrome Manual, Part 2). Provided that the airplane ACN is not more than 10 percent above the published pavement PCN, the bearing strength of the pavement can be sufficient for unlimited use by the aircraft. A figure of 10 percent has been chosen as representing the lowest degree of variation that is significant. When the ACN is above 10 percent of the PCN and is a low-frequency occurrence, the airport authorities may permit operation on this runway. By using the ACN/PCN system, port authorities are able to avoid the overuse of runways and the deterioration of runway surfaces.

In some extreme cases, airport authorities may curtail the frequency of operations, limit the maximum weight on the runway, or completely close the runway for certain types of aircraft. At a multiple-runway airport, this would not cause problems because operation can be switched to a higher PCN runway.

There are a few methods available related to pavement stresses. They were developed by FAA, U.S. Corps of Engineers, Portland Cement Association etc.

For airplanes operating on gravel runways, a few devices were developed to minimize damage. Engine-mounted vortex dissipators prevent the formation of air vortices below the engine inlets, where stones and debris could be sucked into the engine. Table 16.5 lists the elements of a gravel runway kit for the B737 airplane. There are certain procedures associated with unpaved runway operation:

1. The antiskid system should be on for takeoff and landing.

2. The engine vortex dissipator must be operational and turned on.

3. Taxiing should be kept short and at slow speed.

TABLE 16.5   Airplane Protection On Gravel Runway

Main Gear Deflector
Nose Gear Deflector
Inboard Flap Protection: Left
Inboard Flap Protection: Right
Metal Edge Band on Elephant Ear Fairing
Vortex Discipator For Each Engine
Nacelle
Special Protective Finish
Low Pressure Tires on Low Strength Runways

Note: Items are shown in millions of pounds.

16.23   Airport Soil Management

Another environmental concern is water and soil contamination from activities carried out at airports. This could be the result of fuel and/or maintenance chemical spillage and release. Underground tank storage may have some fuel spillage, and the use of herbicides and pesticides can contribute to environment contamination.

Many airports have cleanup programs to remove chemicals left behind by past operations. Preventive measures have now been introduced to improve airport environments.

16.24   Airport Waste Management

With increased traffic, the volume of waste is increasing. To date, little attention has been given to waste separation, recycling, transport, and disposal. Special attention has to be focused on the treatment of hazardous and regular wastes and the disposal facilities. Numerous airports have already begun simple recycling programs for bottles, cans, papers, and other materials. These subjects must be addressed and the problems associated with them solved. A partnership has to be developed between the airlines, airports, local communities, and government. This should be a friendly partnership in order to solve these problems for the benefit of every participant.

16.25   Volcanic Ash

In 1980, the eruption of Mount St. Helen in Washington State brought attention to the problem of volcanic ash. Ash particles cause damage to machinery where moving parts are involved. The range of particles is between 0.5 and 100 μm in size, and most aircraft filters are effective for materials as small as 15 μm. Some of the ash can attack critical aircraft equipment. Aircraft manufacturers recommend maintenance programs to replace filters and other maintenance measures based on the duration of exposure and the density of the ash. This additional maintenance measure is expensive, so airline operators should avoid as much as possible affected areas.

Crazing is partly an environmental problem that affects acrylic windows on airplanes. It is caused by stress due to airplane pressurization and partly by chemicals such as deicing fluids, cleaning materials, waxes, etc. The environmental effects are acid rain and ozone contamination.

Further studies by Boeing disclosed strong evidence that premature crazing also is induced by volcanic material, such as sulfuric acid and solid particles. In the volcanic eruption that happened in April 1982 in Mexico, material was distributed at a relatively low level in the neighborhood of 40,000 ft, which is the range of airline operations. In order to find a satisfactory solution to crazing, further studies are needed to completely understand the problem of ozone contamination.

16.26   Wind Shear

Airlines developed wind-shear detection systems and installed them in their aircraft. These instruments provide warnings during the takeoff roll, takeoff, approach, and go-arounds. These are sensitive phases of flight around the airport vicinity, and it is important that pilots should be alerted about wind-shear phenomena. A graphic presentation is shown in Figure 16.10.

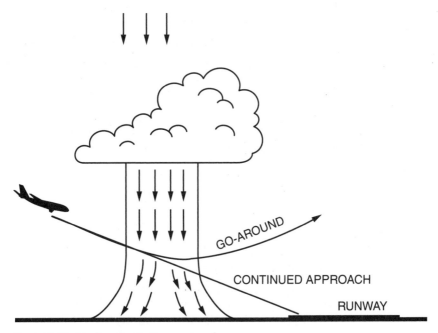

Figure 16.10   Airplane protection on gravel runway.

Short Descriptions of Selected Types of Aircraft

17.1 The B747-400 Aircraft

17.1.1 Introduction

As shown in Fig. 17.1, the first presentation is the layout of the B747 aircraft in three views. It is a four-engine, wide-body, double-aisled aircraft.

17.1.2 Weight statements

Table 17.1 contains the weight statements for the B747-400 aircraft. The engines are manufactured by Pratt-Whitney, Rolls-Royce, and General Electric. In this table, the airplane is presented with various certified maximum takeoff weights between 800,000 and 875,000 lb. The engines are powered with Pratt-Whitney Model PW 4056 engines. Similar tables are available for airplanes powered with Rolls-Royce and General Electric engines.

An operator with a short-range route structure would select an airplane with a lower takeoff weight for less operating cost. Meanwhile, an operator having long-range operations, such as Qantas Airlines, would select an airplane with a higher takeoff weight.

17.1.3 Aircraft configuration

Various airplane configurations include passenger, cargo, convertible, and combi.

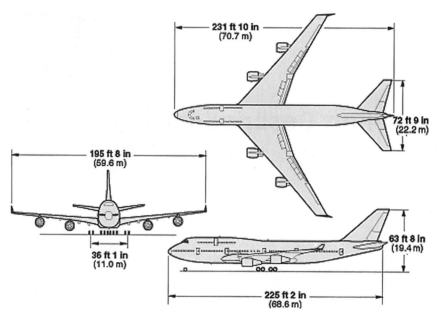

Figure 17.1   Aircraft in three views. (*Courtesy of Boeing.*)

Passenger configuration.   There are 10 doors in the passenger cabin. A certain number of lavatories should be available, as defined by regulations. Galleys also have to be provided.

Passenger seat arrangements vary based on the positioning of the seats. A high-density seating arrangement is shown in Fig. 17.2 that is all economy class with 539 passengers seated on the main deck and 85 passengers on the upper deck.

Two, triclass arrangements are shown in Fig. 17.3. The first-class seats on the main deck have a 61-in pitch; in the business class the pitch is 39 in; and in economy class it is 32 in. Business-class seats are located on the upper deck and have a 38-in pitch. A spiral staircase connects the main deck with the upper deck.

Baggage is stored in the lower compartment. There are three compartments below the main deck, namely, the forward, aft, and bulk compartments. The height of pallets and/or containers is limited to 64 in in the lower lobe. A variety of pallets and/or containers can be used here:

1. Thirty half-width containers (16 in forward and 14 in aft compartments)

2. Fifteen full-width containers (8 in forward and 7 in aft compartments)

TABLE 17.1   B747-400 Weight Statements (Engine: Pratt-Whitney PW 4056)

Characteristics	Units	PW 4056 Engines				
Max design taxi weight	Pounds	803,000	836,000	853,000	873,000	877,000
	Kilograms	364,234	379,207	386,913	395,985	397,800
Max design takeoff	Pounds	800,000	833,000	850,000	870,000	875,000
weight	Kilograms	362,873	377,846	385,553	394,625	396,893
Max design landing	Pounds	574,000	574,000	630,000	630,000	630,000
weight (1)	Kilograms	260,361	260,361	285,763	285,763	285,763
Max design zero-fuel	Pounds	535,000	535,000	535,000	542,500	542,500
weight (2)	Kilograms	242,671	242,671	242,671	246,073	246,073
Spec operating empty	Pounds	394,660	394,660	394,660	394,660	394,660
weight (3)	Kilograms	178,984	178,984	178,984	178,984	178,984
Max structural payload	Pounds	140,340	140,340	140,340	147,840	147,840
	Kilograms	63,687	63,687	63,687	67,089	67,089
Typical seating capacity	Upper deck	42 business class				
(includes upper deck)	Main deck	24 first, 32 business, 302 economy				
Max cargo, lower deck	Cubic feet	5,250	5,250	5,250	5,250	5,250
containers (LD-1)	Cubic meters	149	149	149	149	149
Max cargo, lower	Cubic feet	845	845	845	845	845
deck bulk cargo	Cubic meters	24	24	24	24	24
Usable fuel	U.S. gallons	53,985	53,985	53,985	57,285	57,285
capacity (4)	Liters	204,333	204,333	204,333	216,824	216,824
	Pounds	360,226	360,226	360,226	382,336	382,336
	Kilograms	163,368	163,368	163,368	173,395	173,395

Notes:
1. 630,000 lb landing weight is optional.
2. 542,500 lb zero-fuel weight is optional.
3. Spec operating empty weight reflects three-class 400-passenger arrangement and standard item allowances. Actual OEW will vary with airline configuration and optional equipment. Consult using airline for actual OEW.
4. Optional tail fuel of 3300 U.S. gal is reflected in the higher fuel capacity.
Source: Boeing.

3. Nine pallets, 96 × 125 × 64 in (5 in forward and 4 in aft compartments)

4. Four pallets, 88 × 108 × 64 in or 88 × 125 × 64 in and two half-width containers

5. Five pallets in forward and four pallets and two half-width containers in aft compartments

The containers are

LD-1, 92 × 60.4 × 64 in and a base of 61.5 in

LD-2, full width 186 × 60.4 × 64 in and a base of 125 in

The volume of the bulk cargo compartment is 845 ft³. Figure 17.4 shows the container arrangements for the lower cargo compartment,

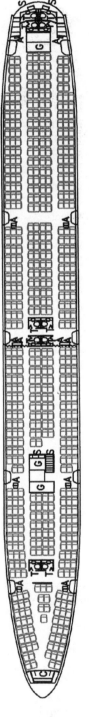

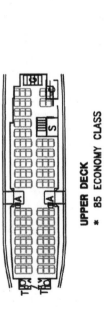

A ATTENDANT
T TOILET
S STORAGE
G GALLEY

UPPER DECK
* 85 ECONOMY CLASS

MAIN DECK
* 539 ECONOMY CLASS

Figure 17.2 High-density seating configuration. (*Courtesy of Boeing.*)

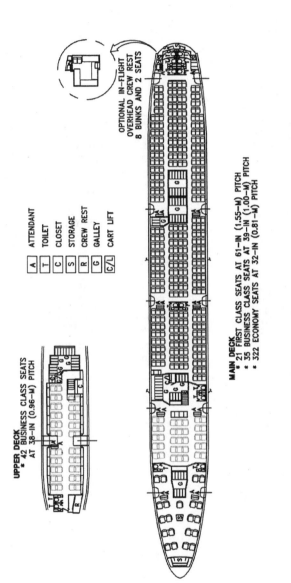

UPPER DECK
* 42 BUSINESS CLASS SEATS
  AT 38-IN (0.96-M) PITCH

OPTIONAL IN-FLIGHT
OVERHEAD CREW REST
8 BUNKS AND 2 SEATS

A	ATTENDANT
T	TOILET
C	CLOSET
S	STORAGE
R	CREW REST
G	GALLEY
C/L	CART LIFT

MAIN DECK
* 21 FIRST CLASS SEATS AT 61-IN (1.55-M) PITCH
* 35 BUSINESS CLASS SEATS AT 39-IN (1.00-M) PITCH
* 322 ECONOMY SEATS AT 32-IN (0.81-M) PITCH

MAIN DECK
* 32 FIRST CLASS SEATS AT 61-IN (1.55-M) PITCH
* 34 BUSINESS CLASS SEATS AT 39-IN (1.00-M) PITCH
* 309 ECONOMY CLASS SEATS AT 32-IN (0.81-M) PITCH

Figure 17.3 Triclass configuration. (*Courtesy of Boeing.*)

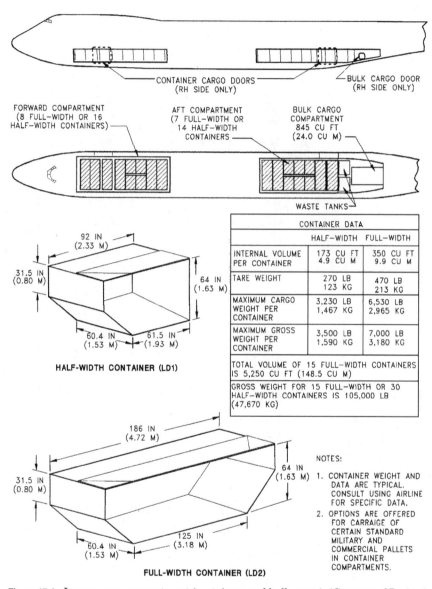

Figure 17.4 Lower cargo compartment (containers and bulk cargo). (*Courtesy of Boeing.*)

where passenger baggage is carried. The lower compartment's total volume is 5250 ft³.

To calculate the lower compartment's revenue potential , the following steps should be taken:

1. Define the average density of the planned freight.

2. Define passenger number times 35 lb for baggage weight per passenger.

3. Define total weight, based on density, carried in lower cargo compartment.

4. Subtract passenger baggage weight.

5. Obtain available revenue-generating weight.

Note: Passenger baggage weight is considered as an average of 35 lb per passenger. This average weight can be changed based on the experience of the airline.

An example would shed more light on this calculation as follows: For a given flight, there are 286 passengers. The density of the planned freight is 9.6 lb/ft³. Assuming 286 passengers, total baggage weight at 35 lb each would amount to 10,010 lb. Total weight based on density is (9.6 lb × 5250 ft³) = 50,400 lb. Subtract passenger baggage weight, that is, 50,400 lb − 10,010 lb, gives 40,390 lb, which represents the total available revenue-generating weight.

The revenue generated by the flight in this example is based on carrying 286 passengers on the main deck. There is also the potential of generating additional revenue through the availability of this 40,390 lb on the lower deck.

Freighter configuration. The B747F freighter's main deck can be loaded with several combinations of pallets and/or containers. Palletized configuration is able to carry, for example, 29 pallets of size 96 × 125 in, or 31 pallets of size 96 × 117.5 in.

Figure 17.5 presents the container arrangements on the main deck of the airplane. Here, twenty-nine 10-ft containers are shown. The length of containers may vary up to 20, 30, or 40 ft. Figure 17.6 presents a typical main deck random intermix. It contains various sizes of pallets and containers in virtually any combination.

The airplane can be loaded through either the nose door and the side door. The nose door permits the loading of long pipes, whereas the side door permits the loading of a 10-ft container. A short list of pallets and containers shows the variety of unit load devices.

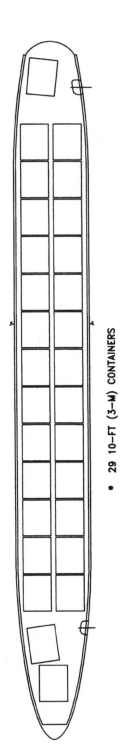

* 29 10—FT (3—M) CONTAINERS

Figure 17.5 Cargo configuration. (*Courtesy of Boeing.*)

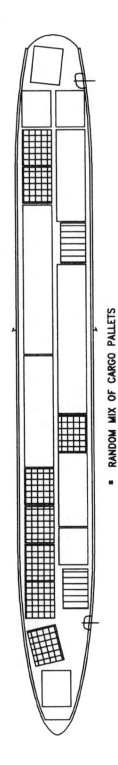

* RANDOM MIX OF CARGO PALLETS AND CONTAINERS UP TO 40 FT (12 M)

Figure 17.6 Mixed cargo configuration (pallets and containers). (*Courtesy of Boeing.*)

Pallets			Containers		
W	L	Ht	W	L	Ht
88	125	87	96	125	96
88	125	96	96	240	96
88	108	96			
96	125	96			
96	117.7	96			

Side door only (containers):

W	L	Ht
96	125	118

Nose door only (containers):

W	L	Ht
96	238	96
96	360	96
96	480	96

This airplane has a large volume capacity, permitting the loading of airplane engines, heavy machinery, outsized vehicles, small aircraft, etc. A controlled environment (air conditioning or heating) permits transportation of fragile objects such as cut flowers, perishable food, and live animals on the airplane.

Table 17.2 lists the cargo capacity and tare-weight summary for selected pallet and container data.

TABLE 17.2   Loading Pallets and Containers

Item	Size, in	Location	Weight, lb	Volume, ft^3	No. of items	Total weight lb	Total volume ft^3
Pallet	96 × 125	Upper	283	630	29	8,207	18,270
Pallet (64 in high)	96 × 125	Lower	283	415	9	2,547	3,735
Total						10,754	22,005
Container	96 × 125	Upper	1,050	605	29	30,450	17,545
Container (64 in high)	60.4 × 186	Lower	470	350	15	7,050	5,250
Total						37,500	22,795

From the total load in palletized configuration, 10,754 lb and in container configuration, 37,500 lb, are assigned the tare weight. When making the decision to palletize or containerize freight, the following also should be considered: Changing from a completely palletized to a completely containerized configuration, the tare weight could increase to (37,500–10,754) 26,746 lb. Considering a random mix of palletized and containerized configuration, a tare weight of 13,373 lb can be assumed over a completely palletized configuration. To carry this extra (tare) weight over a trans-Atlantic trip (3000 nautical miles), an additional fuel burn of 3343 lb (498 gal) is necessary. Thus amounts to $300 per trip at 60 cents/gal. This 13,373 lb of load is lost in potential revenue.

The cost of additional fuel to carry heavier tare weight and the loss in potential cargo revenue would define the cost difference between palletized configuration and a mix of palletized and/or containerized cargo. This loss could be projected on a monthly or yearly basis.

Convertible configuration.   The purpose of the convertible airplane is to satisfy both passenger and freight demands on an airline's route system. It is economically valuable for various reasons. For example, because of seasonal fluctuations, in summer, there is a greater demand for passenger operations (Figure 17.7) and less or no demand at all during the winter, when freight operations could replace the diminishing demand of passengers. Furthermore, on weekdays, for example, the airplane could operate as a cargo airplane, and on weekends, it could operate as a passenger airplane (Fig. 17.8).

There is a small penalty to be paid; namely, the operating weight empty of a convertible airplane is higher than that of a pure freighter,

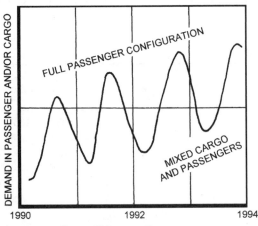

Figure 17.7   Convertible operations.

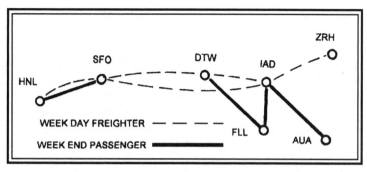

Figure 17.8 All cargo on weekdays and all passengers on weekend operations.

and for this reason, its maximum payload is slightly lower. The multiple benefits of the convertible airplane justify its use. Listed below are the features and benefits of a convertible airplane:

- *Easy conversion.* As traffic dictates, airlines can change the B747 convertible from an all-passenger to an all-cargo configuration. Conversion takes up to 10 hours, based on the configuration.

- *Usable functions.* All seating arrangements of the passenger airplane and all loading arrangements of the cargo airplane are available in the convertible.

- *Reasons for conversion.*
  - Dense seasonal markets (see Fig. 17.7)
  - Passenger plus cargo market development

- *Full convertibility:*
  - Matching weekly demand fluctuations (see Fig. 17.8)
  - Charter flexibility

- *Potential additional utilization:*
  - Capability for passenger and/or cargo configuration arrangements provide for maximum operating flexibility.
  - Direct main deck nose portal loading and unloading are used for the cargo-handling system.
  - Convertible component sizes are selected for ease of handling.
  - Convertible component sizes are consistent with cargo-handling system capabilities.
  - Minimal use of ground service equipment and minimal use of special hand tools are required.

Combi configurations. Configuration is used on routes where passenger and cargo traffic combined could generate a profitable operation but each traffic alone cannot produce satisfactory revenue. In this configuration, cargo is carried on the main deck and in the belly of the airplane. The airplane is designed in such a fashion that the passenger and cargo hardware can be interchanged; the cargo-handling system components replace the passenger interior components, and vice versa.

Figure 17.9 shows a typical combi load. Passenger seats are in the front of the main deck, and containers and/or pallets are carried in the back. At the time of conversion, e.g., from passenger to cargo, first the seats and carpets are removed, followed by the passenger conversion components.

Table 17.3 shows passenger conversion components. The conversion list covers about 565 pieces weighing some 30,000 lb. Only a few items from this list are presented in the table.

After all passenger items are removed, the cargo conversion components are installed (Table 17.4). These components include sill rollers, caster trays, center and side guides and restraints, and so on. It is necessary to establish the amount of time necessary to complete the conversion for scheduling purposes. Total conversion involves approximately 560 pieces weighing about 12,000 lb.

17.1.4   Airplane conversion time

Table 17.5 shows the approximate conversion time in hours for various configurations. The times indicated are based on averages and may be different for each airline.

TABLE 17.3   Passenger Conversion Components

Attendant seat assembly	Lavatory modules
Carpeting	Life-raft container
Center stowage beams	Partitions
Drop-ceiling components	Seat assembly
Escape slides	Side stowage bins
Fixed portion of galley	

TABLE 17.4   Cargo Conversion Components

Adjustable guide installation	Guide installation shock mount
Basic end lock installation	Lock roller tray assembly
Center guide restraints	Outboard side guide
End lock	Power unit drive
End lock and tray installation	Roller tray assembly, 200 ft
End stops	Side restraints
Ganged side restraints	Splitter guide
Guide assembly	Sill rollers

T	TOILET
S	STORAGE
G	GALLEY
R	CREW REST
C/L	CART LIFT

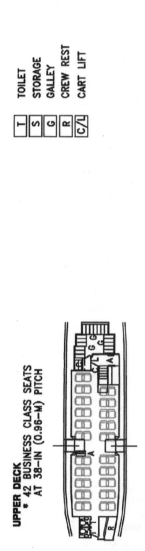

UPPER DECK
* 42 BUSINESS CLASS SEATS
  AT 38-IN (0.96-M) PITCH

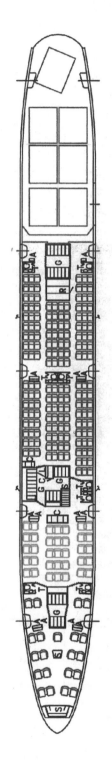

MAIN DECK - COMBI LOAD
* 21 FIRST CLASS SEATS AT 61-IN (1.55-M) PITCH
* 35 BUSINESS CLASS SEATS AT 38-IN (0.96-M) PITCH
* 168 ECONOMY CLASS SEATS AT 32-IN (0.81-M) PITCH
* 7 PALLETS

Figure 17.9 Cargo-passenger partition. (*Courtesy of Boeing.*)

TABLE 17.5   Conversion Time Estimates

	From				
To	All pass.	¼ cargo ¾ pass.	½ cargo ½ pass.	¾ cargo ¼ pass.	All cargo
All pass.	—	5	6	8	10
¾ pass. ¼ cargo	4	—	4	5	6
½ pass. ½ cargo	5	2	—	3	5
¼ pass. ¾ cargo	6	4	2	—	4
All cargo	8	5	4	3	—

17.1.5   Airplane ground time

Certain services have to be performed on the airplane during turnaround. Table 17.6 shows a typical turnaround servicing for a B747 airplane. Since scheduling has to incorporate terminal servicing when establishing ground time during turnaround and/or enroute station, it is essential to know the time to allocate for such services. The times shown may vary from airline to airline and from station to station because of procedural differences, resources available, and management efficiency.

TABLE 17.6   Estimated Ground Time, Domestic Turnaround

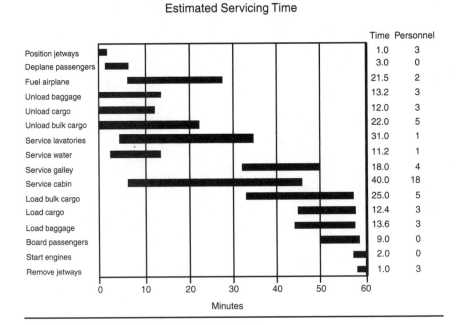

Domestic Turnaround
Estimated Servicing Time

	Time	Personnel
Position jetways	1.0	3
Deplane passengers	3.0	0
Fuel airplane	21.5	2
Unload baggage	13.2	3
Unload cargo	12.0	3
Unload bulk cargo	22.0	5
Service lavatories	31.0	1
Service water	11.2	1
Service galley	18.0	4
Service cabin	40.0	18
Load bulk cargo	25.0	5
Load cargo	12.4	3
Load baggage	13.6	3
Board passengers	9.0	0
Start engines	2.0	0
Remove jetways	1.0	3

Minutes

17.2   The DC-10 Aircraft

17.2.1   Introduction

Figure 17.10 shows the general layout of the DC-10 in three views. There are three versions of the DC-10. The series 10 airplanes are medium-range aircraft that can carry 270 passengers up to 3800 nautical miles;

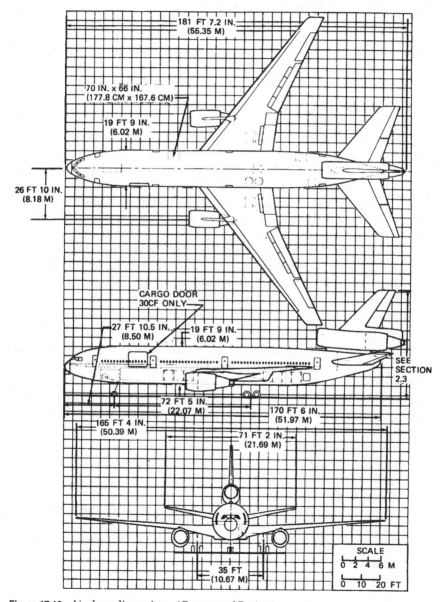

Figure 17.10   Airplane dimensions. (*Courtesy of Boeing.*)

they are powered with GE CF6-6D, -6D1, and 6D1A engines (the D1A engine is limited to 2000 ft airport elevation and has an operating temperature limit of up to 100°F).

The series 30 airplanes are powered with GE CF6-50 engine family, which have a higher thrust rating. The operating range is extended to over 5600 nautical miles. The series 40 airplanes are powered with Pratt-Whitney JT9D-59A engines. The series 30 and 40 airplanes are available in convertible configuration. The freighter derivative is powered with the 50C2 or 59B engine. It has a main cargo door measuring 170 × 120 in, and two belly doors measuring 104 × 66 in.

17.2.2 Passenger configuration

Passenger seats can be arranged (e.g., for mixed classes) for 270 passengers. This could include 48 first-class passenger seats (6 abreast with a 38-in pitch) and 222 economy-class seats (8 abreast with a 34-in pitch). A full economy class could seat 399 to 345 passengers at 10, 9, 8, etc., abreast with 29- to 34-in pitch. Several other configurations are possible, but only the economy class is shown here (Fig. 17.11).

Two types of food service are available: a conventional main-cabin galley system and a lower-level galley system. In the latter, the food is prepared below deck and carried by two elevators to the midcabin center located between first-class and tourist sections. The advantage of a lower-level galley is in the space saved on the main deck, making possible the installation of more (revenue-generating) seats. The position of the galley has a specific effect on revenue. In the case of a lower-level galley, there is a loss of 1600 ft^3 of belly cargo volume. At 8.0 lb/ft^3, this is a loss of 12,800 lb of potential cargo. On the other hand, 15 more passenger seats are available on the upper deck. The decision as to galley position is based on the economics of the revenue-generating 12,800 lb of cargo or the loss of 15 additional passenger seats. The final dollar value will be the answer to the questions of which configuration to use. Each airline has to evaluate its own unique situation and make a decision based on that.

Besides the passengers, the airplane also carries cargo in the lower-level compartment (belly). Taking a high number of passengers, (e.g., 325 passengers and 35 lb of baggage for each passenger), the baggage weight would amount to (325 × 35) 11,375 lb. For an above-floor galley configuration, the total volume of containers is 3792 ft^3 (without bulk). Assuming an 8.7762 lb/ft^3 density, the potential for carrying load in the lower-level compartment is 33,280 lb. Subtracting the passengers baggage weight, the extra potential revenue for carrying cargo in the lower-level compartment would be (33,280 lb − 11,375 lb) 21,905 lb.

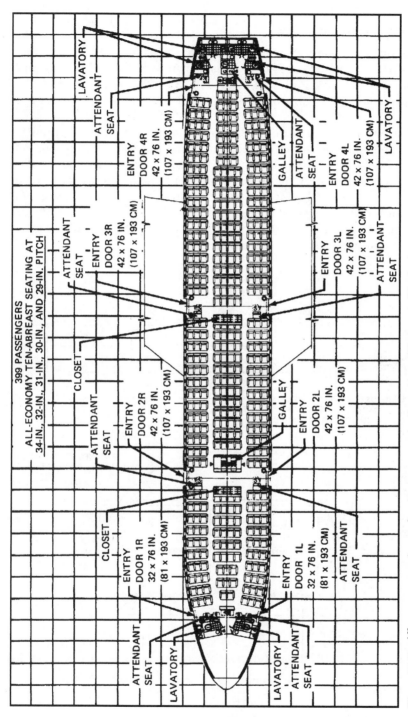

Figure 17.11 All economy maximum-density passenger configurations. (*Courtesy of Boeing.*)

399

17.2.3 Cargo configuration

The airplane's main deck can accommodate pallets and/or containers. Only the palletized configuration is shown in Fig. 17.12 in a symmetrical arrangement. Pallet size is 88 × 125 in. The volume is 10,874 ft³. The airplane is capable of loading a 20-ft container through the main door.

For the lower forward compartment, a 66 × 104 in cargo door is available. For the center compartment, the size of the cargo door is 66 × 70 in. The forward compartment can accommodate five 88 × 125 in pallets or 12 full-width or 24 half-width containers. The center compartment can house 8 half-width or 4 full-width containers. Figure 17.13 shows the containerized configuration, and Fig. 17.14 shows the palletized configuration. In the lower compartments, the heights of pallets and containers cannot exceed 64 in.

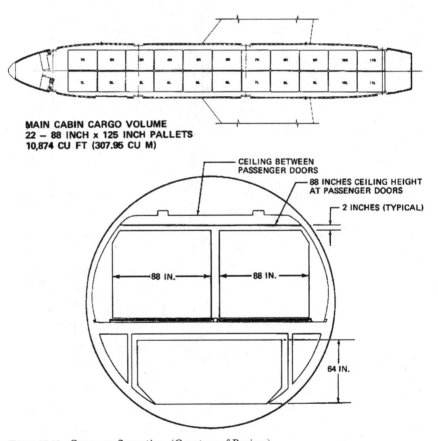

Figure 17.12   Cargo configuration. (*Courtesy of Boeing.*)

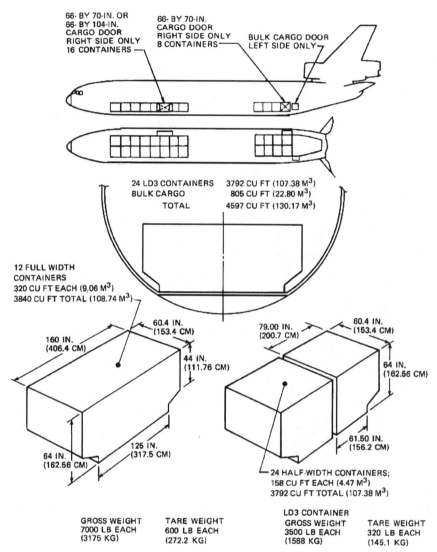

Figure 17.13 Lower compartment (upper galley) containerized configuration. (*Courtesy of Boeing.*)

17.2.4 Mixed cargo-passenger operations

A mixed cargo-passenger arrangement is presented in Fig. 17.15. This operation is effective on routes where the passenger or cargo operation alone would not be able to generate profit. In this configuration, the airplane carries 22 first-class and 193 economy-class passengers. Their baggage is in 14 containers carried on the main deck. This means that the

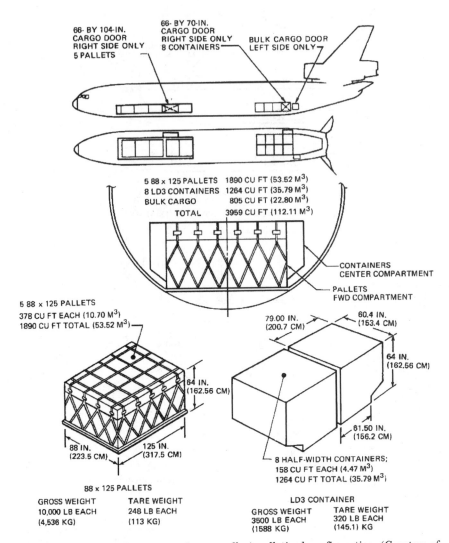

Figure 17.14 Lower compartment (upper galley) palletized configuration. (*Courtesy of Boeing.*)

full cargo capability of the lower deck has the potential to generate profit.

In a mixed cargo-passenger operation, the available number of passengers and the amount of cargo define the passenger seat and pallet and/or container relationship.

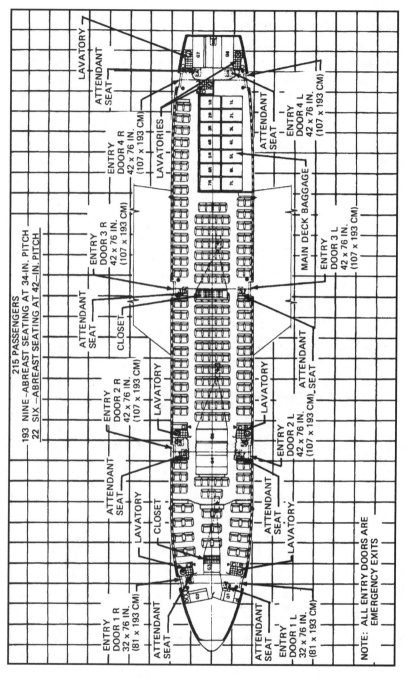

Figure 17.15 Mixed configuration on main deck (passengers and baggage containers). (*Courtesy of Boeing.*)

17.2.5 Payload versus range curve for passenger and freighter operation

Figure 17.16 presents a payload versus range curve for a passenger configuration. The airplane can carry 399 passengers and baggage at a total weight of 81,795 lb. This weight consists of 165 lb per passenger and 40 lb of baggage for each passenger.

Example. The airplane can carry 60,000 lb on a 5000 nautical mile trip with a takeoff weight of 555,000 lb. The question is, how many passengers would the 60,000 lb represent? The average weight of one passenger is 165 lb, and passenger baggage is assumed to weigh 40 lb per passenger. Thus

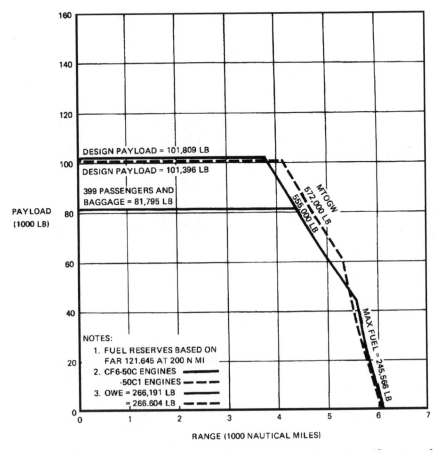

Figure 17.16 Payload versus range curve for passenger configuration. (*Courtesy of Boeing.*)

$$60,000/(165 + 40) = 292 \text{ passengers}$$

The payload versus range curve for cargo operations is presented in Fig. 17.17. The airplane range with maximum payload is 3300 to 3500 nautical miles, and the maximum takeoff weight is 555,000 to 572,000 lb. By carrying 100,000 lb of cargo, a range of 4700 to 5000 nautical miles can be achieved.

17.2.6 Community noise and the environment

Community noise levels of the DC-10 are significantly lower than those of jet freighters equipped with low-bypass-ratio engines. The 90-EPNdB

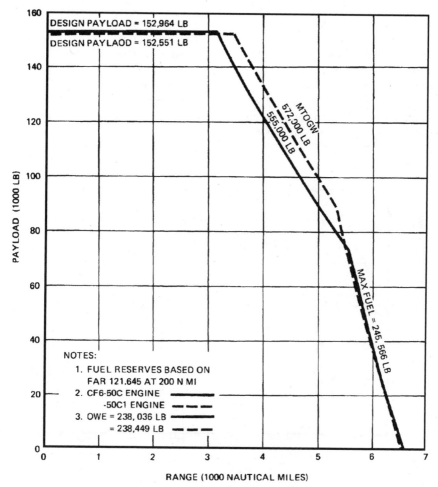

Figure 17.17 Payload versus range curve for cargo configuration. (*Courtesy of Boeing.*)

community impact comparisons of the DC-10 and the DC-8-63 illustrate the significant differences in noise structure.

Low community noise impact is especially important in freighter operations, which usually occur at night and often are subject to noise curfews. The maximum gross weight takeoff and approach noise levels of the DC-10 at FAR 36 measurement locations with either CF6-50C1/C2 or JT8D-59A/B engines are approximately 4 to 7 EPNdB lower for approach and 9 to 11 EPNdB lower for takeoff than those of the DC-8-63 and still satisfy FAA noise requirements. To an observer on the ground, a 10-EPNdB reduction would be heard as an approximately 50 percent decrease in noise because of the logarithmic relationship.

17.2.7 Pavement strength requirements

Each airplane carries an aircraft classification number (ACN). As an example, Fig. 17.18 shows a DC-10 airplane with an ACN presented for flexible pavement. Entering the chart at the airplane's gross weight, move up to the runway subgrade strength line (this may be obtained from the appropriate airport authorities). Turning to the left on the vertical scale, the aircraft classification number can be found. Consequently, this is compared with the pavement classification number (PCN), as described in Chap. 16.

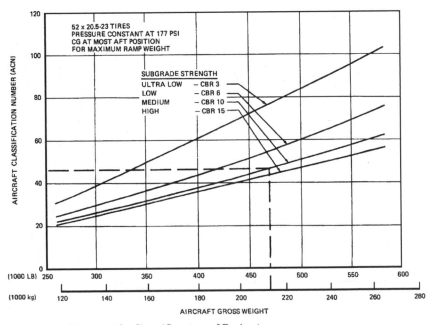

Figure 17.18 Pavement loading. (*Courtesy of Boeing.*)

17.3   MD-11 Aircraft

17.3.1   Introduction

A short description is due for the MD-11 aircraft. This aircraft is practically the further development of a DC-10 airplane. Numerous performance enhancements are built into this aircraft. The aircraft is available in four combinations:

- Passenger
- Freighter
- Convertible freighter
- Combi

17.3.2   Some improved features of the MD-11 versus the DC-10 aircraft

The MD-11 has an increased passenger-carrying capability. It can carry 52 more passengers than the DC-10, and this potential increase is estimated to generate an additional $15.2 million in revenue (Fig. 17.19).

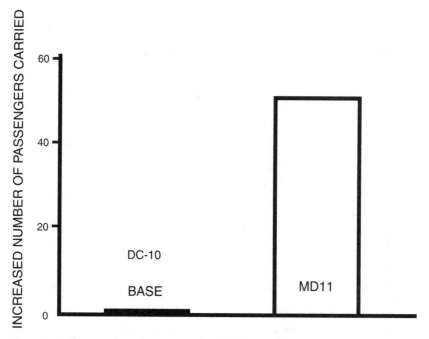

Figure 17.19   Increased passenger capacity of MD-11 versus DC-10.

The advanced aerodynamics and engine technology of the aircraft produce a considerable saving of about 14 percent compared with the DC-10 (Fig. 17.20).

Compared with the DC-10 aircraft, the MD-11 carries a greatly reduced ton-mile cost, as shown in Fig. 17.21. The aircraft's increased capacity with a reduced operating cost naturally brings a better return on investment.

To mention briefly, Lockheed also built trijets in series 1, 100, 200 and 500. These are quite similar to the DC-10 aircraft. The L-10-11 aircraft are not manufactured any more.

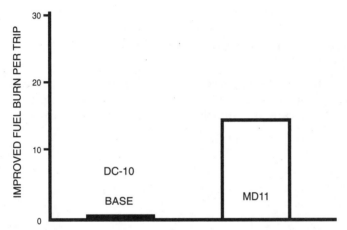

Figure 17.20   Lower fuel burn per trip.

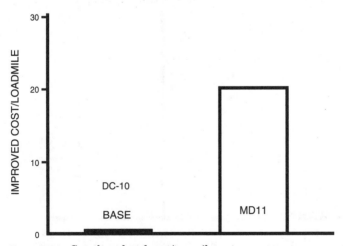

Figure 17.21   Greatly reduced seat/ton-mile cost.

Performance Description and Comparison

18.1 Short Description of the B747-400 Aircraft Performance Capabilities

18.1.1 The B747-400's capabilities

A B747-400 aircraft powered with Pratt-Whitney PW 4056 engines is discussed in this section.

18.1.2 Takeoff field requirements

This aircraft has an excellent runway performance capability. At sea level with maximum takeoff weight on a level runway at standard day conditions, the aircraft requires a 10,500-ft runway only. Compared with other aircraft, e.g., a DC-10-30 (with CF6-50C engines) that requires 11,700 ft of runway under similar conditions, this aircraft's takeoff weight is less frequently limited by runway length (Fig. 18.1).

18.1.3 Payload/range

Figure 18.2 presents a payload versus range chart for a B747-400 airplane with PW 4056 engines, at 0.85 Mach cruise. The vertical scale of this chart indicates the operating empty weight (OEW) plus payload values. In order to obtain payload weight, since the empty operating weight is a known figure of 394,000 lb, subtract the 394,000 lb from the "Operating Empty Weight plus payload", that is,

$$(OEW + payload) - 394,000 \text{ lb} = payload$$

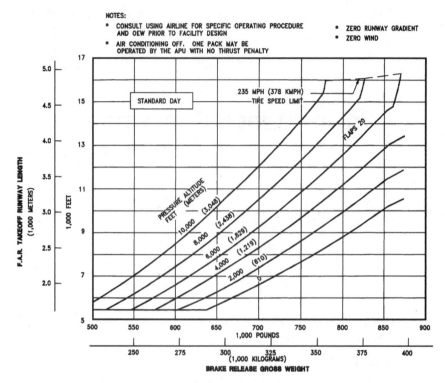

Figure 18.1   FAA takeoff runway length requirement, standard day, PW 4056 engines. (*Courtesy of Boeing.*)

First, the maximum payload should be established. From the chart we read 564,000 lb minus the OEW, that is,

$$(\text{OEW} + \text{payload} = 564,000 \text{ lb}) - 394,000 = 170,000 \text{ lb}$$

of payload for combi, and

$$535,000 - 394,000 = 141,000 \text{ lb}$$

for the passenger configuration.

The airplane is certified for various brake release weights, as shown in Chapter 17, Table 17.1. The next step is to determine the ranges for the certified brake release weights with maximum payload.

A similar tabulation can be developed for the combi configuration. A few examples are shown below for the use of the FAA takeoff runway length requirements chart (Fig 18.1) and payload versus range chart (Figure 18.2).

TABLE 18.1   Various Brake Release Weights and Ranges

Brake release weight, lb	800,000	833,000	850,000	870,000	875,000
Range with max payload, nautical miles	4,750	5,270	5,550	5,900	5,950

Note: For passenger configuration, zero − fuel weight is 535,000 lb, the maximum payload is 141,000 lb, and operating empty weight is 394,000 lb.

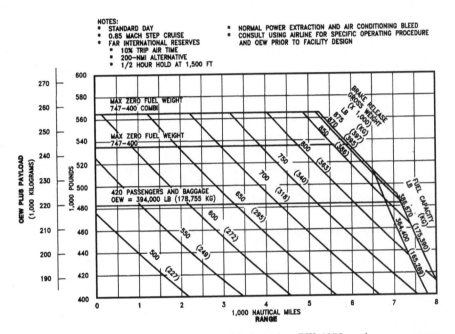

Figure 18.2   Payload versus range for 0.85 Mach cruise, PW 4056 engines, passenger configuration. (*Courtesy of Boeing.*)

Example of baggage-allowance calculation

Number of passengers carried: 420

OEW plus payload (from the chart with 420 passengers): 484,000 lb

Airplane OEW: 394,000 lb

420 passengers at 165 lb each: 69,300 lb

Baggage weight: 20,700 lb

20,700/420 = 49.2 lb baggage per passenger

Example of calculating the range of the airplane

Number of passengers carried: 380

Available runway length: 6700 ft

Sea level, standard day, runway slope: zero

Takeoff weight limiting obstruction under flight path: zero

Wind: zero

Air conditions: off

Takeoff weight based on these conditions (from Fig 18.1): 700,000 lb

380 passengers at 165 lb each: 62,700 lb

Baggage weight at 49.2 lb per passenger: 18,696 lb

Operating empty weight: 394,000 lb

OEW plus payload: 475,396 lb

To determine the range of the trip with an OEW plus payload of 475,376 lb, shown in Figure 18.2, move horizontally to the right to 700,000 lb. The range for this trip is 4500 nautical miles.

Example of calculating required runway length

Payload to be carried up to 5900 nautical miles: 126,000 lb

Airport elevation: 2000 ft

Standard day

Runway slope: zero

Obstruction below flight path limiting takeoff weight: zero

Wind: zero

Air conditions: off

The zero-fuel weight or OEW plus payload is 126,000 + 394,000 lb. Enter Fig. 18.2 on the vertical left side scale at 520,000 lb. Move to the right until the range of 5900 nautical miles is reached. At this point, a brake release weight of 850,000 lb is shown. The conclusion is that with a brake release weight of 850,000 lb over 5900 nautical miles, the airplane can carry a payload of 126,000 lb.

Now enter Fig. 18.1 on the horizontal weight scale at 850,000 lb and move up until the line labeled 2000 ft airport pressure altitude is reached. Moving horizontally from this point, a takeoff runway length of 11,250 ft is indicated.

These charts furnish a very good approximation of the expected performance of the airplane. Nevertheless, to calculate the correct brake release weight for each airport and each runway, the FAA-approved flight manual should be the correct source and should be used in the airline's daily operations.

18.2 Economic Comparison: DC-10 versus DC-8-63

18.2.1 Introduction

This is an economic comparison between two types of aircraft. Specifically, two trips have been selected for a detailed analysis: JFK to LHR and JFK to FRA. Figure 18.3 compares the sizes of the two airplanes.

18.2.2 Aircraft weight statements

The DC-8-63CF aircraft is 5 ft longer than the DC-10-30CF. Meanwhile, the DC-10's tail is 16 ft higher, and its fuselage has a larger diameter. The weight summary of both airplanes is given in Table 18.2.

Figure 18.4 shows the pallet arrangement for the main deck of eighteen 88 × 125 in pallets for the DC-8-63CF. For the DC-10, the pallet configuration is shown in Chapter 17.

18.2.3 Volume and tare weight summary

This is shown in Table 18.3. For both aircraft, the 88 × 108 and 88 × 125 in pallet configurations are tabulated. The maximum number of pallets for the DC-8-63CF is 18 pallets on the upper deck and bulk on the lower deck. For the DC-10-30CF, 30 pallets of the 108-in size and 22 pallets of the 125-in size can be accommodated on the upper deck, with 8 pallets on the lower deck. Based on the available size of pallets and containers, a large number of variations can be obtained.

COMPARISON OF DC-10 AND DC-8-63

DC10...		
WINGSPAN	6.5 FT	GREATER
TAIL HEIGHT	16 FT	HIGHER
LENGTH	5 FT	SHORTER

Figure 18.3 Comparison of DC-10 and DC-8-63 aircraft.

TABLE 18.2   Weight Summary

	DC-8-63CF*	DC-10-30CF
Engine type	JT3D-7	CF6-50C
Max. takeoff weight, lb	355,000	555,000
Max. landing weight, lb	275,000	411,000
Max. zero-fuel weight, lb.	261,000	391,000
Operators empty weight, lb	150,000	236,175
Max. fuel capacity at 6.7 lb/gal, lb	162,649	244,700

*CF = convertible freighter.

18.2.4   Economic assumptions

For economic evaluation, the assumptions should be stated first. They are shown in Table 18.4, which is aimed at comparing these two airplanes for a selected city-pair and conducting an economic evaluation. Landing fees are tabulated in Table 18.5.

Based on revenue, cost, and loads, the breakeven figure can be obtained.

18.2.5   Direct operating cost comparison

The direct operating cost is presented in dollars per statute mile for the range of up to 6000 statute miles. As shown in Fig. 18.5, the cost of the DC-8-63CF at 500 statute miles is $2.6 per statute mile, and the cost decreases as a function of range and reaches the value of $1.99 per statute mile at around 2000 miles. From here on up to 6000 miles, there is hardly any change in this value. The DC-10-30CF dollars per statute mile is about $0.55 per statute mile higher and runs parallel with the DC-8-63CF cost curve (see Fig. 18.5).

18.2.6   Comparison of selected trips

Table 18.6 contains cost figures for selected trips for both airplanes. Based on revenue, cost, and loads, the breakeven figure can be obtained. The comparison is conducted here for JFK-LHR-JFK and JFK-FRA-JFK round trips.

The average trip cost (JFK-LHR and JFK-FRA), including landing fees, for the DC-8-63CF is $16,851 and for the DC-10-30CF is $23,351, or about 38.5 percent higher. The average payload for the DC-8-63CF is 164,476 lb as opposed to 303,143 lb (84 percent higher) for the DC-10-30CF.

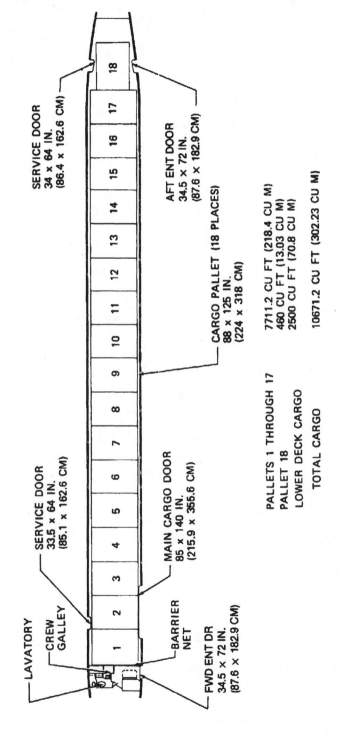

Figure 18.4 DC-8-63 in cargo configuration. (*Courtesy of Boeing*)

415

TABLE 18.3 DC-8-63CF and DC-10-30CF palletized cargo configuration

	DC-8-63CF	DC-8-63CF	DC-10-30CF	DC-10-30CF
Pallet size	108 in	125 in	108 in	125 in
Number of pallets:				
Above floor	18	18	30	22
Below floor (88 × 125)	—	—	8	8
Volume (ft^3):				
Above floor	7,373	8,244	15,462	14,621
Below floor (including bulk)	2,500	2,500	3,534	3,534
Total volume	9,873	10,744	18,996	18,155
Tare weight (lb):				
Above floor	3,870	4,500	6,450	5,500
Below floor			2,000	2,000
Total tare weight	3,870	4,500	8,450	7,500
Density* (lb/ft^3)	10.85	9.91	9.02	10.08

*Based on weight-limited payload.

TABLE 18.4 Economic Assumptions

Date of evaluation	19xx
Airplane study price	DC-8-63CF: $12 million
	DC-10-30CF: $24 million
Year of delivery	19xx
Depreciation	14 years
Residual value	10%
Insurance	1.5%
Utilization	4,380 h/yr
Maintenance	Use manufacturer's estimate
Maintenance allocation:	
Airframe	CAB-reported cost
Engines	80% contracted
Maintenance burden	1.80 × in-house maintenance labor
Labor rates	19xx
In-house	$6.52/mmh
Contract	$15.65/mmh
Contract material premium	1.30
Spares	10% airframe; 40% engine

Note: DOC-IOC relationship is 100%.

TABLE 18.5 Landing Fees ($)

	DC-8-63CF	DC-10-30CF
Frankfurt	569	949
London	705	1,208
New York	588	971

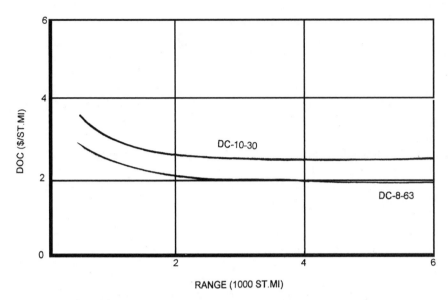

Figure 18.5 Direct operating cost versus range.

TABLE 18.6 Cost figures for DC-10-30CF and DC-8-63CF aircraft

Trip	JFK-LHR-JFK	JFK-LHR-JFK	JFK-FRA-JFK	JFK-FRA-JFK
Airplane type	DC-10-30-CF	DC-8-63CF	DC-10-30CF	DC-8-63CF
JFK-LHR, lb	154,825	92,454	154,825	84,088
LHR-JFK, lb	154,825	81,120	141,811	71,290
R.T. loads, lb	309,650	173,574	296,636	155,378
R.T. loads, ton	140.75	78.80	134.83	70.62
Operating cost, $/round trip	20,435	14,866	22,168	16,386
Landing fees, $/round trip	2,179	1,293	1,920	1,157
Total cost rel. to acft., $/round trip	22,614	16,159	24,088	17,543
Average cost/ton delivered, $	161	205	179	248
Difference, %	Base	+27	Base	+39
Pounds to beakeven	@ 27¢/lb	@ 27¢/lb	@ 29¢/lb	@ 29¢/lb
lb/round trip-cost	83,755	59,848	83,062	60,493
Total operating cost/lb, cents/lb	7.30	9.30	8.12	11.29

18.2.7 Conclusion and profit picture

The conclusion of this study is compiled in Table 18.7, where all pertinent cost information for both airplanes are tabulated side by side. As a conclusion of this study, the DC-10-30CF is the preferred airplane type (for JFK-LHR and JFK-FRA) because it can carry the full load with the exception of the FRA-JFK leg, where the payload is about 13,000 lb below the maximum capacity. This is due to head wind and long range.

The DC-8-63CF cannot carry the maximum load because the length of the trip is beyond its maximum payload range point. As shown in Table 18.7, the economic indicators for the DC-10-30CF are better than those for the DC-8-63CF.

18.3 Economic comparison of a large and medium-sized cargo aircraft

18.3.1 Introduction

The economics of two airplanes are compared in this study. One type of airplane, called X, is medium-sized, and the other, called Y, is a large cargo aircraft. This study was prepared several years ago. The numbers have changed, but the methods and procedures are still applicable in today's practice. The most significant changes are

- Fuel prices

- Depreciation schedule

TABLE 18.7 Cost and profit figures

	DC-8-63CF	DC-10-30CF	Changes, %
Total average cost for round trip	$16,851	$23,351	+38
Average cost per lb, round trip	10.2 ¢/lb	7.7 ¢/lb	+24
Breakeven load @ 28¢/lb, round trip	60,182 lb	83,390 lb	+38
Revenue @ 28¢/lb	$46,053	$84,960	+84
DOC/round trip	$16,851	$23,351	+38
IOC/round trip	$16,851	$23,351	+38
Total operating cost/round trip	$33,702	$46,702	+38
Profit before taxes/round trip	$12,351	$38,258	+209.7
Profit @ 50% load	$6,175	$19,129	+209.7
Total profit of 150 round trips/year @ 50% load factor	$926,250	$2869,350	+209.7*

*The high percentage of difference between the load capabilities of the two aircraft is caused by the long range of the analyzed trips because the length of these two trips extends beyond the maximum payload range point of the DC-8-63CF aircraft. On a shorter range, this difference would be around 50 percent only.

- Spare parts policies
- Cost of labor and materials
- Cost of airplanes

This study conducts a detailed profit and cost analysis for the JFK-LHR round trip.

18.3.2 Ground rules and economic assumptions

Cruise: 31000 ft at long-range cruise

Enroute winds: 75 percent probability

Temperature: Standard

Study route: JFK to LHR and LHR to JFK

Great-circle distance: 3442 statute miles

	Type X	Type Y
With full load	64,092 lb	198,006 lb
With 70% load factor:	44,800 lb	139,297 lb
Crew salaries:		
Dollar per block-hour	351.49	467
Dollar per mile	0.6892	0.849
Block-hours to flight-hours ratio	1.08	1.10
Block speed, mi/h	510	550
Yield considered $0.156 per revenue-ton-mile		

Note: Dollars per block-hours/miles per hour = dollar per mile.

18.3.3 Fuel analysis

Round trip: JFK-LHR-JFK

Wind probability: 75 percent

	Type X	Type Y
Block time, h	15.23	14.36
Block fuel, lb	205,837	372,051
Trip fuel cost*	$3139	$5675
Dollars per mile	0.456	0.8244

*Fuel @$0.1022/gal.

18.3.4   Block to flight time ratio (JFK-LHR-JFK)

Type X airplane:

$$\text{Flight time} = 14.10 \text{ flight hours}$$

$$\text{Block time} = 1.08 \times 14.10 = 15.23$$

This provides 34 minutes of ground time at each end of the mission for taxi and ground maneuvers.

Type Y airplane:

$$\text{Flight time} = 13.05 \text{ flight hours}$$

$$\text{Block time} = 1.10 \times 13.05 = 14.36$$

This provides 39 minutes of ground time at each end of the mission for taxi and ground maneuvers.

18.3.5   Maintenance cost

This includes burden.

Type X airplane: $175.44 dollars/block-hour

$$\frac{175.44}{510} = 0.344 \text{ dollars/mile}$$

Type Y airplane: $347 dollars/block-hour

$$\frac{347.93}{550} = 0.6326 \text{ dollars/mile}$$

18.3.6   Depreciation schedule

Type X airplane:

  Price: $4.75 million

  Straight line: 8 years to 10 percent residual

  Annual utilization: 4000 block-hours per year

Type Y airplane:

  Price: $24.75 million

  Straight line: 16 years to 10 percent residual

  Annual utilization: 4000 block-hours per year

Depreciation cost
Type X airplane:

$$\frac{4.75 - \$0.475}{8 \text{ years}} = 534{,}375 \text{ dollars/year}$$

$$\frac{534{,}375}{4000} = 133.59 \text{ dollars/block-hour}$$

$$\frac{133.59}{510} = 0.2619 \text{ dollars/mile}$$

Type Y airplane:

$$\frac{\$24.75 - \$2.48}{16 \text{ years}} = 1.392 \text{ million dollars/year}$$

$$\frac{\$1.392}{4000 \text{ hours}} = \$348 \text{ dollars/block-hour}$$

$$\frac{\$348}{550} = 0.632 \text{ dollars/mile}$$

18.3.7 Insurance cost

At 1.5 percent purchase price:
Type X airplane:

$$1.5 \times \$4.75 = 71{,}250 \text{ dollars/year}$$

$$\frac{71{,}250}{4000} = 17.80 \text{ dollars/block-hour}$$

$$\frac{17.80}{510} = 0.0349 \text{ dollars/mile}$$

Type Y airplane:

$$1.5 \times \$24.75 = 371{,}200 \text{ dollars/year}$$

$$\frac{371,200}{4000} = 92.80 \text{ dollars/block-hour}$$

$$\frac{92.80}{550} = 0.1687 \text{ dollars/mile}$$

18.3.8 Direct and indirect operating costs and revenues

Direct operating cost (based on JFK-LHR-JFK system):
Type X airplane:

Crew	$0.6892
Fuel	0.456
Depreciation	0.2619
Insurance	0.0349
Maintenance	0.344
Total$/mile	1.786

Type Y airplane:

Crew	$0.849
Fuel	0.8244
Depreciation	0.632
Insurance	0.1687
Maintenance	0.6326
Total $/mile	3.1067

Type X JFK-LHR-JFK systems:

Revenue $-$ 22.4 tons $\times$ 6884 miles = 154,201 revenue-ton-miles

$0.156 $\times$ 154,201 = $24,055 revenue

(Yield = 15.6 cents per ton-mile.)

Total revenue = $24,055

Direct operating cost:

$1.78625 per mile $\times$ 6884 miles = $12,296

Indirect operating cost:

Aircraft and traffic servicing, two departures	$4486
Sales and promotion, 10 percent of revenue	$2405
Ground property equipment, $12.88/block-hour × 15.23 hours	$196
General and administrative, 8% of all IOC	$567
Total indirect operating costs	$7654

Type Y JFK-LHR-JFK systems:

$$\text{Revenue} - 69.6 \text{ tons} \times 6884 \text{ miles} = 479{,}126 \text{ revenue-ton-miles}$$

$$\$0.156 \times 479{,}126 = \$74{,}743$$

$$\text{Total revenue} = \$74{,}743$$

Direct operating cost:

$$\$3.107 \text{ per mile} \times 6884 \text{ miles} = \$21{,}388$$

Indirect operating cost:

Aircraft and traffic servicing, two departures	$11,904
Sales and promotion, 10 percent of revenue	$7,474
Ground property equipment, $39.86/block-hour × 14.36/hour	$572
General and administrative, 8 percent of all IOC	$1,596
Total indirect operating cost:	$21,546

18.3.9 Economic comparison

	Type X	Type Y
Total revenue	$24,055	$74,743
Total operating cost	$19,950	$42,934
Operating profit	$4,105	$31,809
Percent operating profit to revenue	17.06%	42.55%
Operating profit per ton	$183.2	$456.7
Total operating cost/rtm	$0.129	$0.0896

Note: For the Type Y, the revenue is about three times that of the type X, while the cost is only two times more. This is the advantage of a larger airplane.

Operating cost breakdown (70 percent load factor):

	Type X	Type Y
Direct operating cost	7.97 cents/revenue-ton-mile	4.463 cents/revenue-ton-mile
Indirect operating cost	4.96 cents/revenue-ton-mile	4.496 cents/revenue-ton-mile
Total operating cost	12.93 cents/revenue-ton-mile	8.96 cents/revenue-ton-mile
IOC percent of DOC	62.2%	100.7%

18.4 Economic Comparison for a Group of Different-Sized Airplanes and Equal Lifts Comparison

18.4.1 Introduction

In the following description, a small airline's fleet plans will be evaluated. The forecast will furnish the expected load for a North Atlantic operation. Two types of aircraft will be compared for ALFA Airlines, namely, a large airplane of type ABC-999 and a smaller airplane of type XYZ-111.

18.4.2 Economic ground rules and assumptions

A market share will be established and a schedule will be prepared for this operation. The ground rules are listed in Table 18.8.

18.4.3 Defining direct operating cost

A direct operating cost summary and scheduled cargo system is shown in Table 18.9.

Annual direct operating cost is listed in Table 18.10. Total miles based on the schedule and are multiplied by the dollars per mile to furnish the annual cost.

18.4.4 Defining indirect operating cost

Indirect operating cost assumptions are listed in Table 18.11 and a summary is given in Table 18.12.

18.4.5 Total system revenue, costs, and profit

Total system revenue-ton-miles is presented in Tables 18.13 and 18.14.

This study indicates that the three ABC-999 airplanes generate about twice as much profit as the XYZ-111 airplanes. The smaller XYZ-111 airplanes individually carry less payload than the larger ABC-999 airplanes, but the smaller airplanes combined carry more load than the larger aircraft. Since certain servicing costs, such as landing fees and traffic servicing costs, are related to the frequency of the flights and the smaller airplanes have higher frequency, operating expenses are not proportional. These higher operating expenses thus decrease the profit.

TABLE 18.8   Direct Operating Cost Ground Rules

	ABC-999	XYZ-111
Study price	23.14 million	11.70 million
Engine price (each)	0.76 million	.323 million
Annual utilization	3894 hours/year	4091 hours/year
Depreciation	16 years to 10%	14 years to 10%
Maintenance	$330/flight-hour	$220/flight-hour
Crew	$325/block-hour	$215/block-hour
Insurance	$84/block-hour	$48.75/block-hour
No. of aircraft considered	3	11

Note: Load factor of 65 percent is considered for both aircraft.

TABLE 18.9   Direct operating cost and cruise type

Model	Average range, statute mile	Cruise procedure	DOC, $/mile	DOC, $/ton-mile at 65% LF
ABC-999	1450	LRC @ 31000 ft	3.78	0.058
XYZ-111	1247	LRC @ 31000 ft	2.22	0.076

TABLE 18.10   Annual costs for both airplanes

Aircraft	Annual cost
ABC-999	5,630,924 miles × 3.78 $/mile = $21,284,892
XYZ-111	20,228,052 miles × 2.22 $/mile = $44,906,275

TABLE 18.11   Indirect operating costs element

Expense type	ABC-999	XYZ-111
Traffic servicing	$40 per ton	$50 per ton
Aircraft servicing	$1000 per departure	$700 per departure
Sales and promotion	10% of commercial sales	10% of commercial sales
General and administration	2.5% of other indirect costs	2.5% other indirect costs

TABLE 18.12   Indirect Operating Cost Summary

	Annual cost	
3 ABC-999		
Traffic servicing	$40 × 151,892 tons	$6,075,680
Aircraft servicing	$1000 × 3900 departures	$3,900,000
Sales and promotion	10% × $50,634,258	$5,063,425
General and administration	2.5% × $15,039,105	$375,977
	TOTAL	$15,415,082
11 XYZ-111		
Traffic servicing	$50 × 116,324 tons	$5,816,200
Aircraft servicing	$700 × 16,016 departure	$11,211,200
Sales and promotion	10% × $78,212,891	$7,821,289
General and administration	2.5% × $24,848,689	$621,217
	TOTAL	$25,469,906

Note: Traffic servicing cost is based on tonnage handled. Sales and promotion cost is based on revenues, while general and administration cost is based on indirect operating cost.

TABLE 18.13   Total system revenue-ton-miles

Regions	ABC-999	XYZ-111	Total revenue-ton-miles
Domestic	24,113,544	42,411,148	66,524,692
International	285,711,275	437,289,881	723,001,156
International charter	71,427,819	109,322,471	180,750,290
TOTAL			970,276,138

TABLE 18.14   Total System Revenue and Cost

	ABC-999	XYZ-111
Revenue		
Domestic	$3,134,760	$5,513,449
(13 cents/revenue-ton-mile)		
International	$39,999,578	$61,220,583
(14 cents/revenue-ton-mile)		
International charter	$7,499,920	$11,478,859
(10 cents/revenue-ton-mile)		
TOTAL	$50,634,258	$78,212,891
Costs		
DOC*	$21,284,892	$44,906,275
IOC**	$15,415,082	$25,469,906
TOTAL	$36,699,974	$70,376,181
Profit	$13,934,284	$7,836,710

*See Table 18.10
**See Table 18.11 and Table 18.12

The advantage of using the smaller airplanes is the easy adjustment to a low season when only a partial fleet is in operation.

Note: Indirect operating cost could be expressed in percentage of direct operating cost. Traffic servicing can be considered as a function of tons, and its cost could be based on the degree of automation (e.g., mechanized cargo system). In passenger operation, the volume of passengers has an effect on cost. Aircraft servicing can be considered per departure or as a percentage of direct operating cost. Sales and promotions are generally 8 to 10 percent of commercial sales volume. General and administrative expenses are in the magnitude of 2 to 3 percent of indirect operating costs.

18.4.6   Equal lift study

An equal lift study is conducted for the 3 ABC-999 airplanes versus 6.6 XYZ-111 airplanes. This study compares expenses and return on investment values when the same lift is applied for both types of aircraft and the revenues in this case are identical (Table 18.15).

The cost of spare parts for the ABC-999 is $3.47 million. A three-airplane fleet is considered, and the spares constitute 15 percent of airplane price. For the XYZ-111, the spare parts amount to $820,000, which represents 7 percent of the airplane price.

The schedule for the ABC-999 airplane's prepayment is 48 percent of the total cost and 52 percent at the time of delivery. For the XYZ-111 airplane, the prepayment is 35 percent of the total cost 1½ years prior to delivery and 65 percent at time of delivery. Table 18.16 contains the total investment figures.

18.4.7   Operating cost

For 6.6 XYZ-111 airplanes, the direct operating cost is as follows: Annual:

2.12 $/mile × 5,630,924 miles × 2.2 airplanes

System average range 1450 statute miles @ $2.12/mile: $26,262,629

Indirect operating cost:

Traffic servicing (151,892 tons) @ $50/ton	$7,594,600
Aircraft servicing (2.2 × 3900 departures) @ $700/departure	$6,006,000
Sales and promotion (10 percent of commercial sales)	$5,063,425
General and administration (2-½ of all other indirect costs)	$466,600
Total IOC	$19,130,625
Total operating cost	$45,399,254

TABLE 18.15   Determination of Investment, Equal Lift ABC-999 and XYZ-111

Airplane type	ABC-999	XYZ-111
Airframe study price	$20.099 million	$10.410 million
Engines	$3.040 million	$1.290 million
Aircraft study price	$23.139 million	$11.700 million

TABLE 18.16   Total Investment

Airplane type	ABC-999	XYZ-111
Aircraft	$69.417 million	$77.22 million
Spares	$10.410 million	$5.41 million
Total	$79.817 million	$82.63 million

Three ABC-999 airplanes carry the load of 6.6 type XYZ-111 airplanes, or 2.2 type XYZ-111 airplanes carry the load of one ABC-999.

18.4.8   Equal lift summary

Three ABC-999 airplanes are compared with an equal lift with 6.6 XYZ-111 airplanes in Table 18.17.

18.4.9   Profit and rate of return on investment

The annual profit summary (equal lift/equal revenue) of the aircraft is shown in Table 18.18. These figures clearly indicate the disadvantages of the smaller airplanes. The revenues are identical (equal lift capability). The operating costs are much higher for the XYZ-111 airplanes (more landing), causing lower profits (Tables 18.19 and 18.20). Return on investment (will be discussed later) summary (equal lift):

The following assumptions are established for this analysis:

- 100 percent of investment borrowed @ 10 percent rate
- 10-year loan period
- Residual value based on 10 percent of investment price

Note: Once the basic figures are changed, this study could alter the final result.

TABLE 18.17   Investment and Operating Costs

Airplane type	(3) ABC-999	(6.6) XYZ-111
Total investment	$79,817,000	$82,630,000
Operating costs		
Direct operating cost	$21,284,892	$26,262,629
Indirect operating cost	$15,415,082	$19,130,625
Total operating cost	$36,699,974	$45,393,254

TABLE 18.18   Revenue and Profit

Airplane type	(3) ABC-999	(6.6) XYZ-111
Revenue	$50,634,258	$50,634,258
Expenses	$36,699,974	$45,393,254
Profit	$13,934,284	$5,241,004

TABLE 18.19   Return on Investment

Airplane type	(3) ABC-999	(6.6) XYZ-111
Return on investment	+6.68%	5.46%

TABLE 18.20   Residual Values

Airplane type	(3) ABC-999	(6.6) XYZ-111
Residual value (A/C and spares)	$7.98 million	$8.26 million

Evaluation of Investment

19.1 Introduction

To evaluate the advisability of an investment in aircraft for a commercial airline, the following approach has proven successful: First, one must select the desired type of aircraft and identify its performance and economic characteristics. The next steps are to design a market expressed in a schedule format and establish conditions on which the proposed schedule should be based. In the case of a corporate aircraft or a nationalized airline, this would not have a primary importance. Then one must establish a target rate of return for this investment. To compute the return on investment, one must consider the relevant data, some of which are listed below:

1. Airplane(s) delivery date
2. Economic life
3. Purchase price (A, S, G)
4. Depreciable life (A, S, G)
5. Direct and indirect operating costs
6. Depreciation method (A, S, G)
7. Income tax rate
8. Residual value (A, S, G)
9. Airplane residual value at the end of economic life

10. Predelivery payment schedule

11. Miscellaneous additional information

Note: A, S, and G = aircraft, spare, and ground equipment, respectively.

The return on investment is a measure by which various alternate investments can be compared and evaluated. It can predict earnings for a particular fleet, the present fleet, the present fleet with additional aircraft, or a planned new fleet. Cash divestment (disposal of equipment) must reduce investment. Cash flow has to be time accountable because of the time value of money. Cash interest expenses must not be included in total operating cost. Interest is included in the cost of capital.

The reason most of the various methods are not realistic is that some deal with averages and are not concerned with the time shape of the investment. They take all the proceeds by their nominal value regardless of their timing, and this disqualifies each of them from being the only true methods for evaluating a given investment. For example, someone receives $100 today and another $100 in the year 2050. The second $100 does not have the same value as the first $100 received today. The basic element of time makes an evaluation model true and correct. For this reason, discounted cash flow is the best method because it takes the time value of the money into consideration.

19.2   Time Value of Money

In a free economy, funds are invested in projects to generate profit and/or to obtain capital appreciation. There are three distinct periods for each investment project. First, the investment is tied up in the project and is not available for other purposes. Second, cash flow is generated as the project is in full operation. Third, the residual value, if any, should be considered as the project is concluded. When projects run parallel, they may be in various stages. This means that while one project may be tying up funds, another may be generating cash flow to be invested into new projects. Timing is essential, and it can greatly affect the magnitude of the profit.

In general there are a few items that have to be discussed before the various evaluation methods are presented.

19.3   Revenues and Costs

All cash revenues must be accounted for. All cash expenses including taxes on additional earnings resulting only from the investment must be deducted from revenues.

19.4   Tax Shield

All noncash expenses, usually in the form of amortization, must be treated properly to shield earnings from the full impact of corporate income taxes.

19.5   Investments and Resale Value

All cash investments in the equipment must be considered. All cash divestments when equipment disposal takes place must reduce the investment accordingly.

19.6   Timing

As a result of investment, all cash flow must be strictly time accountable because of time value of the money.

19.7   Interest

Interest expense related to borrowed funds cannot be included in direct or indirect operating costs. This would be double counting, and for this reason, it cannot be included in the cash flow calculation.

19.8   Various Evaluation Methods

There are several other methods of calculation that can assist in the evaluation process. A few are listed below.

19.8.1   Payback period

This is measured in years. It is the length of time required to recoup the original full cash outlay for the investment. In other words, how long does it takes for the investment to pay for itself?

All proceeds before the payback period are treated and counted as equal, and all proceeds after the payback period are ignored. This method does not consider the time value of money. To illustrate this, let us compare two alternate investments (Table 19.1).

TABLE 19.1   Comparison of two investment proposals

	Proposal A	Proposal A	Proposal B	Proposal B
Initial investment	$10,000	accumul.	$14,000	accumul.
Year	Amount	Amount	Amount	Amount
1	1000	1,000	500	500
2	2000	3,000	2,000	2,500
3	3000	6,000	3,500	6,000
4	4000	10,000	5,000	11,000
5	5000	15,000	6,000	17,000
6	5500	20,500	7,500	24,500
7	6000	26,500	10,000	34,500
8	6000	32,500	10,000	44,500
9	6000	38,500	10,000	54,500
Etc.				
Payback period		4 years		4.5 years

The yearly cash flow and the accumulated amounts are shown for both investments. It takes generally 3 to 6 years to pay back the investment. All the proceeds received before the payback period are counted as equal, and proceeds received after the payback period are ignored.

19.8.2   Lifetime profit generated

It is not realistic to compare two alternate investments in which one generates $120 million and the other $180 million of profit during the life of the investment. Such a comparison ignores the timing of the cash flows and the size of investments, not to mention that the duration of the investments could be drastically different. In addition, the time value of the money is ignored.

19.8.3   Per dollar investment

This is a simple calculation that divides the lifetime profit by the initial investment. It does not consider the time value of the money. Furthermore, the initial investment is retrieved over the life of the project through annual depreciation charges. The cash proceeds during earlier years are more valuable than the proceeds received during the later years because they can be reinvested earlier and used to generate additional profit.

19.8.4   Average annual proceeds

Here is another simple division. The lifetime cash proceeds are divided by the number of years. The time value of money is again ignored.

19.8.5 Average annual proceeds per dollar invested

This method divides the average annual profit after taxes into the average or original investment. It is based on accounting income rather than cash flow. The time value of money is ignored.

19.8.6 Average annual return on average book value

This approach is applicable only for a short time period. Its failure is in the timing of proceeds expenses. It is calculated by dividing the average income for a given year by the average book value of the airplane during the same period. The time value of money is ignored.

Before continuing the subject of various investment methods, the concept of present value has to be discussed in more details.

19.9 Concept of Present Value of Future Sum and the Future Value of Present Sum

By using the process of discounting, we can determine the present value of a future sum, and by applying compounding, we can determine the future value of a present sum (Fig. 19.1) and show it graphically in a simple example.

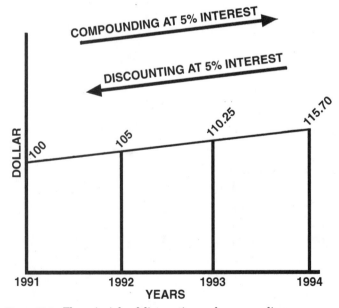

Figure 19.1 The principle of discounting and compounding.

There is another way to present this relationship (Table 19.2) using the same figures as in the preceding example.

This indicates that financially they are equal (Fig. 19.2). For example, $100 in 1991 is equivalent to $105 in 1992. This is called *financially equal,* and it is compounded. Similarly, $105 in 1992 is equal to $100 in 1991, and it was discounted. The figure shows these amounts as a function of time. It is a very common way for illustrating cash flows.

There is another presentation frequently used where all the elements are shown and their relationship is clearly demonstrated (Fig 19.3). For example, 100/0.9090 = 110.0 or 110.00 × 0.9090 = 100. This means that $100 at 5 percent interest compounded annually will be $115.75 in 3 years. The reverse of this situation is that $115.75 discounted at 5 percent interest has a present value of $100 in the year 1991.

Tables are available to calculate discount or compounded rates. For example, in the discount table under 5 percent interest rate at a period of 3 years, the discount factor is 0.8638. Thus, 0.8638 × $115.75 = $100.

The discount table provides rates for different interest rates. As shown earlier, $100 is financially equivalent to $115.75 after 3 years at 5 percent annual compounded interest. That is, a person would be receiving equal value if he or she had $100 today or were to receive $115.75 in 3

TABLE 19.2   Discounting and Compounding

1991	1992
$100	$105
$100	$105

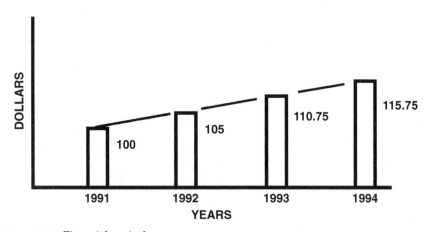

Figure 19.2   Financial equivalency.

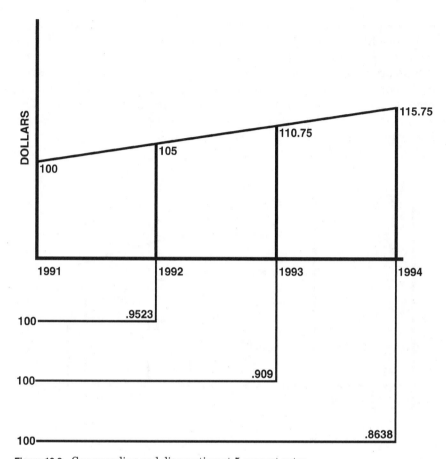

Figure 19.3 Compounding and discounting at 5 percent rates.

years, assuming a discount rate of 5 percent. This is the concept used for cash flows; all cash flows are referenced to a given point of time. By applying their present values, regardless of their time schedule, we are able to compare, at the selected time, the cash inflow and outflow and draw conclusions about the value of the investment. We reference all cash input or output to a given point of time and make these financially equivalent regardless of the timing of their receipt. This is the correct way to compare cash flows.

To illustrate the concept, a graphic presentation shows the situation in Fig. 19.4. Cash outflow has to be referenced to a base year. (*Note:* An arbitrarily selected point in time is generally the first year of operation.) Similarly, cash inflow has to be referenced to the base year. Consequently, one can clearly see the investment and cash inflow relationship. The future cashflow is discounted to the base year or generates present value to the base year. Since every amount of money is referenced to the base year, it is easy to evaluate the project. It will answer the question: Is it worthwhile to invest a certain amount money in this project, or is it better to place it in a bank? If this project brings in the same percentages that a bank would pay for a savings account, the project should not be considered, not to mention the risk involved. The success of the project can be measured by the rate of return on the investment (ROI). As shown in Fig. 19.5, the smaller the time between compounding, the higher the amount becomes.

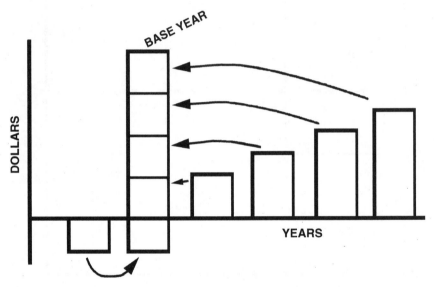

Figure 19.4 Reference time for cash flows.

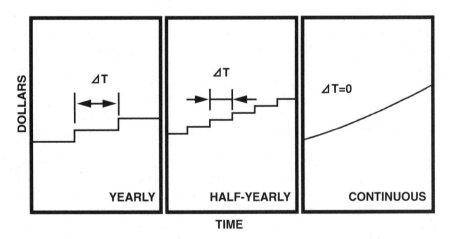

Figure 19.5 Time effect on compounding.

19.10   Cash Flow Analysis

The first step is to subtract the expenses—direct and indirect operating costs—from the total revenue to obtain profit before taxes. With the known tax rate, we establish profit after taxes.

In passenger operations, cargo revenues are additional profits, since passenger operation expenses are already accounted for. Before we proceed, depreciation has to be considered. Depreciation is not a cash expense. It is a bookkeeping entry, and airlines retain it without paying taxes on it. It is for the replacement of equipment. The straight-line depreciation method allows a constant amount of expenses annually, which permits lower costs and higher profits in the early years of an investment. The annual report shows this because stockholders like to see profits. For income tax purposes, the opposite is applicable, such as higher expenses and lower profit. The idea is to pay lower taxes today and higher taxes tomorrow. Here, the time value of money comes into play. In direct operating cost, straight-line depreciation is replaced with accelerated depreciation to reduce the annual book operating profit before taxes in early years. Then the accelerated depreciation is eliminated because it is a noncash expense. This would increase the annual cash operating profit before taxes.

19.11   Internal Rate of Return (IRR) Method

The internal rate of return for an investment proposal is the discount rate that equates the present value of the expected cash outflows with the present value of the expected inflows. The best way to show the method is through an example:

Investment	$1,995
Yearly cash flow	Variable
Duration of project	5 years

$$\$1995 = \$505/(1 + IRR^1) + \$505/(1 + IRR^2) + \cdot + \$445/(1 + IRR^5)$$

where we have to solve it for IRR, and IRR is the internal rate of return.

The exponent of this expression starts with one and is incremented by one with each additional expression. There is a fast way to estimate the value of IRR. The next example shows the method of approximation (Table 19.3).

Divide $677 by the number of years:

$$677/5 = 135$$

TABLE 19.3   Cash flow

Investment	$1990
Duration	5 years
Cash flow	
Year 1	$505
Year 2	$505
Year 3	$606
Year 4	$606
Year 5	$445
Total cash	$2667
Less investment	$1990
Remains	$677

Multiply by 100 and divide by half the amount of the investment. This will be

$$135 \times 100/995 = 13.5\%$$

Let us consider 12 percent as a first try (Table 19.4). The 12 percent rate generates less than the value of the investment; therefore, try 10 percent. Let us set up the same layout (Table 19.5). Interpolation will furnish a 10.40 percent rate of return. With two easy steps, we calculated the rate of return.

19.12   Present-Value Method

With this method, all cash flows are discounted to present value using the required rate of return. If the sum of this cash flow is equal to or greater than zero, the project is acceptable; otherwise, it is rejected.

To show this in the next example, assume a required rate (RR) (where $RR = (1 \times Rx)$ and Rx = rate of return) of return of 10 percent after taxes. Cash flow of $5590 is the cash flow for a given time period.

$$\text{Net present value} = -17{,}990 + [5590/(RR^1) + 5590/(RR^2) + 5590/(RR^3)] + \cdot]$$

$$= -17{,}990 + 21{,}200 = \$3210$$

Note: The exponent of RR is incremented from one, and +1 is added for each subsequent expression of RR.

The result is a positive number, so the project could be accepted. The profitability index (PRFIX) is

$$\text{PRFFIX} = \text{present value of cash inflow} \div \text{present value of cash outflow}$$

TABLE 19.4   First approach: 12%

Year	Cash flow	Present value @ 12%	Present value
First	$505	0.893	$451
Second	$505	0.797	$402
Third	$606	0.712	$431
Fourth	$606	0.636	$385
Fifth	$445	0.567	$252
Total			$1921

TABLE 19.5   Second approach: 10%

Year	Cash flow	Present value @ 10%	Present value
First	$505	0.909	$459
Second	$505	0.826	$417
Third	$606	0.751	$455
Fourth	$606	0.683	$412
Fifth	$445	0.621	$276
Total			$2020

In this example:

$$PRFIX = \$21,200/\$17,990 = 1.18$$

This is greater than 1.00, so the project could be accepted.

There are some additional considerations that should be mentioned very briefly in relation to calculation of the return on investment.

19.13   Graphic Presentation of Internal Rate and Present-Value Methods

It is worthwhile to plot the net present value versus discount rate. When the discount rate is zero, the net present value shows the cash flow of the project (not discounted). Calculating for a few percents (e.g., 5, 10, 15, and 20 percent), a graph can be plotted, as seen in Fig. 19.6.

Two projects, called project A and project B, are shown in Fig. 19.6. The amount of investment for both projects could be read at zero discount rate. For the sake of argument, let us assume that IRR(A) = 15 percent and IRR(B) = 10 percent. If a loan is taken from a financial institution at 15 percent interest rate and the IRR(A) = 15 percent, then we are better off placing the funds in a bank because it is relatively risk-free. If the internal rate of return is less than the cost of the

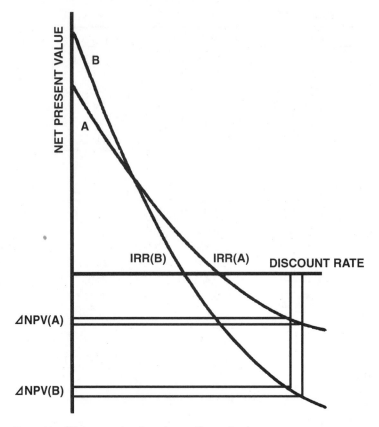

Figure 19.6   Net present value versus discount rate.

moneys used to finance this project, the project should be rejected. If the return is higher than the cost of funds for the project, a surplus will be generated, and the value of the stock will climb.

This example indicates clearly that the internal rate of return should be above the cost of funds.

Taking a second look at the net present value versus discount rate graph, the following can be concluded:

1. As the discount rate increases, the net present value decreases.

2. At low discount rates, project B has higher net present value until the crossover point. Above the crossover point, project A has a higher net present value.

3. Both projects are acceptable if the cost of capital is below the crossover point.

4. Project B is more sensitive to discount rate changes above the crossover point. By inspecting the graph, this relationship becomes obvious. At high discount rates, a small increment in discount rate generates a higher change in net present value for project B.

5. When selecting airplanes, projects are mutually exclusive. In this case, we may reject project A or B.

6. The internal rate of return method assumes that the cash flow is being invested at the rate of IRR, while the net present value method assumes reinvestment at the rate of the cost of capital. This is a more realistic approach. If both methods are analyzed, the project with the higher net present value should be chosen.

There are other considerations, such as inflation, capital rationing, risk of the project, etc., that are beyond the scope of our discussion.

We have presented two methods to evaluate investment, namely, the net present value method and the internal rate of return method. The question arises as to which method would be more feasible to apply. It has been discussed in this text that any method used should satisfy the following criteria:

- The cash flow should be considered for the entire life of the project.
- It should always consider the time value of money.

The project with the higher return should be chosen.

By now we know that the payback period method does not satisfy these criteria. The internal rate of return method satisfies the two requirements; namely, it considers the cash flow for the duration of the project and the time value of money. Sometimes there is more than one solution or no solution for the internal rate of return method. Occasionally, in the cash flow stream, a number of reversals of signs are experienced. If, for example, there are two reversals of signs in the cash flow (Fig. 19.7), we can have two internal rates of return. A multiple reversal of signs is a necessary condition for multiple internal rates of return. This is not the case with the net present value method.

19.14 Nominal Rate and Effective Rate of Interest

The nominal rate of interest does not reflect the actual rate at which the interest is earned. The effective rate is the rate that, when compounded annually, yields the same amount as the nominal rate compounded n times a year. The interest actually earned on an investment depends on the compounding frequency (Table 19.6). For example,

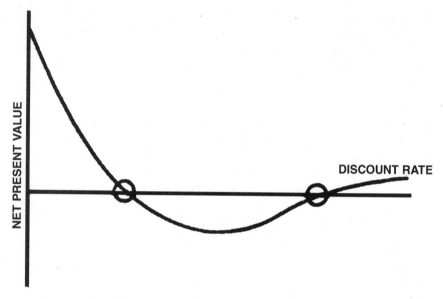

Figure 19.7 Reversal of sign in cash flow.

nominal rate is 8 percent and the time period is 3 years. The effective rate (*EFFRATE*) is calculated as follows:

$$EFFRATE = ((1 + NR/N)^N - 1) *100$$

where *EFFRATE* is the effective rate of interest, *NR* is the nominal rate, and *N* is the frequency of payments.

19.15 Financial Model: Return on Investment

19.15.1 Introduction

After the introduction of some techniques for evaluating an airplane's economic value, based on revenues and expenses, a few models should be examined so that their return on investment can be calculated. The economic worth of an investment is influenced not only by the amount of the investment and its economic life but also by the time, shape, or pattern of earnings. The use of the present-value or discounted-cash-flow concept is necessary to accurately compare the total investment required and the revenues produced. The cash flow generated in an airline approximates the situation where uniform flows are generated at the end of each month and compounded at the end of the year.

The return-on-investment calculation is the best method for defining the profitability of the investment. It is an improvement over the

TABLE 19.6   Frequency of payment vs. final amount

Investment	Frequency	Interest rate	Final amount
1000	Annual	8%	1259.71
	Semiannual	4%	1265.82
	Quarterly	2%	1268.24
	Monthly	0.66%	1271.75

annual cost or any other method, e.g., payback years, etc., because it applies the present-value factors. The return-on-investment method discounts the cash flow back after taxes over the years until it equals the investment. This method requires two steps:

Step 1: Set up a yearly timetable of all revenues and costs predicted to occur over the life of the investment.

Step 2: By a successive iteration process, extend these entries by the continuous compound interest rates until the present value of the receipt equals the investment. Present value discount rate yields the rate of return.

Using this method, it makes no difference when money is invested, either before or after airplane delivery. The amount of expenditures and receipts also can vary. Capitalization, amortization, and salvage value also can fluctuate. This is the one method that evaluates differences among the proposals in a precise time shape of future earnings. It takes into consideration the full life characteristics of each airplane and does not rely on averages or predetermined factors of interest.

This method has the flexibility of absorbing expenditures and revenue as they occur and handling any amount of money expended before the projects begin, as well as the money spent and returned afterwards. It yields a rate of return showing immediately what the airplane operation will cost at a constant rate of cash profit. It answers the questions, Is this investment worthwhile, and will it contribute real profit?

Therefore, the major elements of revenues and expenses will be discussed in more detail, such as operating expenses and revenues, nonoperating income and expenses, tax considerations, amortization, and depreciation.

19.15.2   Source and application of funds

The airline should have the information that summarizes its cash position and needs. This is determined using a source and application of fund statement. This statement summarizes the internal sources of funds and the commitments for use of those funds, allowing the airline to determine the need for external sources of financing. Internal sources are profits, depreciation, and the proceeds from sale of the aircraft. External sources are the sale of stocks, bonds, convertible debentures, and new borrowing. The fund is applied for dividends, loan repayments, training and development, and capital expenditures.

Sources:

Depreciation (internal)

Profits (internal)

Net income (internal)

Deferred tax (internal)

Sale of equipment (internal)

New borrowing (external)

Sale of stocks, bonds, convertible debentures (external)

Applications:

New equipment

Loan repayment

Training and development

Dividends

Others

19.15.3   Airline debt structure

The determination of the magnitude and timing of debt repayment is important in the generation of a financial forecast. It is important to realize the magnitude of payments on existing debt and associated debt before purchasing a new airplane.

19.15.4   Predelivery payments

Purchase of an aircraft requires a predelivery payment to be paid during its construction. The example in the case history in Section 19.16 illustrates the magnitude and timing of these payments. Naturally the scheduling of these payments could affect the decision-making process of the airline. Consequently, the airline strives to include the most representative prices and terms.

There is a need to account for the opportunity costs in arriving at the total investment required for the aircraft. The predelivery payments could earn revenue for the airline in some other use (bonds, bank, etc.). These lost revenues are capitalized from the date of delivery and considered as part of the cost of the aircraft.

19.15.5 Project financing

The percentage of total cost financed can be entirely internal or external or both in any combination, for example, 40 percent internally and 60 percent externally. A computer simulation model is able to handle all the situations related to airplane operation. For example, the predelivery payment is programmed in such a way that it can accept any kind of payment schedule (in the computer model; see Chap. 21).

The interest rate on a loan varies, and so the appropriate rate can be applied. Similarly, the time period of the loan can vary.

19.15.6 Capitalized interest

This is interest on the predelivery payment. The total amount of capitalized interest is added to the cash cost and is depreciated over the aircraft economic life. For discussion purposes, interest rate is considered the interest rate on borrowed funds, and it handles the percentage of borrowed money as a variable. This permits the application of a policy of each borrower when defining the cost of money (in the computer model).

19.15.7 Tax considerations

For tax purposes only, the method of depreciation is the double declining balance and half + 1 life linear, whereas in bookkeeping the straight line method is used.

19.15.8 Investment

The investment in a specific operating asset—an airplane or a fleet—generally consists of more than the price paid for the aircraft. Where interest-free equipment deposits and progress payments are made, an implicit rate of interest should be considered for such money, until delivery of the operating asset is taken. These cash amounts, at times very substantial, are withdrawn from current working capital and quite often are tied up for lengthy periods.

Crew training, development, and other preoperation expenses, if capitalized, may be considered as part of the investment. Support and auxiliary equipment peculiar to specific operating assets also may be considered part of the investment.

Thus the investment in a specific operating asset may consist of

1. Equipment deposit
2. Progress payment
3. Implicit cost on equipment deposit
4. Implicit cost progress payment
5. Final payment
6. Capitalized preoperation expenses
7. Investment in support and ancillary equipment

$$\text{Total investment} = 1 + 2 + 3 + 4 + 5 + 6 + 7$$

19.15.9 Operating revenues

This item can be further divided into

- Passenger revenues from transporting passengers, scheduled or chartered and variable classes (coach, business, first class, etc.)
- Cargo revenues from transportation of cargo, mail, or any other kind of shipment, scheduled or chartered
- Leasing income from property and equipment
- Incidental revenues from services performed in connection with air transportation, such as food, hotels, and miscellaneous services
- Other operating revenues (any other revenue sources not mentioned earlier)

All these items constitute total operating revenue.

19.15.10 Income calculations for a return on investment

Knowing now the method of the return on investment, the income has to be defined. There are two ways to calculate it: the top-down and bottom-up approaches.

19.15.11 Top-down approach to defining revenue

The best way to present the revenue is by applying an example:

- Daily block time: 12.0 block-hours
- Yearly operations: 324 days

$$324 \times 12 = 3888 \text{ block-hours/year}$$

- Block speed: 500 miles/block-hour

$$\text{Total miles: } 3888 \times 500 = 1,944,000 \text{ miles}$$
$$(\text{block-hours} \times \text{miles/block-hour} = \text{miles})$$

- Load = 40,000 lb (or 20 tons)

$$\text{Miles} \times \text{tons: } 1,944,000 \times 20 = 38,880,000 \text{ ton-miles}$$
$$(\text{miles} \times \text{ton} = \text{ton-miles})$$

- Yield: 15 cents/ton-mile

$$\text{Revenue} = 38,880,000 \times 0.15 = \$5,832,000$$
$$(\text{ton-miles} \times \text{dollars/ton-miles} = \text{dollars})$$

This is the potential revenue, if and when the airplane is full (load factor of 100 percent). It is more realistic to choose a 60 percent load factor. In this case,

- Load = 24,000 lb (or 12 tons)

$$1,944,000 \times 12 = 23,328,000 \text{ (miles} \times \text{tons)}$$

- Yield: 15 cents/ton-mile

$$\text{Revenue} = 23,378,000 \times 0.15 = \$3,499,200$$
$$(\text{ton-miles} \times \text{yield} = \text{dollar})$$

For passenger operations, the proper yield should be used. To obtain the revenue, multiply yield expressed in cents per passenger-mile by trip distance and number of passengers.

19.15.12 Bottom-up approach to defining revenue

Trip analysis should be calculated for each segment of the schedule with the help of a simulation program. With this method, in addition to the figures shown in the preceding example, operational limitations are also observed. For example, weight (load) limitations are calculated when the runway is too short or temperature limitations (e.g., at 100°F) are curtailing the takeoff weight in the second segment. Takeoff weight also could also be limited by obstructions. Anytime the takeoff weight is limited, there is a good chance that the load or the required number of passengers is below the maximum, and this may lessen the expected revenues. Using this method, the revenues generated from each trip segment are added up. This is the bottom-up buildup of revenues. (This approach is more realistic because it observes operational limitations, if any.)

19.15.13 Operating expenses

These are expenses incurred while conducting air transportation services. Direct and indirect operations costs can be established with two different approaches: the top-to-bottom and bottom-up approaches.

19.15.14 Top-down approach to defining operating cost

- Daily block times: 12.0 block-hours
- Yearly operations: 324 days/year

$$324 \times 12 = 3888 \text{ block-hours/year}$$

- Daily operating cost: 470 dollars/block-hour (no depreciation included)
- Indirect operating cost: 392 dollars/block-hour
- Total direct and indirect operating cost times block-hours:

$$\text{Total operating cost} = \$3,351,456 = 862 \times 3888$$
$$(\text{dollars} = \text{dollars/block-hours} \times \text{block-hours})$$

19.15.15 Bottom-up approach to defining operating cost

First, generate trip analysis for each trip segment and define direct operating cost. Fuel cost should be calculated for each individual trip, and the actual fuel cost should be applied at each station. Landing fees and transit costs should be considered individually for each station. These cost elements are prestored and could be overridden by updated values on the selected trip. This approach is much closer to actual costs because the calculations are based on a simulation of the real world rather than on average data.

19.15.16 Direct operating cost

This category consists of such flying operations as salaries and expenses, fuel costs, aircraft maintenance, and burden.

19.15.17   Depreciation

Special attention is paid to the airplane depreciation method. It has a significant effect on profit and taxes. The double declining method, with switchover to linear, is applied. It shows higher expenses and lower income for tax purposes. This method enables management to have more money (working capital) available in the early years of operations. This fact is again an indication of the time value of money because the larger the amount available, the more investments can be made for additional profits.

The linear depreciation method shows higher profit to keep stockholders satisfied. Meanwhile, it strengthens the airline's credit rating toward banks and creditors and the stock exchange, improving also the rating of the company's shares.

Depreciation reduces income reported to the government and stockholders. As a result, it is a systematic way to provide for future aircraft (equipment), a way to accumulate tax-free funds for replacement. Two factors determine the depreciation:

1. *Service life.* This is affected by physical deterioration, inadequacy, obsolescence, and changes in art, legal requirements, diminished or increased demand, or casualties.

2. *Net residual or salvage value.* This is affected by future service demands, technological developments, the possibility of reuse, and resale or junk value.

19.15.18   Method of depreciation

This is a double declining balance, which uses a constant percentage factor to double the straight-line rate up to the half-life plus 1 year. It changes to the straight-line rate considering the salvage value of the airplane, has a tax advantage in the early years, and makes a greater return contribution to the working capital during the early years.

Depreciation assumes an accurate prediction on life and salvage. Profits after taxes and dividends largely determine the amount that can be spent on modernization and expansion of the airplane fleet. The depreciation period must not necessarily coincide with the time period of the loan.

The depreciation residual may go from 0 percent up to any specified value. The yield can vary yearly based on economic circumstances. This complete control of the financial section allows the airline to specify its own policy and set up its own specification.

19.15.19   Amortization

This applies to various charges attached to air transportation services such as preoperating expenses.

19.15.20   Indirect operating cost

This includes expenses incurred on ground equipment, aircraft ground services, passenger servicing in air, and general and administrative expenses.

19.15.21   Ground and other equipment depreciation

This is the depreciation of any equipment, tools, or property in connection with air transportation. For example, this might include a tow tractor, baggage carts, staircases, and so on.

19.15.22   Ground and other equipment amortization

These are preoperating expenses, lease improvements, and intangibles related to ground equipment and property.

19.15.23   Cash flow

Annual operating revenue is defined by the planned use of the aircraft; direct cash and indirect operating expenses are estimated; direct cash and indirect operating expenses are subtracted; and gross cash operating profit is obtained. Gross cash operating profit is then reduced by the annual depreciation of the investment, other operating income deductions, and income taxes to yield after-tax operating profit. Generally, the proper income tax rate is used for U.S.-registered operators (the present tax rates always should be applied).

The after-tax operating profit is then increased by the amount depreciated and the investment tax credit (where and when applicable if any) to show the annual cash flow or cash income. Annual cash flow consists of

1. Annual operating revenues

2. Annual direct cash operating expenses

3. Annual indirect operating expenses

4. Annual depreciation

5. Other deduction from operating revenues

6. The expression of 1 − corporate tax rate

7. Tax credit, if any

$$\text{Annual cash flow} = [1 - (2 + 3) - 4 - 5] \times 6 + 4 + 7$$

19.15.24 Cash stream

The annual cash flow at the end of each month is considered to accrue in 12 equal streams. These monthly streams consist of two factors:

1. A part of it to liquidate or recoup the investment

2. Earnings or interest on investment

Monthly earnings, in turn, earn simple interest to the end of the year. At that time they are discounted and deducted from the balance of the investment.

$$\text{Cash stream} = \text{cash flow}/12$$

19.15.25 Leasing

The cost of leasing any equipment from outside is not subject to depreciation.

19.15.26 Operating income

This is the pretax income from operations. It represents the difference between total operating revenue and total operating expenses.

19.15.27 Interest expense

This is the interest required on the amount of loans.

19.16 Investment: Case History

19.16.1 Introduction

In the following example, two airplanes are purchased for $18,166,000, and the total investment amounts to $22,000,000. Opportunity costs, depreciation, and discounted cash flows are tabulated, and the rate of return on this investment is calculated.

19.16.2 Computation of implicit cost

Purchase price of two aircraft	$18,166,000
Implicit costs on advance payments	$694,900
Capitalized preoperation costs	$1,000,000

Peculiar support and ancillary equipment, including $2,139,100
spares (same life as aircraft)

Investment: $22,000,000

The prepayment cost has to be analyzed. Assuming the cost of two airplanes at $18,166,000, we have the prepayment schedule shown in Table 19.7 (opportunity cost at 10 percent per annum).

The first prepayment is given in 1983, and the airplanes' delivery takes place in 1986. Assuming a 10 percent interest rate on $826,000, in the first year $82,600 would be earned, in the second year the same amount plus another 10 percent on the $82,600 (total of $90,900), and in the third year by the same calculation $90,860 plus 10 percent added to it for a total of $99,900, making the total prepayment $1,099,400 (some figures are rounded up to the next 100). Continuing the same for the following years, an additional $694,900 is paid beyond the $18,166,000.

The prepayment generates additional revenue for the manufacturer; however, for the buyer, this amount represents a lost opportunity cost because the funds, being placed in escrow, are not usable. The lost opportunity cost is amortized over the economic life of the airplane.

19.16.3 Depreciation

Depreciation policy: double declining balance for the first 7 years ($0.5 \times N + 1$) and linear up to the twelfth year at 15 percent of aircraft purchase price (Table 19.8).

19.16.4 Discounted cash flow

Table 19.9 presents discounted cash flow. The years between 1988 and 1996 are not shown due to limited space.

Interpolation for the rate of return on the investment would result in about 23 percent for 12 years. Selecting points on the rate of return graph, Table 19.10 results. This information about the rate of return

TABLE 19.7 Prepayment schedule

Year	Payment	1983	1984	1985	1986	Total
1983	826,000(1)	82,600	90,900	99,900(4)	—	1,099,400
1984	826,000(2)	—	82,600	90,900(5)	—	999,500
1985	2,480,000(2)	—	—	248,000(5)	—	2,728,000
1986	14,034,000(3)	—	—	—	—	14,034,000
Total	18,166,000					18,860,900

Note: 1 = equipment deposit; 2 = progress payment; 3 = final payment; 4 = implicit cost on equipment deposit; 5 = implicit cost on progress payment.

TABLE 19.8 Depreciation schedule investment: $22,000,000

Annual depreciation	Balance to depreciate	Year of service
3667	18,333	1
3055	15,278	2
2546	12,731	3
2122	10,609	4
1768	8,841	5
1473	7,368	6
1228	6,140	7
683	5,457	8
683	4,774	9
683	4,091	10
683	3,408	11
683	2,725 residual	12

TABLE 19.9 Discounted cashflow

	1986	1987	1997
Cash operating profit	6.810	6.810	6.810
Depreciation	3.667	3.056	0.683
Other deduction	0.100	—	—
Pretax profit	3.043	3.754	6.127
Posttax profit	1.582	1.952	3.186
Annual cash flow	6.670	5.008	3.869
Cum. cash flow	6.670	11.678	53.166
Disc. factor, 6%	0.969	0.914	0.511
Annual cash flow disc.	6.463	4.577	1.977
Cum. cash flow	6.463	11.040	39.261
Disc. factor, 10%	0.950	0.864	0.333
Annual cash flow disc.	6.336	4.327	1.288
Cum. cash flow	6.336	10.663	33.093
Disc. factor, 20%	0.907	0.756	0.102
Annual cash flow	6.050	3.786	0.395
Cum. cash flow	6.050	9.836	23.397
Disc. factor, 30%	0.780	0.669	0.048
Annual cash flow	5.803	3.350	0.186
Cum. cash flow	5.803	9.153	18.188

(years between 1988 and 1996 are not shown)

TABLE 19.10   Recovery period

Discount	Recovery period
0	4.20
6	4.98
10	5.66
20	9.47
25	14.00

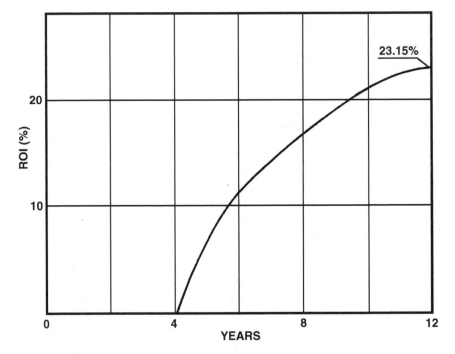

Figure 19.8   Return on investment.

is shown in Fig. 19.8. Here, the rate of return is presented in graph form as a function of time.

19.16.5   "True" average rate of return

The next step is to interpolate the cumulative cash flows in order to obtain the "true" average rate of investment.

Total investment	22,000,000
Cumulative cash flow at 20% discount	23,397,000
30% discount	18,188,000

Residual   2,724,000
Single-payment present value factor at 20%   0.1122
                                          at 30%   0.0429

Present-value factor $= 1/(1 + i)^N$, where i = interest rate and N = number of years.

$$\frac{23,397,000 + 2,724,900\ (0.1122) - 22,000,000}{23,397,000 + 2,794,000\ (0.1122) - 18,188,000 - 2,724,900\ (0.0429)}$$

$$= 0.315$$

$$X = [(0.315\ (0.30 - 0.20)] + 0.20 \qquad X = 23.15\%$$

23.15% = true average annual return

Note: Tax laws are in a constant change. In this example, the then-applicable tax laws were used, e.g., investment tax credit. The method of calculation should always be adjusted to the existing new tax laws.

19.17 Sensitivity Study of Return on Investment

19.17.1 Introduction

The reliability of the rate of return figure depends on the estimates and forecasts of the user. Therefore, a systematic analysis was made to determine the sensitivity of return on investment (ROI) to selected variables. All variables, however, work together in determining return on investment (ROI). The analysis was conducted to measure the change in ROI in relation to change in a variable. The sensitivity of one variable is determined by changing it and holding all other variables constant. The change noted in the rate of return is attributed to the sensitivity of the variables under analysis.

The following variables are examined for sensitivity: load factor, airplane price, and revenue yield.

19.17.2 Load factor

The load factor's effect on the rate of return is significant. In Fig. 19.9, the load factor at the airplane price of $16.5 million increases from 40 to 70 percent, which is a 75 percent increase in tons flown, and the rate of return increases from 24.5 to 60.6 percent. The percentage increase in the rate of return is nearly 150 percent, or double the percentage change in the load factor. As expected, the percentage increase in the return rate diminishes as the load factor approaches a higher value (Table 19.11).

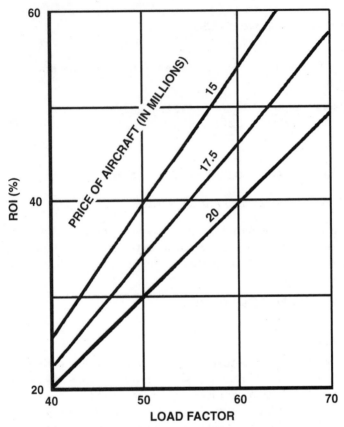

Figure 19.9 Return on investment versus load factor.

TABLE 19.11   Effect of load factor on Return on Investment (ROI)

Aircraft price (in millions)	Load factor, 40%	Load factor, 50%	Load factor, 60%	Load factor, 70%
13.2	30.6	45.9	61.4	78.0
15.1	27.1	39.8	53.0	66.8
16.5	24.5	36.3	48.2	60.6
18.0	22.5	33.1	43.9	55.1
20.0	20.0	29.7	39.4	49.2

Note: Yield = 17.5 cents/revenue-ton-mile.

19.17.3 Price of aircraft

The relationship of an airplane's price to its rate of return is presented in Figure 19.10.

The price of the investment is inversely related to the rate; the price increase causes the ROI to decrease when all other variables are held constant. The amount of the ROI decrease is shown in the "ROI change" column in Table 19.12. The percentage increase in the price of the airplane is from $13 million to $20 million. The relationship of ROI to price change appears to be inverse.

19.17.4 Revenue and yield

The effect of the revenue yield on ROI is shown in Table 19.13. A revenue yield increase from 15 cents per revenue-ton-mile to 17.5 cents per revenue-ton-mile caused the ROI to increase an average of 13 percentage points on the aircraft tested. In other words, the 2.5 per revenue-ton-mile increase in yield, a 17 percent jump, is responsible for creating the increase in the rate of return.

The relationship between revenue, price of aircraft, and ROI is presented in a graph in which the relationship is linear (Fig. 19.11).

19.17.5 Most common errors committed
when calculating return on investment

Calculating direct operating cost. For purposes of ROI, the direct operating cost (DOC) must not include depreciation but must include fuel cost.

Calculating indirect operating cost. When assuming indirect operating cost (IOC) as a percentage of DOC, which incidentally is a rather crude method, DOC does not include B depreciation. Therefore, IOC must be adjusted accordingly.

TABLE 19.12   Effect of Airplane Price on ROI

Load factor, %	Aircraft price, $13.1 mil, % ROI	Aircraft price, $20.0 mil, % ROI	ROI change, %
50	45.9	29.7	−16.2
60	61.4	39.4	−22.0
70	78.0	49.2	−28.8
80	96.7	59.5	−36.2

Note: Yield = 17.5 cents/revenue-ton-mile.

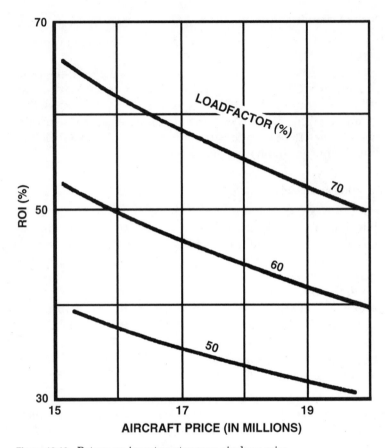

Figure 19.10 Return on investment versus airplane price.

TABLE 19.13 Effect of Revenue Yield on ROI

Aircraft price, $ in millions	Revenue yield, cents/revenue-ton-mile	Revenue yield, cents/revenue-ton-mile	ROI change, %
	15.0	17.5	
13.2	43.5	61.4	17.9
15.1	37.7	53.0	15.3
16.5	34.4	48.2	13.8
18.0	31.4	43.9	12.5
20.0	28.1	39.4	11.3

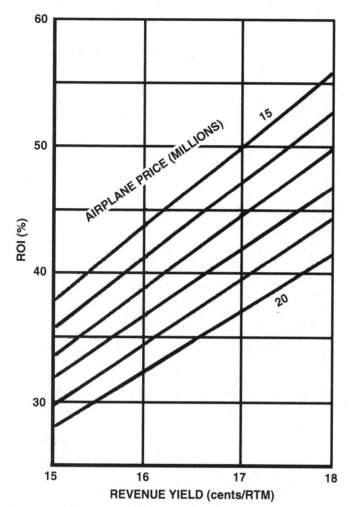

Figure 19.11 Return on investment versus yield.

Total investment. The total investment should include all capital investments involved, including spares and additional ground equipment.

New equipment versus old equipment. When comparing the purchase of new equipment to the continued use of old equipment, the salvage value of the old equipment should be considered as a capital credit, which reduces the total initial investment of the new equipment.

Case Histories: Aircraft Investment Evaluations

20.1 Return on Investment for Two Groups of Airplanes

20.1.1 Introduction

The purpose of this study is to determine the return on investment figures for two groups of different-sized airplanes. There are four airplanes in group A and two larger airplanes in group B.

20.1.2 Basic economic assumptions

To evaluate these two types of airplanes in group A and B, the investment methodology is used. This method is based on the discounted cash flow concept; it determines the return on the average aircraft investment. The depreciation applied is based on the double declining balance method that changes to straight-line depreciation at a year beyond the half-life of the investment.

The payment schedule for group A is as follows: It requires 31 percent of aircraft price 1 year before delivery and 69 percent of aircraft price during the year of delivery. For group B, 20 percent of aircraft price 2 years before, 31 percent 1 year before, and 49 percent during the year of delivery are required. Group A delivery is 1 year ahead of group B (Table 20.1).

20.1.3 Discounted cash flow

Tables 20.2 and 20.3 contain the discounted cash flow presentation.

TABLE 20.1   Basic data for the calculation of Return On Investment (ROI)

	Group A	Group B
Total investment	47,300,000	55,610,000
Depreciation period (years)	12	12
Residual value (%)	10	10
Economic life (years)	12	12
Delivery date	1990	1991
Corporate tax rate (%)	48	48
Interest (%)	8	8
Load factor	60	63
Average load at load factor	134 tons	134 tons
Number of aircraft	4	2
Average load/airplane, tons	33.5	67
DOC ($/mi)	2.01	3.64
Cash DOC	1.58	2.74
IOC to DOC (%)	75	75
Cash IOC	1.12	2.05
Yield (cents/ton-mile)	15	15
System distance (mi)	2000	2000

Note: The purpose of this case history is to demonstrate how to apply the return on investment method for aircraft evaluation. Reviewing the cash flow summaries is a good way to get familiar with this method.

20.1.4   Return on investment for both airplane groups

Taking the yearly average return on investment and presenting it on a graph would indicate the shapes of the return on investments (Fig. 20.1). For group A, a higher return is indicated based on this financial information. Group A should be considered. However, there are some other operational factors that could have some influence when the final decision is made.

20.2   Return on Investment: Per Civil Aeronautical Board (CAB)

20.2.1   Introduction

The General Passenger Investigation began in 1956 and evaluated domestic fare changes that would enable the proper financial development of airlines. Reasonable levels of rates on investments were established in 1960 for the four largest airlines at 10.25 percent return and 11.25 percent return for other carriers. In 1965, large carriers earned

TABLE 20.2 Group A

	1990	1991	1992	1993	1994	1995	1996	1997	1998	1999	2000	2001
Revenue	43,249	43,249	43,249	43,249	43,249	43,249	43,249	43,249	43,249	43,249	43,249	43,249
Cash costs	23,898	23,898	23,898	23,898	23,898	23,898	23,898	23,898	23,898	23,898	23,898	23,898
Depreciation	7,883	6,569	5,475	4,562	3,802	3,168	2,640	1,694	1,694	1,694	1,694	1,694
Interest changes	3,784	3,469	3,153	2,838	2,523	2,207	1,892	1,577	1,261	0,946	0,631	0,315
Startup costs	0.0	0.0	0.0	0.0	0.0	0.0	0.0	0.0	0.0	0.0	0.0	0.0
Pretax profit	7,683	9,313	10,723	11,951	13,026	13,975	14,819	16,080	16,395	16,711	17,026	17,341
Income taxes	3,688	4,470	5,147	5,736	6,253	6,708	7,113	7,718	7,870	8,021	8,172	9,324
Tax credits	3,311	0.0	0.0	0.0	0.0	0.0	0.0	0.0	0.0	0.0	0.0	0.0
Posttax profit	7,306	4,843	5,576	6,214	6,774	7,267	7,706	8,362	8,525	8,689	8,853	9,017
Annual cash flow	15,190	11,412	11,050	10,776	10,575	10,435	10,346	10,056	10,220	10,384	10,548	10,712
Cumulative cash flow	15,190	26,602	37,652	48,428	59,004	69,439	79,785	89,840	100,060	110,444	120,991	131,703
Return on average investment over the economic life of the investment is 0.2555												
Average return on investment	0.0	0.020	0.040	0.060	0.080	0.100	0.120	0.140	0.160	0.180	0.200	0.220
Year	4,001	4,185	4,389	4,614	4,862	5,149	5,488	5,871	6,350	6,925	7,720	8,806

TABLE 20.3 Group B

	1991	1992	1993	1994	1995	1996	1997	1998	1999	2000	2001	2002
Revenue	43,197	43,197	43,197	43,197	43,197	43,197	43,197	43,197	43,197	43,197	43,197	43,197
Cash costs	20,675	20,675	20,675	20,675	20,675	20,675	20,675	20,675	20,675	20,675	20,675	20,675
Depreciation	9,435	7,863	6,553	5,461	4,550	3,792	3,160	2,028	2,028	2,028	2,028	2,028
Interest charges	4,529	4,152	3,774	3,397	3,019	2,642	2,265	1,887	1,510	1,132	0,755	0,377
Startup costs	0.0	0.0	0.0	0.0	0.0	0.0	0.0	0.0	0.0	0.0	0.0	0.0
Pretax profit	8,558	10,508	12,196	13,665	14,953	16,089	17,098	18,608	18,985	19,363	19,740	20,118
Income taxes	4,106	5,044	5,854	6,559	7,177	7,723	8,207	8,932	9,113	9,294	9,475	9,656
Tax credits	3,963	0.0	0.0	0.0	0.0	0.0	0.0	0.0	0.0	0.0	0.0	0.0
Posttax profit	8,413	5,464	6,342	7,106	7,775	8,366	8,891	9,676	9,872	10,069	10,265	10,461
Annual cash flow	17,849	13,327	12,894	12,567	12,326	12,158	12,051	11,704	11,900	12,096	12,293	12,489
Cumulative cash flow	17,849	31,176	44,071	56,637	68,963	81,121	93,172	104,876	116,776	128,872	141,165	153,654
Return on average investment over the economic life of the investment is 0.2352												
Average return on investment	0.0	0.020	0.040	0.060	0.080	0.100	0.120	0.140	0.160	0.180	0.200	0.220
Year	4,320	4,534	4,771	5,033	5,345	5,699	6,109	6,625	7,258	8,101	9,294	11,172

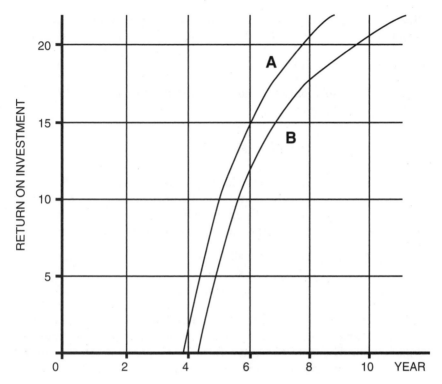

Figure 20.1 Return on investment versus years.

11.20 percent and smaller carriers earned 15 percent return on their investment. From this point in time, the ROI began to decrease slowly. After reviewing the status of the industry again in 1969, the CAB increased the rate of return on investment to 12.0 and 12.35 percent, respectively. Similarly, the CAB made a decision on the debt and equity relationship in airline operation, but this would be beyond the scope of our discussion.

20.2.2 CAB definition of return on investment calculations

In the following, the return on investment will be discussed according to the Civil Aeronautics Board rules and is presented here for purely historical interest. These rules for determination of return on investment differ from the set method of calculating discounted cash flow. They differ with respect to the handling of depreciation of flight equipment and interest payments. Furthermore, they differ in the inclusion of interest of prepayment. The annual depreciation of flight equipment here is considered linear.

For buildings (at that time) and for ground equipment, a 20- and a 10-year linear depreciation with zero residual is applied, respectively.

ROI = profit after taxes + interest ÷ average investment

The time value of moneys is not considered here. This is the disadvantage of this method.

20.2.3   Example presenting CAB's return on investment calculations

An example will show the procedure for this presentation. The value of aircraft totals $230 million, and this includes aircraft price, interest on prepayment, and spares. Depreciation period is 12 years, and linear depreciation is applied. A residual of 15 percent is considered.

Aircraft total investment	230,000,000
Residual, 15%	−34,500,000
Base for depreciation	195,500,000

Yearly depreciation is $16,300,000 (195,500,000/12). Total investment, including startup cost is $265,000,000. The average investment is $172,250,000. The debt-to-equity ratio is 65 percent. At an annual interest rate of 9.1 percent, this amounts to

$$\$10,188,587 = (172,250,000 \times 0.65) \times 0.091 = \text{interest}$$

The yearly operating cost is $90,000,000 during the economic life of the aircraft.

Profit-before-taxes is $20,000,000. This is the difference between revenue and operating costs. However, both are constant from year to year.

A corporate tax rate (at that time) was 48 percent.

$$\$9,600,000 = \$20,000,000 \times 0.48 = \text{taxes}$$

Profit-after-taxes is

$$\$20,000,000 - \$9,600,000 = \$10,400,000$$

The average investment at 65 percent of the original cost is $172,250,000. Applying the formula,

$$11.95\% = (10,400,000 + 10,188,587)/172,250,000 = \text{ROI defined by CAB}$$

The return on investment in this case was 11.95 percent per CAB definition.

21

Airline Simulation Study

21.1 Introduction

Airline simulation is an effective tool for planning purposes. It furnishes management with the answers about economics, effective planning, and many other questions concerning their planned operations. It indicates the relationship between many operational and economic factors. It is an effective management tool. The high cost of investment makes it necessary for management to be able to apply simulation studies to evaluate its various operational plans.

The following steps are taken for a simulation study:

1. Establish a market. Design a schedule and a matching airplane type that should satisfy the needs of this market.

2. Establish the economic ground rules for this type of aircraft. Design a budget and estimate the aircraft's direct and indirect operating costs based on industry averages or actual figures.

3. Evaluate the performance and cost figures for each trip, and project them for a 1-year period and project it into the future.

4. Generate indices and unit costs and compare them with industry averages.

5. Conduct in-depth analyses and try to improve schedule and/or select another type of aircraft. Repeat this process if necessary until a satisfactory result is achieved.

A few flowcharts will be presented for various parts of this simulation to indicate the logic applied here. These charts do not show the nitty-gritty details, but they do furnish a good idea about the structure of the simulation model (see Fig. 21.1).

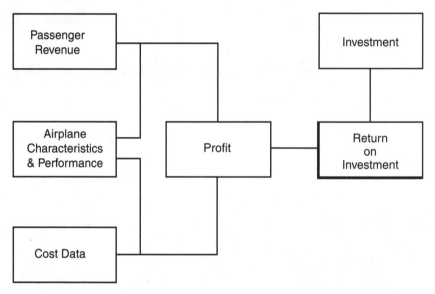

Figure 21.1   Network simulation model: General logic.

Knowing passenger revenues, cost data, aircraft characteristics, and performance, a profit is defined. Considering the characteristics of the investment as a final answer, the return on investment is established.

21.2   Discussion of Simulation Studies

A good simulation study should have enough flexibility and capability built in so that it can simulate the realistic operational and economic conditions of an airline. It should contain the basic data, and for each individual trip, it should have the capabilities to change the data based on the real-world requirements.

In the following, let us assume that the management of a new airline generated a schedule based on its marketing research. Management plans to fly five used B747-200 airplanes in passenger configuration over this proposed schedule. By applying a simulation program, the first question management asks is how profitable the operation is and what is the magnitude of profit.

The schedule is planned for flights between the United States and Europe. It is known ahead of time that the long trans-Atlantic flights will be profitable and the inter-European flights may lose money. The question is, Do the long flights generate enough money to cover the losses of the short flights, and could this arrangement generate satisfactory profit for this operation?

Subsequent questions would concentrate on improvements of the proposed schedule to obtain the best operational and economic results. The accepted three-letter codes are listed for the airports in the schedule:

JFK   J. F. Kennedy Airport, New York

LHR   Heathrow, London

BOS   Logan, Boston

BRU   Zaventem, Brussels

ZRH   Cloten, Zurich

FCO   Fiumicino, Rome

A schedule is presented in Table 21.1. The odd flight numbers are to be read from top to bottom on the left-hand side and from bottom to top on the right-hand side. For example, flight Z501 is a flight from JFK through London, Brussels, and Zurich. The return flight number is Z502 and should be read up on right-hand side.

Flight frequencies per week:

Z501-2   7

Z503-4   7

Z505-6   7

Z507-8   1

Z509-10   1

Z111-12   2

21.3 Economic Assumptions

The selected type of airplane is a used B747-200 powered with JT9D-70A engines (Table 21.2).

Total investment figures are shown in Table 21.3.

Opportunity cost is tabulated in Table 21.4. Depreciation is shown in Table 21.5.

Insurance data are tabulated in Table 21.6.

Interest figures are presented in Table 21.7.

21.4 Establishment of Operating Costs

Direct operating cost elements

Flying operations   1450 dollars/block-hour

Insurance   15 dollars/block-hour

TABLE 21.1 Proposed Table of Schedule

Trip	Z501	Z503	Z505	Z507	Z509	Z511	Trip	Z502	Z504	Z506	Z508	Z510	Z512
JFK	18:00	20:00	22:00				JFK	18:12	21:56	14:14			
BOS				22:00	22:00	22:00	BOS				21:40	22:19	18:36
LHR	6:40	8:40	10:40	10:21	10:21	10:21	LHR	16:31	20:16	13:30	20:22	21:01	17:18
	7:40	9:40		11:21	11:21	11:21		15:31	19:16		19:22	20:01	16:18
BRU	8:40	10:40		12:21	12:21	12:21	BRU	14:29	18:13		18:19	18:58	15:16
	9:25	11:25		13:06	13:06			13:44	17:13		17:34	18:13	
ZRH	10:39						ZRH	12:29			16:19		
FCO		13:26		14:19	15:07		FCO		15:06			16:07	

Note: Trips with odd flight numbers read down; trips with even flight numbers read up.

472

TABLE 21.2 Initial Conditions

Price of five aircraft ($20 million each)	$100,000,000
Aircraft conditions	Used
Depreciation	10 years
Residual	10 percent
Insurance rate	0.35 percent
Interest rate	10 percent
Loan term	Depreciable life

TABLE 21.3 Total investments in dollars

Five used aircraft ($20 million each)	$100,000,000
Opportunity cost	$2,302,000
Spares, startup cost, misc. expenses*	$24,622,000
Total	$126,924,000

*Spares, 17% of aircraft price.

TABLE 21.4 Opportunity Cost at 10 Percent per Annum

Year	Payments	1990	1991	1992	1993	Total
1990	2,000,000 (1)	200,000	220,000	242,000 (4)	—	2,662,000
1991	4,000,000 (2)	—	400,000	440,000 (5)	—	4,840,000
1992	8,000,000 (2)	—	—	800,000	—	8,800,000
1993	86,000,000 (3)	—	—	—	—	86,000,000
Total	100,000,000					102,302,000

Note: 1 = equipment deposit; 2 = progress payment; 3 = final payment; 4 = implicit cost on equipment deposit; 5 = implicit cost on progress payment.

TABLE 21.5 Depreciation

Aircraft and spares	$117,000,000
Depreciation residual, 10% of aircraft + spares	$11,700,000
Amount to be depreciated	$105,300,000
Depreciated for	10 years
Depreciation per year	$10,530,000
Total block-hours per year (5 aircraft)	22233.5
Dollar per block-hour	473.6

TABLE 21.6 Insurance

Aircraft price	$100,000,000
Insurance rate	0.35%
Insurance amount per year	$350,000
Total block-hours per year (5 aircraft)	22233.5
Dollar per block-hour	15.7

TABLE 21.7 Interest

Interest considered for	$126,924,000
Interest rate	10%
Interest per year	$12,692,400
Total block-hours per year (5 aircraft)	22233.5
Dollar per block-hour	570.8

Maintenance and burden	2920 dollars/block-hour
Ownership (depr/interest)	1044 dollars/block-hour

Fuel is based on fuel price at each departure station, and fuel cost is a function of block fuel.

Note: Flying operations and maintenance costs are in line with industry averages for this type of airplane.

Indirect operating cost. This consists of four categories: landing fees, traffic costs, aircraft handling costs, and overhead. Overhead consists of cost elements that do not fit in any of the previously mentioned categories. The costs are shown for each station the airplane flew into during the simulation study.

A simulation report consists of three parts:

1. The first part describes the airplane and lists its weights, characteristics, and cost figures. They are needed to conduct this study.

2. The second part contains trip analyses. It contains a set of trips indicating payloads, schedules, fuel requirement per trip, block and trip times, etc.

3. For each trip, a financial evaluation is presented. Management is interested in profit, revenues, and costs.

The first part presents aircraft weights and cost data (Table 21.8).

In the second part of this presentation, trip data are shown. The Flight Department and Sales, Marketing, and Maintenance Departments are involved and are able to use the presentation for their own planning. The

TABLE 21.8   B747-200, JT9D-70A, Weight Summary. Simulation report part 1

Ramp wt.	823,000 lb
Takeoff wt.	820,000 lb
Landing wt.	630,000 lb
Zero-fuel wt.	560,000 lb
Empty opert. wt.	375,000 lb
Payload wt.	185,000 lb
No. of pass	500
Passenger wt.	165 lb
Baggage wt.	35 lb
Pass. + bag. wt.	200 lb
TTL pas. + bag. wt	100,000 lb
Addnl. load wt.	83,600 lb
Tare wt.	1400 lb
Fuel tank cap.	348,300 lb
Configuration	Passenger

Cost Summary	

Direct Operating Cost	
Flying operations, $/block-hour	1450
Insurance, $/block-hour	15
Maintenance and burden, $/block-hour	2920
Depr./rent/interest, $/block-hour	1044
DOC less fuel cost, $/block-hour	5429

*Fuel cost is calculated at each departure station based on fuel price and required fuel. Add fuel cost to DOC less fuel cost to obtain direct operating cost.
Note: Passenger yield used (when not specified): 11.90 cents/passenger-mile; Addnl. cargo yield used (when not specified): 17.3 cents/ton-mile. Overhead may be considered: 5310 $/block-hour.

Flight Department plans trips according to schedule and prepares all the administrative paperwork for smooth operations. Crew scheduling and fuel purchases have to be planned and implemented, and all other components associated with a trouble free operation have to be completed. The Sales Department has to know the arrival and departure times, passenger ticket prices, cargo rates, weights, volumes, and range (for partial load, too) for each type of airplane. The Maintenance Department has to know the schedule to efficiently plan its maintenance work at the main base and at certain stations where maintenance is conducted.

21.5   Discussion of Trip Analysis

The simulation should have the flexibility to accommodate operational variables—in this study—for each individual trip. A few of these variables are listed below:

1. *Maximum takeoff weight.* This can be any weight based on airport runway length, temperature, obstacles, elevation, and any other limitation found in the FAA-approved flight manual.

2. *Season.* Yearly, seasonally, monthly.

3. *Cruise.* Any Mach number, long-range cruise, etc. Published in the airplane operational manuals.

4. *Cruise altitude.* Step-climb cruise or constant-altitude cruise.

5. *Landing weight.* Could be limited by landing runway length.

6. *Alternate station.* Any suitable alternate based on the operator's choice.

7. *Reserve fuel.* Domestic, international, or island reserve fuel policy.

8. *Zero-fuel weight.* Could be limited by center of gravity (seldom).

9. *Payload.* Could be selected from 1 lb up to the maximum or, when left blank, will be calculated by the program.

10. *Number of passengers.* Could be selected from one passenger to the maximum or, when left blank, will be calculated by the program.

11a. *Additional cargo (passenger operation).* Operator's choice.

11b. *Freight (cargo operation).* Could be selected from 1 lb up to the maximum, or when left blank, will be calculated by the program.

12a. *Tare weight (passenger operation).* Related to item 11a. Operator's choice.

12b. *Tare weight (cargo operation).* Operator's choice.

13. *Airplane operating weight empty.* Actual individual operating weight can be specified, indicated by its tail number.

14. *Wind/reliability.* Enroute wind and reliability can be selected, e.g., 85 percent.

15. *Taxi-in, taxi-out, and ground time.* For each airport based on the operator's experience, various times can be assigned; when left blank, prestored data are applied.

16. *Distance.* Great-circle or airway miles.

17. *Yields.* Based on each operator's experience.

This short description presents some of the features of this simulation study. When operators do not select their choice, prestored industry values are applied (Tables 21.9a and 21.9b).

Airplane operating weight empty is based on tail number, registration number, or any other identification number. In this simulation study, an operator can design the proper ground time for the operation—for each

TABLE 21.9a Trip Analysis Program. Simulation report part 2

	JFK-LHR	LHR-BRU	BRU-ZRH	ZRH-BRU	BRU-LHR	LHR-JFK
Equipment	B747-200	B747-200	B747-200	B747-200	B747-200	B747-200
Engine	JT9D-70A	JT9D-70A	JT9D-70A	JT9D-70A	JT9D-70A	JT9D-70A
Season	Yearly	Yearly	Yearly	Yearly	Yearly	Yearly
Max. takeoff wt., lb	820,000	820,000	820,000	820,000	820,000	820,000
At airport temp., °F	70	64	65	66	65	64
Cruise	M.84	M.84	M.84	M.84	M.84	M.84
Cruise altitude, ft	33,000	15,000	25,000	25,000	15,000	33,000
Takeoff wt. req., lb	632,546	487,564	488,085	494,541	489,187	661,441
Trip fuel, lb	145,905	12,696	18,413	19,195	13,830	172,450
Landing weight, lb	486,641	474,867	469,671	475,346	475,356	488,991
Alternate station	AMS	ORY	STR	ORY	AMS	BOS
Alternate dist., N.M	161	153	70	153	161	166
Fuel to alternate, lb	12,823	12,432	6,826	12,432	12,822	13,061
Holding (etc.) fuel, lb	22,818	11,435	11,844	11,913	11,533	24,929
Reserve fuel, lb	35,641 I	23,867 I	18,671 I	24,346 I	24,356 I	37,991 I
Zero-fuel weight, wt.	451,000	451,000	451,000	451,000	451,000	451,000
Gross payload, lb	76,001	76,001	76,001	76,001	76,001	76,001
TTL pass. + bag. wt., lb	60,000	60,000	60,000	60,000	60,000	60,000
Number of passenger	300	300	300	300	300	300
Additional cargo, lb	14,600	14,600	14,600	14,600	14,600	14,600
Tare weight, lb	1,400	1,400	1,400	1,400	1,400	1,400
Operating wt. emp., lb	375,000	375,000	375,000	375,000	375,000	375,000
OWE is based on	NXXX	NXXX	NXXX	NXXX	NXXX	NXXX
Configuration	Passenger	Passenger	Passenger	Passenger	Passenger	Passenger

TABLE 21.9b Trip Analysis Program. Simulation report part 2

	JFK-LHR	LHR-BRU	BRU-ZRH	ZRH-BRU	BRU-LHR	LHR-JFK
Trip time, h:m	6:10	0:30	0:43	0:45	0:32	7:10
Block time, h:m	6:40	1:0	1:13	1:15	1:02	7:40
Block speed, mph	520	179	243	238	172	452
Block fuel, lb	149,505	16,296	22,013	22,795	17,430	176,050
Total fuel, lb	183,346	38,364	38,885	45,341	39,987	212,241
Distance, n.m.	3025 G	156 G	260 G	260 G	156 G	3025 G
Wind/reliab., kts/pct	19/85	−6/85	−11/85	−31/85	−51/85	−54/85
Day of departure, GMT	Sun	Mon	Mon	Mon	Mon	Mon
Local departure, h:m	18:0	7:40	9:25	12:29	14:29	16:31
GMT departure, h:m	23:0	6:41	8:26	11:29	13:29	15:32
Day of arrival, GMT	Mon	Mon	Mon	Mon	Mon	Mon
Local arrival, h:m	6:40	8:40	10:39	13:44	15:31	18:12
GMT arrival, h:m	5:41	7:41	9:39	12:44	14:32	23:13
Ground time, min	60	45	110	45	60	60

Note: Maximum number of passengers is 500; applying a 60 percent load factor; 300 passengers are considered.

individual trip as necessary—and it has the choice to change it. The advantage of a simulation study is that practically all the operational elements can be changed to conduct a "what-if" analysis and find out the best solution.

Sometimes a shorter runway at a given airport does not permit a required takeoff weight for the trip. Maybe some of the passengers (or, for cargo operations, part of the load) have to be left behind. This, in turn, could decrease profit. There are a few solutions available:

- *Introduce a one-stop operation.* This is more expensive (e.g., two landing fees); however, it could turn profit into losses or vice versa.

- *Select another airport in the vicinity of the airport shown on the schedules.* For example, London Heathrow to Gatwick airport, New York J. F. Kennedy to Newark, and/or La Guardia and/or Newburgh

- *Select another airplane with a better takeoff performance.*

The trip analysis presents a list of operational elements applicable to the selected trips. Every phase of this calculation is shown in detail.

The maximum takeoff weight is based on the parameters of the departure airport and satisfies the regulation described in the FAA-approved flight manual. This is the legal maximum take-off weight for the selected airport and runway. For JFK airport and Runway 31L, 820,000 lb is the maximum takeoff weight for the selected temperature.

The label "Takeoff weight required" indicates that only a takeoff weight of, e.g., 632,546 lb is required to complete this trip.

Fuel information is necessary for an effective fuel purchasing plan. Block times are required for, e.g., crew scheduling, whereas arrival and departure times are required for flight, maintenance, sales, and marketing departments. Payload figures are important to flight, sales, marketing, and accounting and other relevant departments.

Figure 21.2 contains elements of trip data. Airplane performance and economics are generated for each trip to evaluate them for the return on investment.

Passenger revenue is shown in Fig. 21.3. It shows passenger potential, distribution absorption, and the realized route passengers based on the proposed schedule.

Figure 21.4 shows operating cost data. It shows the direct and indirect operating costs and, finally, the total trip cost.

21.6   Financial Evaluation of the Trips Listed on the Schedule

The third part contains cost information (Table 21.10). The Accounting Department would be interested in this section of the report. The

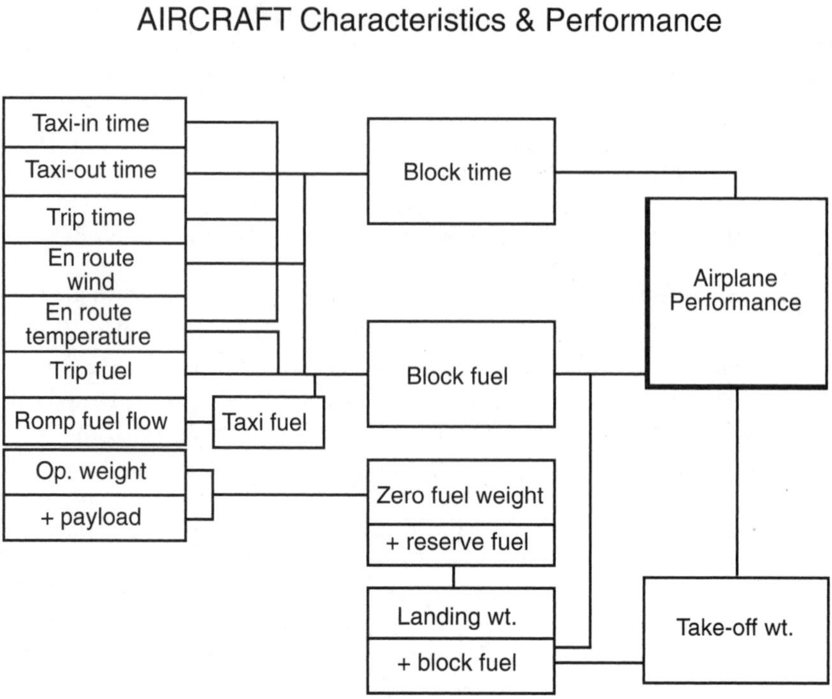

AIRCRAFT Characteristics & Performance

Figure 21.2 Aircraft characteristics and performance.

Finance Department would review revenue, cost, and profit (before taxes) figures.

In these costs, station cost figures are part of indirect operating costs (Table 21.11). Similarly, the landing fees, traffic servicing costs, and handling costs are shown for all stations the airplane covers in this schedule. Fuel prices are listed for each station.

Once the simulation is completed for flights Z501 and Z502, the figures can be projected for yearly operation. Fifty weeks is considered per year due to required maintenance downtime.

The simulation should be conducted for each flight number listed in the schedule (Table 21.12).

Preparing similar tables for all the flight numbers in the schedule (not shown here in detail), a summary could be prepared for each of them (Table 21.13).

Note: The dollars per block-hour here indicate dollars in profit before taxes divided by block-hours. This indicates that each hour, this airplane is generating a certain amount of money. In the case of flight number Z509-10, the profit generated is $841.3 per block-hour. Below

Determination of Passenger Revenue

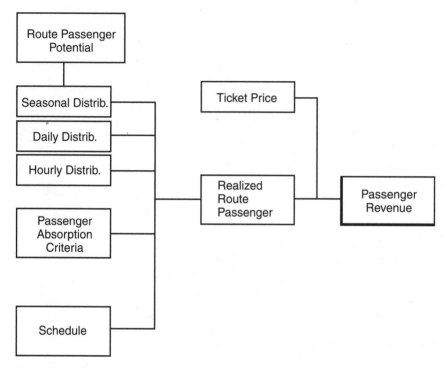

Figure 21.3 Determination of passenger revenue.

are several explanations for the large differences in dollar profit per block-hour:

- Flight number Z501-2 consists of a long trans-Atlantic flight that is profitable and a few inter-European short flights, which reduce the profit, as shown in Table 21.12.

- Flight number Z505-6 consists of a trans-Atlantic flight only, which generates profit, and there are no inter-Europe flights that would reduce the profit.

For the airplane, the following assumptions were applied: The airplane is configured for passenger operations and is able to carry 500 passengers. Applying a 60 percent load factor, this means that only 300 passengers generated a profit (before taxes) of $30,977,950 on a revenue of $348,660,000 or an operating margin of 8.8 percent. The profit after taxes will be $20,135,668, and this produces a net margin of 6.5 percent.

Operating Cost Data

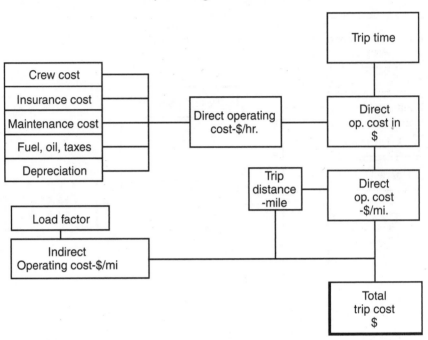

Figure 21.4 Operating cost data.

There are a few unit indicators and ratios that are useful when comparing various scheduling plans and analyzing them for performance and economic reasons:

Operating revenue per dollar investment:

$$\$348,660,700/\$126,924,000 = \$2.74$$

Operating expenses per dollar investment:

$$\$317,682,750/\$126,924,000 = \$2.50$$

Operating profit before taxes per total investment:

$$\$30,977,950/\$126,924,000 = \$0.244$$

Operating profit after taxes per total investment:

$$\$20,135,668/\$126,924,000 = \$0.1586$$

TABLE 21.10 Financial Data

Flying operations, $	9,686	1,450	1,778	1,815	1,508	11,140
Fuel and oil cost, $	12,739	1,948	2,270	3,031	1,798	21,043
Fuel price, cents/gal	57.00	80.00	69.00	89.00	69.00	80.00
Insurance, $	100	15	18	18	15	115
Maintenance + burden, $	19,505	2,920	3,581	3,655	3,038	22,434
Dep/rent/interest, $	6,974	1,044	1,280	1,307	1,086	8,021
DOC, $	49,004	7,377	8,927	9,826	7,445	62,753
IOC, $	39,353	9,393	13,506	11,997	9,185	46,573
Total trip cost (D + I), $	88,357	16,770	22,433	21,823	16,630	109,326
Revenue-passengers, $	124,223	6,406	10,675	10,675	6,406	124,223
Total pass. + bag. wt., lb	60,000	60,000	60,000	60,000	60,000	60,000
No. of passgrs carried	300	300	300	300	300	300
Breakeven no. of pass.	213	785	630	613	778	264
Breakeven pass. LDFCR%	71.13	261.77	210.13	204.41	259.58	88.01
Potential cargo rev., $	4,396	226	377	377	226	4,396
Potential cargo, lb	14,600	14,600	14,600	14,600	14,600	14,600
Tare weight, lb	1,400	1,400	1,400	1,400	1,400	1,400
Unit cost in cents/pass.	8.46	31.15	25.01	24.33	30.89	10.47
Yield pass., cents/pass-mi	11.90	11.90	11.90	11.90	11.90	11.90
Profit potential, $	35,866	-10,364	-11,758	-11,148	-10,224	14,897

Note: Cargo revenue not included in profit figures (17.31 cents/ton-mile). Reserve fuel: I = international; D = domestic; S = 2 hrs fuel. G = great circle; A = airway miles; Total fuel = ramp + trip + reserve fuel; block fuel includes taxi-out/in fuel; taxi-in fuel taken from reserve fuel. Each column is associated with a certain trip, as shown in Table 21.9. The first column is for JFK-London, etc.

TABLE 21.11   Fuel and Indirect Operating Cost elements. Simulation report part 3

Station	Fuel price, cents/gal	Indirect operating cost Landing fee, $	Traffic svc., $	Handling cost, $
Paris, ORY	0.91	3227	2479	2669
New York, JFK	0.57	2624	1743	1995
Amstrdm, AMS	0.82	3370	1388	755
Boston, BOS	0.56	1866	1882	1921
London, LHR	0.80	729	1670	900
Brussels, BRU	0.69	1151	1383	1911
Sttgrt., STR	0.78	378	359	361
Zurich, ZRH	0.89	2795	2756	2348

*Fuel cost is part of direct operations cost.

TABLE 21.12   Flights of Z501 and Z502: Daily and Yearly Figures

Trip	Revenue	Total cost	Profit before taxes	Cargo revenue, potent.	Block time
JFK-LHR	$124,223	$88,357	$35866	$4396	6:40
LHR-BRU	6,406	16,770	−10364	226	1:00
BRU-ZRH	10,675	22,433	−11758	377	1:13
ZRH-BRU	10,675	21,823	−11148	377	1:15
BRU-LHR	6,406	16,630	−10224	226	1:02
LHR-JFK	124,223	109,326	14897	4396	7:40
TTL/TRIP	$282,608	275,339	7269	9998	18:49
Total/year	$98,912,800	$96,368,650	$2,544,150	$3,499,300	6591:18

Note: This is a daily trip, and 50 weeks is considered per year. Two weeks is planned for maintenance.

TABLE 21.13   Summary for Each Flight Number: Revenue, Costs, Profit before Taxes, Cargo Revenues, Block-Hours, and Profit Index

Flight no.	Revenue	Total cost	Profit before taxes	Cargo potential	Block-hours	$/block-hour
Z501-2	$98,912,800	$96,368,650	$2,544,150	$3,499,300	6,591.55	385.9
Z503-4	109,609,800	102,678,800	6,930,000	3,878,000	7,163.45	967.41
Z505-6	86,956,100	69,189,050	17,767,050	3,077,200	5,016.9	3541.75
Z507-8	13,447,500	13,242,600	204,900	475,800	906.6	226.0
Z509-10	14,975,500	14,144,050	831,450	529,900	988.3	841.3
Z511-12	4,760,000	22,059,600	2,700,400	876,200	1,566.6	1,723.6
Total/year	348,660,700	317,682,750	30,977,950	12,336,400	22233.50	1393.3

A few other indices are mentioned here:

- Productivity
- Average available seat-mile per airborne hour
- Average available ton-mile per airborne hour

Note: Airborne time is the same as trip time; no taxi and ground times are involved because this could distort the figures.

Fuel:

Gallons of fuel consumed per block-hour

Cost of fuel per gallon

Aircraft capacity:

Average available seats per aircraft mile

Average available tons per aircraft mile

In a similar fashion, utilization and traffic figures can be expressed per miles and per hours, to mention a few. They are always helpful for comparative reasons.

In the last column of Table 21.13 and 21.14 an index is introduced. This is the profit index expressed in dollars per block-hour and dollars per mile. This is the relationship between profit (before taxes) and the block time or the miles necessary to fly this flight number. For example, for flight number Z501-2, the profit before taxes generated for a 1-year period was $2,544,150, and this required a total block time of 6591.55 hours or a mileage of 2,408,700 nautical miles or 2,770,000 statute miles. This means that each hour flown produced $385.9, and each mile generated $1.056 per nautical mile or $0.918 per statute mile. This index answers the question of how to rank each flight from the profit standpoint. Ranking each flight number by its profit index is a good indication of the economy of each trip.

TABLE 21.14   Profit Indices per Block-Hour and per Mile

Rank	Flight number	Profit (B.T.) per block-hour	No. of segments
1	Z505-6	3541.75	2
2	Z511-12	1723.60	4
3	Z503-4	967.41	6
4	Z509-10	841.30	6
5	Z501-2	385.90	6
6	Z507-8	226.00	6

The ideal schedule should contain only trips that are at least around 2000 miles or longer for this type of airplane, such as JFK-London, because they are profitable. According to the Marketing Department, this is not feasible, as was mentioned before, because the airplanes are filled with passengers whose destinations are beyond London, such as Brussels, Zurich, and Rome (Table 21.15).

A compromise is necessary to plan the least number of short trips and still satisfy the demands of management for a profitable operation. Another solution could be to plan the long trips with Boeing 747 airplanes while the market for trips from London to Brussels or Zurich, etc. is serviced with a B757 or a smaller airplane. This would make the short trips more economical. One drawback, though, is that passengers have to change airplanes at London.

Another source of revenue is the cargo carried in the belly of the airplane. Assuming a conservative load of 16,000 lb (14,600 lbs of payload and 1400 lb tare weight), a yearly profit potential (before taxes) of $12,336,400 could be expected (based on 14,600 lb). Naturally, a certain amount from this revenue has to cover cargo handling costs and insurance.

Reviewing all information of this simulation study, each trip should be evaluated for its operational and economic characteristics, and trips not generating profit should be deleted. Certain trips that have a growing potential should be followed closely. Other types of airplanes should be evaluated. The whole simulation study should be repeated until a satisfactory schedule and its matching airplane type have been established.

21.7   Profit (B.T.) per mile and per block hr.

This schedule generates a certain amount of profit, but would this amount of profit be satisfactory related to the size of the investment? For this reason, this investment should be evaluated. The internal rate of investment method should be applied, and the discounted cash flow should be used. Every cash flow should be time accountable, which means that each cash input or outflow of money must be con-

TABLE 21.15   Profit (B.T.) per mile and per block-hour.

Total miles	Profit (B.T.) /mile
2,408,700 nautical miles	1.286 $profit/nautical mile
2,770,000 statute miles	1.118 $profit/statute mile
Total flying time	Profit (B.T.)/block-hours
22233 block-hours	1393 $profit/block-hours

B.T. = Before taxes

sidered according to its time. Figure 21.5 is a flowchart indicating the main elements of this simulation model. This model answers the question. For a given schedule and airplane type, what return on investment can be expected?

Table 21.16 shows the assumptions that are made in the simulation. This is followed by financial assumptions such as amount financed, period of loan, rate of interest, and payment schedule (Table 21.17).

Cash flow summary contains revenue, cash cost, various depreciations, interests, profits, and annual and cumulative cash flows (Table 21.18). Based on these figures, the average rate of return is presented and its distribution as a function of time is listed.

21.8 Conclusion

Based on the preceding figures for this selected schedule and aircraft, the return is 10 percent. If this return were not satisfactory, the schedule and the airplane combination would have to be changed until a satisfactory return were produced that would be acceptable to management. This requires a recycling of the data until a satisfactory result is established.

Financial Calculation for Return on Investment

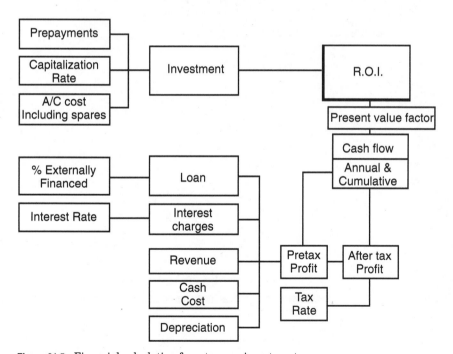

Figure 21.5 Financial calculation for return on investment.

TABLE 21.16   Return on Investment, B747-200-70A, Assumptions. Simulation report

Block distance, statute miles	3,000	Daily utilization, hours	12
Block speed, MHP	500	Payload, tons	50
Number of aircraft	5	Operating factor, day/year	320
Depreciation period, years	10	Depreciation residual, %	10
Order date	1990	Delivery date	1993
Cost of airplanes, $	20,000,000	Capitalization rate, %	10
Spare parts cost, $	17,000,000	Other investment cost, $	4,622,000
Ground equipment cost, $	3,000,000	Total investment, $	124,622,000

Depreciation aircraft; double declining balance; ground equip., linear 7 yrs; others, linear depreciation

TABLE 21.17   Payment schedule. Simulation report.

Amount financed, %	99
Period of loan, years	10
Rate of interest, %	10.0
Payment schedule:	
Payment 1	$2,000,000 1990
Payment 2	$4,000,000 1991
Payment 3	$8,000,000 1992

TABLE 21.18 Return on Investment. Simulation report

	Cash Flow Summary Millions of Dollars									
	1993	1994	1995	1996	1997	1998	1999	2000	2001	2002
Revenue	348.66	348.66	348.66	348.66	348.66	348.66	348.66	348.66	348.66	348.66
Cash costs	317.68	317.68	317.68	317.68	317.68	317.68	317.68	317.68	317.68	317.68
Depreciation A/C	23.40	18.72	14.98	11.98	9.58	7.67	7.17	7.17	7.17	7.17
Depreciation others	0.46	0.46	0.46	0.46	0.46	0.46	0.46	0.46	0.46	0.46
Depr. ground equip.	0.43	0.43	0.43	0.43	0.43	0.43	0.43	0.00	0.00	0.00
Interest charges	12.34	11.10	9.87	8.64	7.40	6.17	4.94	3.70	2.47	1.23
Pretax profit	−5.65	0.26	5.24	9.47	13.10	16.25	17.98	19.65	20.88	22.11
Income taxes	−0.85	0.09	1.78	3.22	4.48	5.59	6.19	6.78	7.21	7.64
Posttax profit	−4.80	0.18	3.46	6.25	8.61	10.66	11.79	12.87	13.67	14.47
Annual cash flow	19.49	19.79	19.33	19.12	19.09	19.22	19.85	20.50	21.30	22.10
Cum. cash flow	19.49	39.28	58.60	77.72	96.81	116.04	135.88	156.38	177.69	199.79

Average return on investment over the economic life of the investment is 10.0444.

Percent	5	8	10
Year	7.85	8.97	9.98

22

Definitions

22.1 The Freedom Rights and Open Sky Definitions

An airline is governed by certain rules when flying and carrying traffic into another country. These rules consist of categories of freedom, as described and illustrated below. Home country is designated with the letter H, while foreign countries are designated as 1 and 2.

Freedom no. 1 is the right to overfly a foreign country (Fig. 22.1).

Freedom no. 2 is the right to land in a foreign country in case of emergency, for a fuel stop, or for repair but not to receive or discharge passengers or freight (Fig. 22.2).

Freedom no. 3 is the right to take passengers and/or freight from the home country and deposit them in a foreign country (Fig. 22.3).

Freedom no. 4 is the reverse of freedom no. 3; it is the right to take on passengers and/or freight in a foreign country and discharge them in the home country (Fig. 22.4).

Freedom no. 5 is the right to transport passengers and/or freight between two foreign countries (Fig. 22.5).

Freedom no. 6 is the right to carry passengers and/or freight between foreign countries through the home country (Fig. 22.6).

Freedom no. 7 is the right to operate entirely outside the territory of the flag state (Fig. 22.7).

The eighth freedom is referred to as *cabotage*.

Cabotage. The right to carry passengers and/or freight within a foreign country (Fig. 22.8).

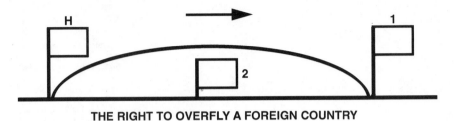

THE RIGHT TO OVERFLY A FOREIGN COUNTRY

Figure 22.1 Freedom no. 1.

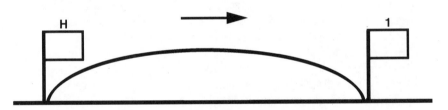

THE RIGHT TO MAKE A MAINTENANCE AND/OR FUEL STOP

Figure 22.2 Freedom no. 2.

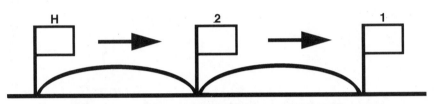

THE RIGHT TO CARRY PASSENGERS IN A FOREIGN COUNTRY

Figure 22.3 Freedom no. 3.

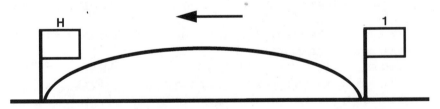

THE RIGHT TO CARRY PASSENGERS FROM A FOREIGN COUNTRY

Figure 22.4 Freedom no. 4.

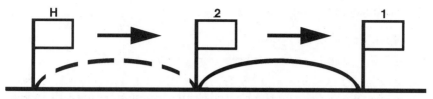

THE RIGHT TO CARRY PASSENGERS BETWEEN 2 FOREIGN COUNTRIES

Figure 22.5   Freedom no. 5.

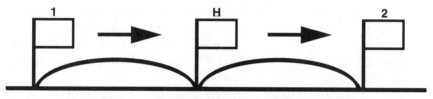

**THE RIGHT TO CARRY PASSENGERS BETWEEN 2
FOREIGN COUNTRIES, THEN THE HOME COUNTRY**

Figure 22.6   Freedom no. 6.

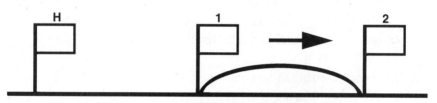

**THE RIGHT TO CARRY PASSENGERS ENTIRELY
OUTSIDE THE HOME COUNTRY**

Figure 22.7   Freedom no. 7.

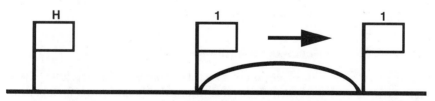

THE RIGHT TO CARRY PASSENGERS WITHIN A FOREIGN COUNTRY

Figure 22.8   Freedom no. 8.

U.S. open sky definition:

1. Open entry on all routes.

2. Unrestricted capacity and frequency on all routes.

3. Unrestricted route and traffic right, the right to operate service between any point in the United States and any point in European countries, including no restrictions as to intermediate and beyond points, change of gauge, routing flexibility, coterminalization, or the right to carry fifth-freedom traffic.

4. Double-disapproval pricing in third- and fourth-freedom markets and (1) in intra-EC markets: price-matching rights in third-country markets, (2) in non-intra-EC markets: price leadership in third-country markets to the extent that the third- and fourth-freedom carriers in those markets have it.

5. Liberal charter arrangement (the least restrictive charter regulations of the two governments would apply, regardless of the origin of the flight).

6. Liberal cargo regime (criteria as comprehensive as those defined for the combination carriers).

7. Conversion and remittance arrangement (carriers would be able to convert earnings and remit in hard currency promptly and without restriction).

8. Open code-sharing opportunities.

9. Self-handling provisions (right of carrier to perform/control its airport functions going to support its operations).

10. Procompetitive provisions on commercial opportunities, user charges, fair competition, and intermodal rights.

11. Explicit commitment for nondiscriminatory operation of and access for computer reservation system.

22.2 Control Surfaces of an Airplane

Figure 22.9 shows the location of control surfaces of an airplane. There are inboard and outboard flaps and ailerons. Spoilers are for spoiling the airflow over the wing. For example, in landing configuration, spoilers are used to shorten the ground run. On the leading edge there are slats that when extended generate more lift at takeoff. At the tail, the following control surfaces are shown: the rudder in a vertical position and the elevator and stabilizer in horizontal positions. The airplane operation manual furnishes detailed descriptions of all control surfaces.

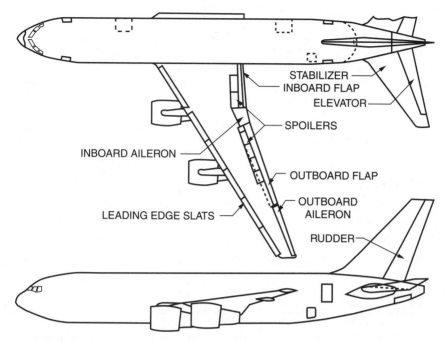

Figure 22.9   Control surfaces of an airplane.

22.3   Current and Constant Dollar Definition

The definition for current dollar or constant dollar is explained in Table 22.1. Sometimes manufacturers express the prices of airplanes in constant dollars and sometimes in current dollars. This is the reason for showing these definitions here. Constant dollars are expressed in a base period; the figure shows changes in quantity only. Current dollars are expressed in the current period; the figure changes in price and quantity.

TABLE 22.1   Current and Constant Dollar

	1990	2000
Number of CD disks	100	200
Price per unit	$10	$15
Gross national product in current dollars	1000	3000
Gross national product in constant 1990 dollars	1000	2000
Implicit price deflator (base is 100)	100	150

Remark: In 1990, 100 CD disks were produced, and in 2000 the amount produced was doubled. In 2000, the gross national product (GNP) was doubled to $2000. This is based on constant 1990 dollars. Considering current dollar values, the GNP in 2000 is $3000. The implicit price deflator for the GNP in 2000 is 150. This explains briefly the difference in current and constant dollars.

22.4   Air Speed Definitions

$V_{ind}.$ The instrument-indicated airspeed corrected for instrument error only. Abbreviation: IAS.

$V_c.$ The calibrated airspeed, which is equal to the indicated airspeed corrected for position error. Abbreviation: CAS.

$V_e.$ The equivalent airspeed, which is equal to the calibrated airspeed corrected for adiabatic compressible flow for a particular altitude. Abbreviation: EAS.

$V_{tas}.$ True airspeed. It is obtained by dividing equivalent airspeed by the square root of density ratio. Abbreviation: TAS.

The Mach number is named after an Austrian scientist Ernst Mach. The ratio of actual airplane speed to the speed of sound is expressed in Mach numbers. Equation (22.1) shows the speed of sound value for various conditions:

$$V_s = \sqrt{\gamma\, g\mathrm{RT}} \tag{22.1}$$

where γ = ratio of specific heat
  g = gravitational constant
  R = gas constant (air)
  T = temperature

23

Recommended Readings

23.1 Periodicals

Air Cargo World

Aircraft Economics

Air Law

Air Transport (ATAA)

Air Transport World

Airline Fleet and Asset Management

Aviation Daily

Aviation Week & Space Technology

ICAO Journal

Journal of Air Law and Commerce

Journal of Aviation / Aerospace Education and Research

Official Airline Guide

Regional Airline World

Transportation Law Journal

Wall Street Journal

World Airline News

23.2 U.S. Government Publications

United States Civil Aeronautics Board:

Air Carrier Financial Statistics

Air Carrier Traffic Statistics

Aircraft Operating Cost and Performance Report

Commuter Air Carrier Traffic Statistics

Handbook of Airline Statistics

Origin-Destination Survey of Airline Passenger Traffic

Quarterly Cargo Review

Trends in the Airline Cost Elements

Trends in Unit Cost

Uniform Systems of Account and Reports for Certified Air Carriers

Other government agencies:

United States General Accounting Office. Airline Competition: Impact of Changing Foreign Investment and Control Limits on U.S. Airlines, 1992

International Aviation: Measures by European Community Could Limit U.S. Airlines' Ability to Compete Abroad, 1993

U.S. Federal Aviation Administration: FAA Aviation Forecast, Fiscal Years 1994–2005, 1994

23.3   List of Aviation-Related Books and Publications

Banfe, Charles F. Airline Management. Englewood Cliffs, NJ: Prentice-Hall, 1992.

Brigham, E. F. D. Fundamentals of Financial Management. New York: Dryden Press, 1986.

Butler, William, Robert Kavesh, and Robert Platt. Methods and Techniques of Business Forecasting. Englewood Cliffs, NJ: Prentice-Hall, 1974.

Davis, R. E. G. The History of Air Express in the United States. 12th International Forum for Air Cargo. Warrendale, PA: SAE, 1984.

De Looff, James L. Commuter Airlines. Hicksville, NY: Exposition Press, 1979.

Doganis, Rigas. Flying Off Course: The Economics of International Airlines, 2d ed. London: Harper Collins Academic, 1991.

Garrison, Paul. The Corporate Aircraft Owners Handbook. Blue Ridge Summit, PA: Tab Books, 1981.

Gialloreto, Louis. Strategic Airline Management. London: Pitman, 1988.

Geselle, Lawrence E. Airline Re-Regulation. Chandler, AZ: Coast Aire Publications, 1990.

Gordon, Robert J. Airline Costs and Managerial Efficiency. National Bureau of Economic Research. New York: Columbia University, 1965.

James, George W. Airline Economics. Lexington, MA: D. C. Heath, 1982.

King, Jack L. Corporate Flying. Glendale, CA: Aviation Book Co., 1987.

Kotler, Philip. Marketing Management: Analysis, Planning and Control, 4th ed. Englewood Cliffs, NJ: Prentice-Hall, 1980.

Morrison, Steven, and Clifford Winston. The Economic Effects of Airline Deregulation. Washington D.C. The Brookings Institution, 1986.

O'Connor, William. An Introduction to Airline Economics, 3d ed. New York: Praeger, 1985.

Radnoti, George. "Application of Electronic Data Processing: Airport Analysis in Airline Operations and for Manufacturers." *IEEE Transactions, Aerospace and Electronic Systems,* 1972.

Radnoti, George. "System Analysis for an Airline Operational Environment Through a Computerized Network Simulation Model." *IEEE Transactions, Aerospace and Electronics Systems,* 1972.

Radnoti, George. "Computer Program, Revenue and Cost." *IEEE Transactions, Aerospace and Electronics Systems,* 1972.

Richardson, J. D. *Essentials of Aviation Management,* 2d ed. Dubuque, IA: Kendal Hunt, 1981.

Shaw, Stephen. *Air Transport: A Marketing Prospective.* London: Pitman, 1982.

Shaw, Stephen. *Airline Marketing and Management.* London: Pitman, 1985.

Spencer, Milton H. *Contemporary Economics,* 5th ed. New York: Worth Publishers, 1983.

Taneja, Nawal K. *Airlines in Transition.* Lexington, MA: D. C. Heath, 1981.

Taneja, Nawal K. *Introduction to Civil Aviation.* Lexington, MA: D. C. Heath, 1987.

Tretheway, Michael W., and Tae H. Oum. *Airline Economics: Foundation for Strategy and Policy.* Vancouver, BC: University of British Columbia Press, 1982.

Index

ABOUT THE AUTHOR

GEORGE RADNOTI holds a doctorate in air transportation economics and a Master's degree in mechanical engineering from the University of Engineering and Economics of Budapest, where upon graduation was teaching as an assistant professor. He has completed Boeing's Jet Performance School. He is presently President of Computerized Aviation Consulting and has previously held positions as senior vice president and major shareholder of Icarus Computerized Consulting Services (CCS), presently operated by Jeppesen & Co.